Modern Construction

Lean Project Delivery and Integrated Practices

Industrial Innovation Series

Series Editor

Adedeji B. Badiru

Department of Systems and Engineering Management
Air Force Institute of Technology (AFIT) – Dayton, Ohio

Industrial Control Systems: Mathematical and Statistical Models and Techniques
  Adedeji B. Badiru, Oye Ibidapo-Obe, & Babatunde J. Ayeni

Learning Curves: Theory, Models, and Applications
  Mohamad Y. Jaber

Project Management: Systems, Principles, and Applications
  Adedeji B. Badiru

Statistical Techniques for Project Control
  Adedeji B. Badiru

Technology Transfer and Commercialization of Environmental Remediation Technology
  Mark N. Goltz

Modern Construction
Lean Project Delivery and Integrated Practices

Lincoln H. Forbes

Syed M. Ahmed

CRC Press
Taylor & Francis Group
Boca Raton   London   New York

CRC Press is an imprint of the
Taylor & Francis Group, an **informa** business

CRC Press
Taylor & Francis Group
6000 Broken Sound Parkway NW, Suite 300
Boca Raton, FL 33487-2742

© 2011 by Taylor and Francis Group, LLC
CRC Press is an imprint of Taylor & Francis Group, an Informa business

Printed in the United States of America on acid-free paper
10 9 8 7 6 5 4 3 2 1

International Standard Book Number: 978-1-4200-6312-7 (Hardback)

Library of Congress Cataloging-in-Publication Data

Forbes, Lincoln H.
    Modern construction : lean project delivery and integrated practices / Lincoln H. Forbes, Syed M. Ahmed.
        p. cm. -- (Industrial innovation series)
    Includes bibliographical references and index.
    ISBN 978-1-4200-6312-7 (hardback)
    1. Building. 2. Lean manufacturing. 3. Construction industry--Management. 4. Construction industry--Cost control. 5. Project management. I. Ahmed, Syed M. (Syed Mahmood), 1960- II. Title.

TH145.F67 2011
690.068--dc22   2010034911

Visit the Taylor & Francis Web site at
http://www.taylorandfrancis.com

and the CRC Press Web site at
http://www.crcpress.com

To

My wife Laurel Forbes

& son Richard Forbes

For their patience and support

From Lincoln H. Forbes

Contents

Comments by Greg Howell, Co-Founder of the Lean Construction Institute

This book, lead-authored by Lincoln Forbes, makes an important contribution to the construction industry. He has been an active participant in the Lean Construction Institute since 2004 and with co-author Syed Ahmed brings a unique background in facilities design, construction management, quality management/assurance, and industrial engineering. He connects emerging lean construction theory and practices with its roots in industrial engineering. In this he saves the baby of lean construction and the bath water of more traditional industrial engineering and productivity improvement practices. The book also connects lean construction with the LEED movement. This makes great sense as lean construction rests on a conceptual foundation and understanding of waste that explains why projects managed on a traditional basis are so often adversarial and difficult to control. Where traditional practice tries to optimize the piece, lean aims to optimize at the project level. This requires a different approach to managing work. Our environmental problems arise from similar practices, locally optimizing at the expense of larger systems. Lean won't save the planet but it does help us understand how to manage larger, more complex systems far more effectively than is possible with practices aimed at local optimization.

Preface

This book was written to address a pressing need for performance improvement in the construction industry. The construction economy is a large sector of many countries, hence its performance directly influences their competitiveness. The book introduces the subject of lean construction as a methodology for improving the entire design and construction supply chain. It presents other tools and techniques for improving performance as well, and links them with lean construction as support activities, although they can enhance the performance of an individual stakeholder in the supply chain, such as a mechanical subcontractor. At the core of the lean philosophy is the aim of "global optimization" where entire project performance is maximized, in contrast with "local optimization" where an individual stakeholder benefits, often at the expense of others. For this very reason, the book advocates an understanding of such techniques as quality management, BIM, ergonomics, and sustainable practices, but that construction professionals should use them to transition from local to global optimization as lean construction is increasingly adopted.

Several members of The Lean Construction Institute shared concepts, experiences, and cases during the development of the manuscript, notably Greg Howell and Glenn Ballard, co-founders of the Lean Construction Institute. They were most generous with their time, ideas, and encouragement, and the book has benefited significantly from their support. Valuable input and sharing of ideas came from several other leading members of the lean construction community, including Will Lichtig, Peter Gwynn and Tariq Abdelhamid. (Their contributions and many others are recognized in the Acknowledgments section.)

During the past several decades, the manufacturing and service industries have greatly increased their levels of productivity, quality, and profitability through the application of process improvement techniques and information technology. A major contributor to the competitiveness of these sectors is their embrace of the lean production philosophy that evolved from the "Just-In-Time" system that was the foundation of Toyota's success in automotive manufacturing.

The construction industry lags far behind the manufacturing and service industries with regard to the application of performance improvement and optimization techniques. According to the Bureau of Labor Statistics (BLS), in the past 40 years the productivity of non-farm industries has increased by over 100%, while that of the construction industry has stagnated. Several research studies have shown that at least 30% of wasted resources are caused by entrenched attitudes in the management of projects. Waste in construction occurs especially in the interaction between trades, and in the handoffs of work from one trade to another. Studies by the Construction Industry Institute have identified that 10% of project costs is typically spent on field work, and that only 43% of the time spent on construction activity may be considered productive.

As global competitiveness increases, so will the expectation of higher levels of productivity in the delivery of constructed facilities. Increasingly, customers in other industries expect their suppliers to be flexible, communicative, and responsive to their needs in today's fast-paced business environment. Construction organizations that fail to utilize the foregoing approaches will find themselves at a competitive disadvantage.

Construction industry practitioners have been continually seeking to apply better technologies and processes to improve project delivery, but there is a lack of unified strategy and there is little incentive to change. Most construction contracts place the parties to construction in adversarial roles, although approaches such as "design–build" and "partnering" have diminished this challenge to a limited extent. Most recently, initiatives such as Lean Project Delivery and Integrated Project Delivery have been adopted successfully on a relatively small number of projects. This success has been largely attributable to the interest of willing project owners who have set the stage for the close integration among all parties that is needed to create a so-called lean environment. Although the parties to a construction project are interdependent, they tend to make decisions that further their own self-interests. Industry specialization has separated designers from the management of the production process, and this is perpetuated by the potential legal ramifications of apportioning project risk.

Given this industry scenario, innovation is still possible—in the internal operations of each party as well as within the combined activities of the project team, depending on the extent to which they are willing to optimize the entire supply chain for delivering the project. This book recognizes that the supply chain is as strong as its weakest link and encourages innovation within the operations of each of the parties, as a prelude to making these parties strong, effective links.

The book provides a dual approach for improving construction processes: (a) several chapters are devoted to describing lean construction and its applications; and (b) process improvement approaches are presented in other chapters that relate to lean construction but may also improve projects in a so-called traditional environment. It is hoped that readers in a traditional construction environment will be encouraged to use the principles described to develop readiness for a lean environment.

It is recognized that true lean construction can best occur when all the construction stakeholders—especially general contractors, construction managers, subcontractors, and material suppliers—are committed to the concept of optimizing the flow of activities holistically. It is also recognized that optimizing the performance of individual actors in the construction supply chain does not ensure that the overall process will be optimized or greatly improved if the handoffs from one trade to another do not observe lean principles. Design activities should address the construction process as well as the finished product. This adoption of lean construction is progressing gradually with the passage of time. It is being applied selectively to projects where a small number of owners, designers and contractors are aware of its benefits and include it in their design and construction contracts. Lean practices enhance the utilization of resources, reducing waste, and overall costs. They also enable practitioners to improve the quality of the built environment, secure higher levels of customer/owner satisfaction, and simultaneously improve their profitability. This book also explains how industrial engineers may position themselves to play a vital role in the construction industry. Readers are encouraged to become actively involved in the lean construction movement by joining a chapter of the Lean Construction Institute (LCI). LCI may be contacted at www.leanconstruction.org/

What's in This Book

This book provides tools and techniques designed to enhance the project management process. The treatment of lean methods provides specific procedures for the modification of planning, and scheduling activities to improve the interaction between the parties, reduce waste, and improve performance.

In order to accommodate a wide variety of readers, the book explains the basis of a number of tools and techniques. It is assumed that, for example, some construction practitioners may not have a detailed knowledge of tools and techniques. By the same token, nonpractitioners of construction are provided with an overview of construction delivery methods as a foundation for applying process improvement techniques that may already be familiar to them. This book should serve as a contemporary reference for construction practitioners and as a supplementary text for students.

Chapter 1 provides an overview of the construction industry and the relationships among its participants—designers, general contractors, construction managers, contractors, subcontractors, material suppliers. It explains the construction sector as a significant portion of a nation's economy, pointing to the consequences of comparatively low productivity. The chapter briefly describes current design processes and construction delivery methods and advances the adoption of lean methods to address industry underperformance.

Chapter 2 describes productivity measurement in construction as a foundation for improving performance regardless of the means used—lean construction or quality-based initiatives. It provides the theoretical background for comparing work produced by construction processes to the resource inputs—labor, materials, energy, etc. This chapter explains productivity concepts and how they relate to performance measurement in construction organizations. It also provides examples of productivity and performance measurement methods.

Chapter 3 explains lean theory and provides a very important foundation for the book by explaining how lean concepts apply to the construction environment. It provides a historical background of the Toyota System and the production management concepts that grew from it. The chapter chronicles the major contributions made by a number of researchers, and lists milestones in the development of the lean construction movement. The principles of reducing waste and adding value, as well as meeting the requirements for a lean environment are presented.

Chapter 4 builds on the previous chapter by describing The Lean Project Delivery System™ as a proven framework for delivering projects on time, within budgets, with high-product quality, and minimal safety incidents. The application of The Last Planner® System is explained in detail, including the process of learning from performance measurement and analysis. The chapter explains the foregoing systems in detail, including related topics such as designing for lean operations and target value design.

Chapter 5 explains lean process measurement including PPC determination, charting of reasons for non-completion (RNC), five-why analysis, rolled PPC and plus/delta analysis. The chapter also introduces readers to lean tools and techniques such as value stream mapping, Kaizen methodology, 5S, and A3 reports.

Chapter 6 explains how to create an environment that is conducive to lean construction. Lean coaching and how it prepares project participants for adopting lean beliefs and lean practices are described. Detailed cases explain the activities of subcontractor organizations.

Chapter 7 devotes considerable detail to the deployment of lean techniques in construction projects. The book discusses relational contracting and its deployment as a proven vehicle for lean construction. It presents case studies of project delivery methodologies that have evolved from The Lean Project Delivery System™, such as Integrated Project Delivery and Lean Production Management.

Chapter 8 describes information technology applications to the management of the design and construction processes for benefits similar to those experienced in the manufacturing and service industries. Information technology–based systems integration and building information modeling among other topics are discussed. Two cases by Professor Salman Azhar of Auburn University, Alabama, illustrate the application of BIM to construction projects.

Chapter 9 addresses quality improvement initiatives in construction and describes total quality as an approach to doing business that attempts to maximize the competitiveness of an organization through the continual improvement of the quality of its products, services, people, processes, and environments. The application of Six Sigma techniques in construction projects is also described.

Chapter 10 first presents sustainability as an important foundation for lean construction, based on principles advanced by the U.S. Green Building Council and the Department of Energy. "Green Construction" and "Sustainability" have become a major concern in the design, construction, and operation of facilities as increasing attention is focusing on global warming and other environmental phenomena. Commissioning is described as serving two important roles, ensuring sustainable building performance through "Green" certification, and supporting project-wide quality control.

Chapter 11 makes available a variety of industrial engineering–related performance improvement tools and techniques. While these tools and techniques have been traditionally used in other industries to improve performance, this chapter explains how the techniques may be used more extensively to improve construction processes, both in a lean environment as well as in a traditional environment.

Chapter 12 addresses construction site safety practices. It is an area of concern for employers of construction workers; this concern for safety is extended to the employees of subcontractors as well. Successful project delivery requires that safety incidents be minimized. The chapter describes several strategies for improving safety, beginning with the design process, especially as a foundation for lean construction.

Chapter 13 presents several topics that relate to human performance in the work environment, such as motivation, diversity, and ergonomics. These topics can have a significant impact on a construction team's ability to maximize its performance, especially with lean construction.

Chapter 14 explains the application of systems integration to design and construction projects through three cases presented by industrial engineers in the construction industry. The strategies described are integrated to improve project performance in a traditional environment and further explain how a transition can be made to lean construction. The chapter also discusses possible roles for IEs in the construction industry, and how they might prepare themselves.

Chapter 15 describes the application of post-occupancy evaluation to construction projects and links the techniques to the Deming cycle for continuous improvement. An important extension of the weekly review that is conducted on lean-based projects is promoted; learning from project outcomes is essential for improving design and construction processes. An enhanced model of the project delivery process is proposed, based on lessons learned in the ongoing evolution of lean construction applications.

Acknowledgments

This book has materialized through the selfless sharing of ideas and experiences by many people in the construction industry, especially a number of members of the lean construction community. Their kind support throughout the process of preparation and publication is most highly appreciated. Lincoln Forbes wishes to recognize and thank the following people.

Greg Howell and Glenn Ballard, co-founders of the Lean Construction Institute, are owed a debt of gratitude for helping to make this book possible. Greg shared ideas in the early stages of the book's development that helped in shaping its direction. He also participated in dialogue on best practices for lean coaching and lean implementation as a roadmap for future lean adopters. Glenn Ballard addressed the historical evolution of lean construction, based on his participation since its inception. He also shared important findings from his many research studies over the years that influenced the content of the book.

William Lichtig, Chief Strategic Officer for McDonough, Holland & Allen, PC, made available his pioneer work with the Integrated Form of Agreement (IFOA). Peter Gwynn, principal, Lean Implementation Services and Dave Koester, Project Manager, Regional Contractors Alliance (RCA), documented a highly successful application of Lean Production Management. Matt Horvat, consultant with Lean Project Consulting, developed illustrative cases with typical scenarios that may be faced by new adopters of lean construction. Hal Macomber, principal, Lean Project Consulting, kindly shared many articles and papers, including those on target value design and study action teams. Robert Blakey, principal, Strategic Equity Associates, made insightful observations on the process of introducing organizational change and coaching management and workers. Professor Tariq Abdelhamid, editor of the *Lean Construction Journal*, graciously read through a number of chapters and provided helpful feedback. Ed Anderson, principal, Anderson Technical Services, reviewed several chapters and assisted with relevant content. Ed also opened the door to a wide array of experiences, garnered from past assignments in lean projects. Professors Fred Aghazadeh of Louisiana State University and Jim Moore of the University of Southern California provided invaluable feedback by reviewing portions of the manuscript.

Dennis Sowards, principal of Quality Support Services (QSS), introduced associates and clients who were willing to share their lean journeys. They included Chris Warren, Director of Risk Management and Continuous Improvement, Tweet/Garot Mechanical Inc.; Ted Angelo, Executive VP, Grunau Company, Inc.; Tracy Dabrowski, CFO, Belair Excavating/Contracting, and Paul Pasqua, former director. Other outstanding lean cases were provided by Owen Matthews, President of Westbrook Air Conditioning and Plumbing in Orlando, Florida, recognized as the originator of the Integrated Project Delivery (IPD) methodology and Jay Berkowitz, President of Superior Window Company and a passionate lean advocate. The following industrial engineers made a noteworthy documentation of lean-related cases drawn from classical IE tools and techniques: Jeff Mason, CEO, Integrated Facility Services (IFS); Al Attah, President of Macval Associates, Dallas, Texas; Jorge Cossio, President, ITN de Mexico, S.A. de C.V.; and Dr. Larry Nabatilan, formerly of Louisiana State University, who provided an ergonomic study.

Lincoln Forbes extends special thanks to Dr. Vincent Omachonu of the University of Miami, in appreciation for paving the way for this book and for his ongoing advice and guidance.

Both authors would like to thank the following supporters of a great team effort: Prof. Salman Azhar of Auburn University, who contributed cases and narrative in Chapter 8 on BIM and ICT. Rizwan Farooqi, PhD scholar in the construction management program at Florida International University, who deserves special recognition. He dedicated many hours to leading a team of graduate students to investigate a number of topics, and actively participated in the process of developing synopses and summaries that represented a significant contribution to the book. Thanks and appreciations are also extended to the following students who were an important part of the research team: Lakshmi Priya, Saraswathi Vadali, Vishval Mehta, and Farhan Saleem. The authors would also like to thank students Virushali Davakhar and Christina Jordan for their support.

Finally, the authors extend sincere thanks to the following individuals for their help with illustrations: Darrlyn Choate, Charlene Thompson, and Richard Forbes.

Authors

Lincoln H. Forbes, PhD, PE, LEED® AP, is an adjunct professor at Florida International University (FIU), Miami, in the College of Engineering and Computing's Construction Management and Engineering Management programs. A registered professional engineer in Florida, Dr. Forbes has over 30 years of experience in various aspects of facilities design, construction, and maintenance as well as quality/performance improvement. He has been administrative head of several facilities-related departments and functions in Miami-Dade County Public Schools, the nation's fourth largest district. These positions included in-house construction, design services, construction quality control, post-occupancy evaluation, project warranty services, and quality management. He has also provided research and technical support on sustainability practices, commissioning, construction systems, and methods and materials. Dr. Forbes has also been involved in strategic planning and performance improvement. Earlier in his career, he worked in consulting as a senior electrical-mechanical design engineer and site engineer with ADeB Consultants (International), and later as Chief, Management Analysis with the Dade County Public Works Dept., Florida.

Dr. Forbes is a member of The Lean Construction Institute, a non-profit organization dedicated to promoting knowledge and awareness of lean construction principles. He is the founding president of the Construction Division of the Institute of Industrial Engineers (IIE). The division promotes IE's active involvement with construction performance improvement. Earlier roles in the IIE included director of the government division and president of the Miami chapter. Dr. Forbes is a senior member of ASQ and the American Society for Healthcare Engineering (ASHE). He has served as an examiner for the Florida Sterling Council's Quality Award.

Dr. Forbes received his PhD from the University of Miami, specializing in the improvement of quality and performance, including healthcare facilities design and construction. Previously, he obtained an MBA and an MS in industrial engineering at the University of Miami. He earned a BSc in electrical engineering at the University of the West Indies.

Dr. Forbes has published and presented many papers internationally on the application of lean techniques and quality and productivity improvement in construction. He authored a first-time chapter on construction in the 2005 *Handbook of Industrial Engineering.* He has chaired several IIE conference tracks on IE applications in construction. He has served as a columnist on continuous improvement in the construction industry for the ASCE publication *Leadership and Management in Engineering,* and as a reviewer for the ASCE *Journal of Construction Engineering and Management.*

Syed M. Ahmed, PhD, is a full professor and chairperson of the Department of Construction Management at East Carolina University (ECU) located in Greenville, North Carolina. Prior to that, he was an associate professor and graduate program director in the Department of Construction Management of the College of Engineering & Computing at Florida International University (FIU) in Miami. He was also an affiliated faculty member of the Department of Civil & Environmental Engineering at FIU. Dr. Ahmed received his PhD in 1993 and his MSCE in 1989 from Georgia Institute of Technology majoring in construction management, with a minor in industrial engineering and management science. He also holds an undergraduate degree (BSCE) in civil engineering from the University of Engineering & Technology in Pakistan (1984).

Dr. Ahmed has over 25 years of international experience in teaching, research, and consulting in the United States, Hong Kong, Pakistan, Mexico, and Jamaica. His areas of interest and expertise are construction scheduling, quality and risk management, project controls, construction safety, construction procurement, and construction education and information technology. He has received over $500,000 from various funding agencies including the U.S. Agency for International Development, U.S. Department of Defense, and Florida Department of Education. He is the author of several books and has published extensively with over 125 publications in international journals and conferences in his areas of expertise.

Dr. Ahmed is currently on the editorial board of more than six international journals and serves as the associate editor of *American Society of Civil Engineers* (ASCE) *Journal of Construction Engineering & Management*. In the past 10 years, Dr. Ahmed has been invited to conduct numerous workshops, and to give lectures and presentations at different international seminars in Hong Kong, Pakistan, Australia, and India. In 2002, Dr. Ahmed conceived, planned, and organized the First International Conference on Construction in the twenty-first century (CITC) (http://www.fiu.edu/~citc). Dr. Ahmed has chaired five CITC conferences with various conference themes including Challenges and Opportunities in Management and Technology; Sustainability and Innovation in Management and Technology; Advancing Engineering, Technology, and Management; Accelerating Innovation in Engineering, Technology, and Management; and Collaboration and Integration in Engineering, Technology, and Management.

Dr. Ahmed also conceived and organized the First International Conference on Construction in Developing Countries (ICCIDC-I) in Karachi, Pakistan in August 2008 and the second conference (ICCIDC-II) in Cairo, Egypt in August 2010 (http://www.fiu.edu/~iccidc).

1

Overview of the Construction Industry

This chapter provides background information on the relationships among the participants in the construction industry. These relationships greatly influence the effectiveness of construction delivery, especially with projects involving lean-based delivery methods. It explains the recent history of the construction sector as a significant portion of a nation's economy, pointing to the consequences of comparatively low productivity. The subsequent chapters provide tools, techniques, and approaches for improving the performance of the construction industry. The term "construction" is used in a general sense that includes both design and construction processes, unless specified otherwise.

Background on Industry Performance

The construction industry has traditionally been one of the largest industries in the United States. As reported by the Bureau of Labor Statistics (BLS), U.S. Department of Labor, the value of construction in 2006 was $1,260.128 billion, representing 8% of the gross domestic product (GDP) (Figure 1.1). The industry employed approximately 7.614 million people in 2007 (Figure 1.2). By its very nature construction activity in the United States has not been subjected to the trend toward offshoring that has plagued both the manufacturing and service industries. The BLS report titled "State of Construction 2002–2012" forecasts that 58.4% of U.S. jobs will be construction-related at the end of that decade. Although other industries have blazed a trail to higher levels of quality and performance, the majority of construction work is still based on long-established techniques.

Construction productivity has been lower than that of other industries. The BLS does not maintain an official productivity index for the construction industry, the only major industry that is not tracked consistently. Productivity in construction is usually expressed in qualitative terms and cannot be analyzed scientifically. A number of independent studies estimated that construction productivity increased between 1966 and 2003, at a rate of 33%, or 0.78% per year.

This productivity growth is less than one-half that of U.S. nonagricultural productivity gains during the same period, which averaged 1.75% annually. An estimate by the National Institute of Science and Technology (NIST) points to a decline in U.S. construction productivity over a 40-year period to 2007 that was −0.6% per year, versus a positive nonfarm productivity growth of 1.8% per year. The increased productivity may be attributable to mechanization, automation, prefabrication, less costly and easier to use materials, and a reduced real cost of labor.

Most of the improvement in construction productivity has been the result of research and development (R&D) work in the manufacturing industry related to construction machinery. Earthmoving equipment has become larger and faster; power saws have replaced handsaws. However, as the use of power tools increases the consumption of energy, their contribution to productivity improvement may be eroded.

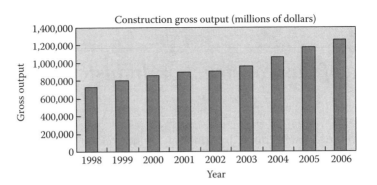

FIGURE 1.1
Construction gross output 1998–2006.

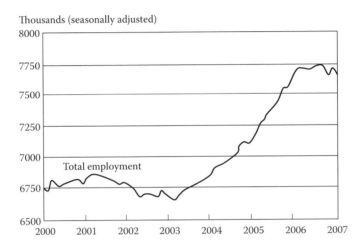

FIGURE 1.2
Construction employment: 2000–2007. (Courtesy of the U.S. Department of Labor, Bureau of Labor Statistics.)

Tilt-up construction is one example of a technology that has improved construction productivity. According to the Tilt-up Concrete Association, tilt-up construction has proven to be a fast and economical construction methodology for small buildings of one to four stories, but is far from new. It became popular in California in the late 1950s, initially for warehouses, then its use subsequently extended to shopping centers, churches, townhouses, and so on. Successful introduction of the technique is credited to one Robert Aiken who constructed several buildings using "lifting jacks" to raise building panels in Illinois and Ohio. It is believed by some that a Roman architect discovered—2000 years earlier—that it would be easier to "cast a slab of concrete on the ground, then hoist (tilt) it up into position, than to build two wood forms. Place concrete between them, and then remove the forms."

Reasons for Low Productivity

There is evidence that improved productivity is related to the levels of R&D in an industry, but the extent of R&D effort in the construction industry has remained very low in the

United States. The overall productivity in construction has been greatly affected by regulatory controls, the environment, climatic effects, cost of energy, and other factors. Productivity improvement has never been the focus of the construction industry, probably due to the lack of a model that can fit together all the fragmented components of the overall construction process.

Productivity in the construction industry is largely unmeasured, and those measures that do exist are contradictory and conflicting. Research has pointed to a significantly high level of wasted resources in the construction industry—both human and material (Figures 1.1 and 1.2). An article titled "Construction and the Internet" in *The Economist*, dated January 15, 2000, noted that up to 30% of construction costs is due to inefficiencies, mistakes, delays, and poor communications.

- The Construction Industry Institute (CII) estimates that in the United States, 10% of project cost is spent in one rework.
- Between 25 and 50% of construction costs is lost to waste and inefficiencies in labor and materials control.
- Losses are incurred in errors in information when translating designs to actual construction. Inadequate interoperability accounts for losses of $17 to 36 billion per year. This problem is due to communication difficulties between software used by different operators in the design and construction supply chain.

Need for New Approaches to Construction

Global competition is increasing rapidly; in 2007 construction contracts won by Chinese companies increased in the United States and Europe by 160%. In the case of the United States, this trend would suggest that the competitiveness of domestic construction organizations is being challenged on its home turf. Construction spending is moving east to China, India, the Middle East, and Africa (CRIMEA). Construction organizations that are the most competitive are likely to benefit from this trend.

Clearly, new approaches are urgently required to address the construction industry's ills. Construction is said to need a greater sense of mission, focus, and industry support. There are several obstacles to its competitiveness in the United States:

- Declining productivity
- A mindset for minimum first cost
- Prescriptive standards and codes that stifle productivity

The industry is said to need more R&D, and to move beyond a reliance on historical surveys to identify new approaches for getting the job done. This problem is not confined to the United States. Sir John Egan's report in 1998 "Rethinking Construction," identified many shortcomings in construction practices in the United Kingdom and advocated radical change. The government in the United Kingdom appointed a construction task force that was comprised of a chairman and nine members representing senior representatives of construction organizations and their clients.

The task force studied several industries, including auto manufacturing and grocery retailing to accomplish the following objective:

- Quantify the scope for improvement of construction efficiency
- Examine current construction practice and assess the potential for its improvement by way of innovation
- Identify specific actions and good practice that would help achieve more efficient construction
- Identify projects that might help to demonstrate the improvements that can be achieved

Sir John Egan's 1998 report, also called "The Report" made positive observations about the UK construction industry, citing its ingenuity and capacity to handle difficult and innovative projects (Egan 1998; Table 1.1). At the same time, the report pointed to many industry weaknesses such as: low profitability, the urgent need to train the work force, inadequate R&D investment, and clients' misunderstanding of cost and value relationships. At the heart of the report was a desire to develop a change of style, culture, and process in the industry as opposed to recommending mechanistic activities. Proverbs, Holt and Cheok (2002) conducted a follow-up analysis of fifteen recommendations from the Egan report, using a sample drawn from the Managing Directors of 100 UK contractors. The study was intended to rank the recommendations with regard to the construction industry based on a) Impact, i.e., potential effectiveness (b) Effectiveness, i.e., the potential to achieve the intended purpose, and (c) Perceived potential for success, computed as the product of scores for impact and effectiveness.

- Impact was correlated most highly
  - with Development of long-term relationships

TABLE 1.1

Construction Initiatives

Number	Recommendation
1	The commitment of leadership from all parties
2	Greater customer focus
3	Application of products and services from other industries
4	Promote best practice in health and safety
5	Provide decent working conditions and promote people as the construction sector's greatest asset
6	Measure and assess own performance, set performance targets, and share information
7	Development of an integrated process instead of reliance on convention
8	Improved quality and fewer defects
9	Use of demonstration projects to develop and illustrate innovations
10	Rethinking (i.e., reengineering) the construction process
11	Adopt lean thinking as a method of sustaining performance improvement
12	More and better training
13	Development of technology (e.g., standardization and prefabrication)
14	Development of long-term relationships
15	Greater use of performance improvement tools and techniques (e.g., value management, benchmarking, etc.)

Source: Adapted from Proverbs, Holt, and Cheok, Summary of J. Egan, Report, *Rethinking Construction*, 2002.

- Commitment of leadership from all parties
- Greater customer focus and More and better training
- Effectiveness was correlated most highly with
  - Development of long-term relationships
  - Commitment of leadership from all parties
  - Improved quality and fewer defects
- Perceived potential for success was correlated most highly with
  - Development of long-term relationships
  - Commitment of leadership from all parties
  - Greater customer focus and More and better training.

These findings supported the recommendations of the Egan Task Force.

In the United States, the need for innovative methods to improve construction performance is underscored by the ongoing and anticipated shortage of skilled labor in an industry that is highly labor-intensive. Studies have shown that the traditional sources of workers—vocational school programs and apprenticeships—have not been keeping up with the demand. In the area of technicians and equipment operators alone there is a projected shortage.

Private and public organizations in the United States have been concerned about construction cost and quality for many years. The Construction Users Anti-Inflation Roundtable was formed to investigate the causes of inflation in the cost of construction. In 1972, this organization joined with a number of others to form the Business Roundtable, represented by the presidents of the 200 largest U.S. companies. This initiative led to the creation of the Construction Industry Cost Effectiveness (CICE) project to promote quality, efficiency, productivity, and cost-effectiveness in the construction industry. The CICE produced a January 1983 report titled "More Construction for the Money." The CICE report observed that construction was a $300 billion activity at that time, and that even a modest application of the teams' recommendations could save the industry at least $10 billion annually. The report revealed deficiencies in the planning, design, and construction of projects and that there was slow acceptance and use of modern management methods to plan/execute projects. In particular, owners were said to be unwilling to bear the costs of modern systems.

The CII was created in 1983 at the University of Texas–Austin. The CII has conducted many research studies on construction productivity during that time to the present.

Causes of Poor Construction Industry Performance

The observations of the CICE project in 1983 are still relevant in 2010 and beyond. The study teams noted that "more than half the time wasted in construction is attributable to poor management practices." They concluded that "if only the owners who pay the bills are willing to take the extra pains and pay the often small extra cost of more sensible methods they would reap the benefit of more construction for their dollars." The waste in construction occurs mostly in the interaction between trades. The self-interest of parties makes them put themselves first.

The construction industry is not unified on the measurement of productivity or overall performance. Many researchers define construction project success in terms of (a) cost:

within budget, (b) schedule: on-time completion within schedule, (c) safety: high safety levels—few or no accidents, and (d) quality: conformance with specifications and few defects. Some view quality as one point of a triangle with cost and schedule; quality is often the first to be sacrificed in favor of cost savings and schedule reductions. Many contractors believe that it is impossible to meet desirable benchmarks in all four factors simultaneously and that there is a zero–sum relationship between them. In other words, an accelerated schedule results in an increased cost and lower levels of safety and quality. Material waste is a significant factor in construction costs—9% by weight in the Dutch construction industry and 20–30% of purchased materials in the Brazilian construction industry. Material wastes are caused by several sources such as design, procurement, material handling, operations, and so on.

At least four main elements are important to overall performance:

1. Productivity: primarily measured by cost, satisfactory productivity is work accomplished at a fair price to the owner and with reasonable profit for the contractor
2. Safety: accident-free projects
3. Timeliness: on schedule and everything on hand when needed
4. Quality: conformance to specifications and satisfying owner's needs

In terms of the prevailing industry practices, the concept of construction performance does not emphasize productivity and quality initiatives. The work of many researchers has revealed an industry tendency to measure performance in terms of the following: completion on time, completion within budget, and meeting construction codes. Very little attention has been directed to owner satisfaction as a performance measure. Although the Malcolm Baldrige National Quality Award was established in 1987, no construction-related organization has won this coveted award during the 20 years of its inception.

There is a growing emergence of subcontracting. In effect, so-called general contractors perform the role of construction manager, with individual contracts, and with specialty subcontractors. The subcontractors are often priced in a manner that does not reflect the general contractor's agreement with the owner; even if the owner pays a high price, the subcontractor may still have to work with inadequate budgets, often compromising quality as a result.

Safety is a major source of waste. In the year 2006 alone, construction accounted for approximately 21% of all deaths and 11% of all disabling injuries/illnesses in the workplace (BLS 2006).

Poor communication inhibits productivity. Communication tends to be via the contract. Essentially, the designer is paid to produce a design expressed in the form of specifications and drawings. The contractor is expected to use these as a means of communication and produce the completed facility. This communication often does not work as well as it should.

There are large gaps between expectations and results as perceived by construction owners. Symbolically, Value (V) = Results (R) – Expectations (E). Consequently, since expectations often outweigh the results, construction owners feel that they receive less value than they should. Studies to quantify the gaps or dissonance zones between the three parties to construction (i.e., owners, designers, and contractors in health care facilities projects), revealed significant differences between them (Forbes 1999) In the area of owner satisfaction factors for example, public owners and designers differed on seven of nine criteria, owners and contractors differed on five of nine criteria, while designers and contractors disagreed on the relative importance of two criteria.

Innovation is adopted very slowly. Small contractors often lack the expertise or financial resources to adopt technological advances—adoption is inhibited further by fear and uncertainty. Roofing contractors, for example, tend to use the same time-honored methods to ensure that supplies and equipment are on-site each day. Expediters, at additional cost, deliver items that are frequently forgotten.

Needed training often does not get to the decision makers in the construction industry. Construction management (CM) programs around the country have been providing higher levels of training for managers; however, this training has not reached the ultimate decision makers in the industry. Efforts to enhance quality and productivity are likely to be frustrating under this scenario.

Owners have not specifically demanded productivity and quality. There is a general lack of productivity/quality awareness in the industry among all parties, including owners. Owners have come to accept industry pricing—they have not been able to influence the productivity of the industry—prices have simply trended upwards with fluctuations based on market conditions. By contrast, manufacturing activities have become cheaper over time on a per unit basis.

Architect/engineer contracts are said to be unclear with respect to professional standards of performance, often leading to unmet expectations. Construction owners feel that typical A/E contracts protect designers at the owner's expense. For example, prevailing contract language relieves designers of any role in the case of a lawsuit or arbitration between an owner and contractor. An outgrowth of this is the practice of "substantial completion" where a facility is usable but may have a small but significant percentage of remaining work in the form of a "punch list". An owner often has a very difficult time in persuading a contractor to finish that work.

Few large design or construction companies, and virtually no small companies, have implemented the concept of a quality or productivity manager—cost cutting trends have resulted in such a position being viewed as an unjustifiable luxury. On the positive side, there is a growing interest in lean construction techniques, and as of 2010, a small number of projects have been carried out based on a lean approach. However, the deployment of these methods is mostly driven by consultants, as "lean" is still in its emerging stages.

There is little, if any, benchmarking—many manufacturers and service organizations have become preeminent by adopting the best practices of benchmarked organizations. Construction has done very little of this, due to distrust, fear of losing competitive advantage, but more likely, simply by being anachronistic.

Mistakes, rework, poor communication, and poor workmanship are part of an ongoing litany of deficiencies that seem to be accepted as being a natural part of construction activity. Safety is a major national concern. Construction has an abysmally poor safety record, worse than virtually all other industries.

Categories of Construction

In order to understand how improved methods can be applied to the construction industry, it is helpful to understand that environment; it is truly diverse, so much so that its participants have found it easy to rely on such clichés as "the industry is like no other" and "no two projects are alike," to maintain the status quo in which long-established management traditions are seen as an arcane art that others cannot fully understand.

The BLS refers to three major headings: General Building Contractors SIC Code 15, Heavy Construction (except building) SIC Code 16, and Special Trade Contractors SIC 17. These are further subdivided into 11 SIC Code headings that include:

- Commercial building construction: offices, shopping malls
- Institutional construction: hospitals, schools, universities, prisons
- Residential: housing construction, including manufactured housing
- Industrial: warehouses, factories, process plants
- Infrastructure: Road and highway construction, bridges, dams

Who Are the Parties Involved in Construction?

1. Owners: They originate the need for projects and determine the locations and purpose of facilities. They arrange for design, financing, and construction.
2. Designers: They are usually architects or engineers who interpret the owner's wishes into drawings and specifications that may be used to guide facility construction. In the design/build concept, they may be a part of the construction team.
3. Constructors: These are contractors and subcontractors who provide the work force, materials, equipment and/or tools, and provide leadership and management to implement the drawings and specifications to furnish a completed facility.
4. The labor force: This is comprised of foremen, craftsmen, or journeymen, and skilled or semiskilled apprentices or helpers. Many different crafts are represented, such as masons, pipefitters, carpenters, electricians, and so on.
5. Major suppliers: Equipment and material manufacturers, transporters
6. Financial institutions: banks, construction financial organizations
7. Lawyers, insurers
8. Federal and local regulators: code enforcement professionals
9. Public services
10. Utilities
11. Safety professionals
12. Quality assurance/quality control professionals
13. Lean facilitators/coaches/mentors: These professionals provide support in the implementation of new lean approaches in design and construction.

Project Delivery Methods

Construction projects are described as having certain characteristics; each project is unique and not repetitive. Projects come in various shapes, sizes, and complexities. A project is said to work against schedules and budgets to produce a specific result. The construction team cuts across many organizational and functional lines that involve virtually every department in the company. The suppliers typically have a contract with the contractor, but not with the owner or designers. Government agencies oversee both the design and construction to ensure compliance with prevailing state or local construction codes.

There are a number of models for the process of designing and constructing facilities. Several of these models have been in existence for many years and have been used with varying degrees of success, depending on the type of project and the skills required. The pros and cons of these delivery methods are discussed below in items 1 through 5. However, the emphasis of this book is on the more recently developed project delivery systems such as: lean project delivery and relational contracting. These systems are described in detail in Chapters 6 and 7.

The project delivery methods include:

1. Design-bid-build (DBB)
2. Design-build (DB)
3. Engineer-procure-construct (EPC)
4. Design-CM contracts
5. Design-agency CM contracts
6. Fast-track construction
7. Partnering
8. Relational contracting/lean project delivery

Design–Bid–Build Contracts

Design-bid-build contracts represent the most frequently used type of project delivery systems for most construction projects, and are considered to be the "traditional" delivery method. DBB projects have the following characteristics (Figure 1.3):

1. The project is conceptualized by the owner.
2. Planning is carried out based on the objectives to be met and on economic and technical feasibility.
3. Programming is carried out to identify the uses and desired sizes of various spaces, followed by a schematic design to identify relationships of these spaces relative to each other. The scope of the project, preliminary budget, and schedule are derived.
4. Detailed design is usually carried out in stages, with intermediate checkpoints for verification by the parties to the project.
5. The design work culminates in the preparation of completed drawings and specifications, representing bid documents as well as detailed cost estimates. The bid documents are used to solicit construction bids, or are otherwise used to negotiate a construction price.
6. Bid analysis is carried out and a legally binding contract is then awarded. The drawings, specifications, and signed documents then become construction documents.
7. The contractor is given access to the site and instructed to proceed, based on legally established time frames. A contract may contain incentives for timely completion, as well as penalties for avoidable delays or cost overruns.

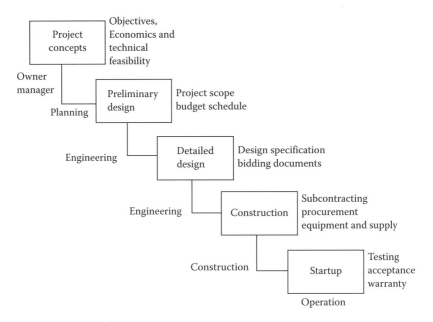

FIGURE 1.3
Traditional design-bid-build construction phases.

8. The owner, or agents of the owner such as architects/engineers or construction managers, monitor(s) the progress of the construction, ensuring that interim payments to the builder/contractor match construction progress.

9. At completion, there are acceptance inspections, leading to the commissioning of the facility for the owner's use. The project is turned over to the owner.

Design-bid-build has many well-known shortcomings: There is a greatly protracted process to do programming, design, bidding and bid award, followed by construction. Oftentimes, delays further extend the project duration and may result in cost inflation as the time extends. Litigation and disputes are very common with this method of construction delivery because of dissonance between the expectations of the three parties—owners, designers, and contractors. In many cases "absence of litigation" is one very important performance indicator. Litigation can be an especially thorny issue for public organizations. An aggressive pursuit of claims against designers and contractors can be politically undesirable and unpleasant.

Advantages:

1. The design team is impartial and looks out for the interests of the owner.

2. The system treats potential bidders fairly and improves the owner's decision-making ability.

3. It assists the owner in establishing a reasonable cost for the project.

4. Completed projects generally meet acceptable quality levels.

Disadvantages:

1. Design-team failures may increase the cost of a DBB project and delay it.

2. There is increased risk for the general contractor and the possibility of a compromise in quality in order to lower the cost of the project.

3. As the general contractor is brought to the team postdesign, there is little opportunity for input on cost-effective alternatives.

4. Pressure may be exerted on the design and construction teams, which may lead to disputes between the architect and the general contractor.

Design–Build Projects

Design-build (DB) projects accelerate delivery through concurrent design and construction activities. As is typical of all types of projects, a DB project is conceptualized by the owner; planning is carried out based on the objectives to be met, and on the economic and technical feasibility of the project. Site acquisition may be implemented at any point before contract award, but is best done as early as possible to ensure that the design will not have to be aborted. Planning and schematic design are carried out by the owner's design professional, and may include infrastructure and foundation details for the project. This information allows construction to start shortly after contract award, while the design builder continues the preliminary design to obtain a final design. Typically, their design professional develops a preliminary design and cost and schedule proposals for the overall project. In some DB projects the owner may review proposals from a number of design builders and enter into a legally binding contract with one that provides the most appropriate proposal. The design builder is given access to the site and instructions to proceed, based on legally established time frames. This type of contract may also contain incentives for timely completion, as well as penalties for avoidable delays or cost overruns. The DB organization initiates construction while finalizing the detailed design. At intermediate checkpoints, verification is done by the parties to the project. The design culminates in the preparation of completed drawings and specifications that are used to complete the project. The owner or agents, such as architects/engineers or construction managers, monitor the progress of the construction, ensuring that approvals for interim payments match the progress of the construction work. At the completion of the construction there are acceptance inspections, leading to the commissioning of the facility for the owner's use (Figure 1.4).

Design-build has the potential to provide better quality, especially with regard to the following factors that are considered to be subsets of quality: (a) communication is simplified and accelerated, as the owner has one point of contact; (b) the adversarial nature of the three-way relationship in DBB projects is avoided; (c) conflicts over the intent of the specifications and their deployment are resolved internally; (d) the accelerated completion of DB projects lends itself to greater owner satisfaction; and (e) cost growth is minimized for the owner. In order to make this potential a reality, public owners have to act in a proactive manner to secure good quality construction from DB projects, as speed and cost containment can also bring unsatisfactory quality, if unchecked. For example, the designer does not have a professional/fiduciary relationship with the owner, and may, in some instances, be required to favor the builder's needs to meet bid price limitations over the owner's needs.

Advantages:

1. Design-build is generally the fastest project delivery system.

2. There is a single entity responsible for design and construction.

3. There is an early cost and scheduling commitment.

4. Conflicts between project professionals are internalized and may not involve the owner.

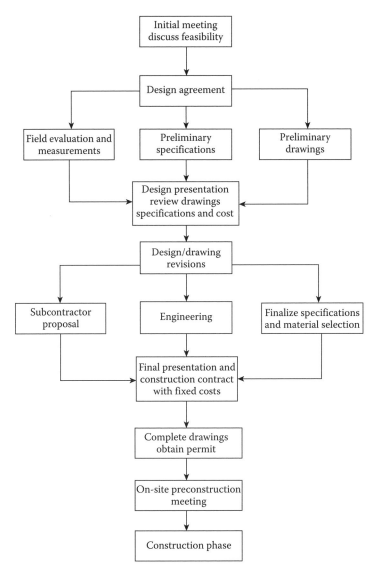

FIGURE 1.4
Process flow for a design-build project.

Disadvantages:

1. A design builder may provide reduced building features in order to protect profit margins.

2. Involvement of the owner is generally limited to the early stages of a project.

3. Hidden reductions in quality are possible when cost savings and design changes are determined by a design builder.

4. The designer does not represent the owner's interest, but is responsible instead to the contractor.

Engineer–Procure–Construct (EPC) Projects

These are configured in a manner very similar to DB projects. Most of the design and construction functions are performed or managed by one organization (Figure 1.5). This model, however, is used primarily for industrial projects that emphasize engineering design, as opposed to architectural design. The EPC projects typically have commissioning and maintenance phases included to allow for a plant to reach its designed operating capacity after acceptance.

Advantages:

1. The system has control over financial expenditures.
2. There is better communication between the owner, designer, and the contractor.

Disadvantages:

1. The system does not eliminate the need for an owner's representative.
2. The designs may be unimaginative, emphasizing cost over quality.
3. The postcontract variations are difficult to bring about.
4. The system is not suited for refurbishment work.

Design–Construction Management (CM) Contracts

The owner typically hires a CM organization, for a fee, to provide professional management services. Trade contractors contract directly with the owner on an individual basis and not through the construction manager, although the CM advises the owner on the formation and conduct of those contracts. The owner also contracts separately with a design concern (i.e., an architectural/engineering (A/E) firm) to obtain the design documents.

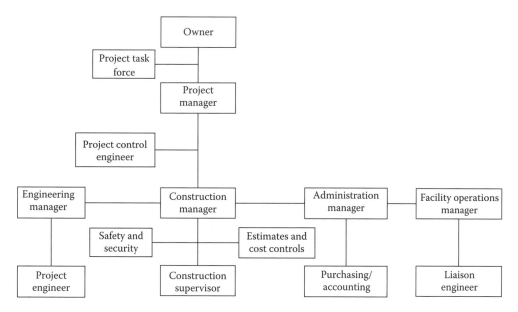

FIGURE 1.5
Organization chart for an engineer-procure-construct (EPC) project.

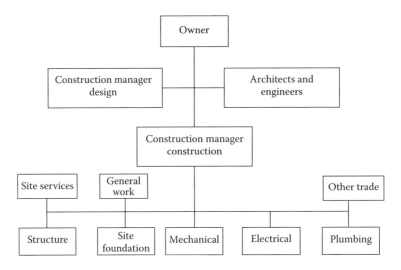

FIGURE 1.6
Organization chart for a design CM contract.

In some instances, the A/E firm may play the role of the CM. This form of contracting places a heavy responsibility on the owner to coordinate the work, as the trade contractors do not have contracts with each other and have no contractual obligation to cooperate (Figure 1.6).

Advantages

1. The CM agent, with construction expertise, provides the owner with an agent in addition to the designer to supervise the project. This reduces the owner's management burden in large or complicated projects.
2. The CM agent's project scheduling and capability to competitively fast-track some items may speed up the process and save money for the owner.
3. The CM agent's cost estimating and construction expertise at the design phase assist in monitoring construction costs.

Disadvantages:

1. The CM agent is an added cost. The owner is at risk for final construction cost; actual construction costs are not known until construction is complete.
2. Multiple prime contracts increase paperwork and administrative time, and increase the potential for construction disputes and claims.
3. The CM agent typically has less clout to resolve design or construction issues than a general contractor and serves only as a mediator.

Design–Agency CM Contracts

In this type of contract the owner hires a design team to prepare project construction documents, and also hires a construction manager (CM) to oversee the construction phase of the project. This is often done on the basis of a lump-sum or fixed-price contract. The

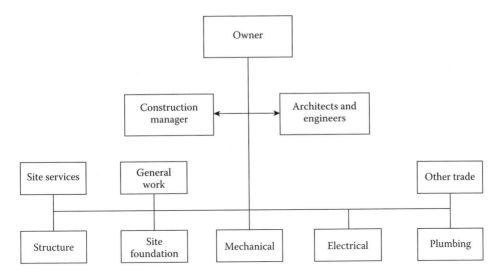

FIGURE 1.7
Organization chart for a design-agency CM contract.

CM may act as an agent of the owner, contracting directly with all the trade contractors. The CM prepares bid packages that are priced competitively by the trade contractors, and reviews these bids to select the most appropriate ones (Figure 1.7).

Advantages:

1. Fewer management resources are needed.
2. Participants' roles are clear and widely accepted.
3. The process is well established and universally understood.
4. This is a fixed-price contract based on complete documents with little room for change orders.

Disadvantages:

1. Contracts are awarded based on the low bid regardless of past performance.
2. Construction starts after design work is complete and approved.
3. Design quality may suffer from a lack of input from contractor(s) and sub(s).
4. The designers' fees may increase as change orders increase.
5. Changes in the scope (i.e., design, unforeseen conditions, timeframe, etc.) will generate change orders.

CM-at-Risk Contracts

The construction manager in this type of contract assumes the risk of pricing, and contracting directly with the respective trade contractors. In general, CM-type contracts are not as amenable to quality initiatives as DBB and DB contracts.

Construction Management at Risk (CM@Risk) is increasingly being used by the U.S. public sector. Like design-build, it facilitates improved quality through a selection process based on factors other than the low bid. A scoring system is used to consider the previous

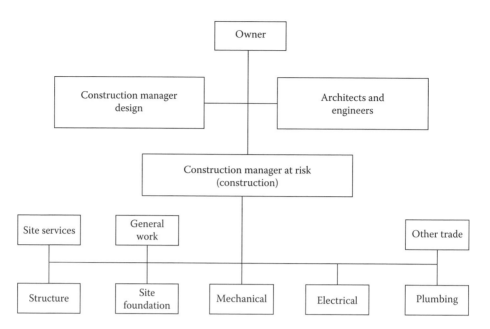

FIGURE 1.8
Organization chart for a CM-at-risk contract.

performance of a contractor, based on various criteria. It is not the cheapest method—it is best used where there is uncertainty, such as renovation projects where the current state of a facility or its infrastructure may not be entirely known. This uncertainty is reduced by having the CM involved in managing the design phases of a project, in the selection of sub or specialty contractors, and then assuming the risk for successful completion. Bidders tend to build in safety cushions for unforeseens, but a guaranteed maximum price (GMP) approach may be used to set a specific limit to the owner's project cost. Consequently, project budgets for CM@Risk are somewhat generous, resulting on less emphasis by contractors on cost reduction, and fewer compromises in the area of quality (Figure 1.8).

Advantages:

1. Early cost-commitment gives the owner project cost security.
2. The at-risk entity is responsible for managing the construction process and has more influence over subcontractors.
3. The CM contractor can reduce the owner's burden in the management of large or complicated projects.
4. The CM contractor reviews projects for constructability, cost, and schedule, potentially reducing change orders and delays.
5. Fast-tracking and multiple prime contracts may speed construction.

Disadvantages:

1. The management role of the CM contractor is an added cost.
2. The CM contractor may provide reduced building features than are available with a bid approach, in order to protect the margin of profit.

3. There is a potential conflict of interest with having one entity serve as both construction manager and contractor.

Fast-Track Construction

Fast-track construction is valuable in meeting accelerated schedules demanded by the owner. It allows a contractor to commence construction immediately after contract award, while a designer simultaneously completes the construction documents. It may be carried out with or without a design builder.

Advantages:

1. The most important advantage of the system is its conformance of the basic design.
2. The system also has better constructability compared to the other systems.
3. The system also requires demanding documentation and minimizes the cost of overruns.
4. The system increases productivity by speeding up the construction process.

Disadvantages:

1. The system can create misunderstanding between the owner, designer, and the contractor.
2. The system is prone to an increased number of errors from the designer.
3. To correct errors or to change to more advantageous designs requires more change orders than would be needed with standard construction scheduling.
4. In the fast-track process, incomplete drawings and specifications are incrementally released for bidding, governmental review, and construction.

Partnering

Partnering is the formation of a project team to deliver a construction project; the team commits to open communications in a spirit of trust, and works to accomplish mutual project goals. While the team members work supportively to meet mutual goals, they also focus on their individual goals. Yet, they recognize that their individual success is linked to the overall success of a project.

First-level partnering involves team formation for an individual project. On the other hand, second-level partnering involves members of a team that have worked together successfully on one or more projects and have come to recognize the benefits of a long-term relationship. Kubal (1994), points out that team members do not all have to be the same in order to have second-level partnering; the team may have a combination of participants from previous teams.

Partnering has resulted in projects that are far less reliant on litigation to ensure that project objectives are met. It significantly reduces the fragmentation that characterizes many construction projects, as a result of which the employees of the involved companies tend to work more on an individual level rather than an organizational level. The reduced reliance on legal resources often lowers project costs, to the benefit of both owners and design and construction providers.

The success of partnering activities has been linked to certain common factors:

1. Selecting a team based on competent member companies that can be mutually respected based on their respective track records.
2. Selecting a team with member companies whose top management will commit to a spirit of teamwork.
3. Conducting partnering sessions with an experienced facilitator who can work in a neutral capacity to conduct team-building exercises.
4. Clarification of project goals and performance requirements, and "buy-in" by the respective companies.
5. Operating procedures: communications, problem resolution, and documentation protocol.
6. Mutually agreed systems for measuring performance.

Project success is further enhanced by having employees of the involved companies view themselves as working with each other as opposed to viewing their interaction as only a relationship between companies.

Advantages:

1. The completed construction is of a higher quality standard.
2. Overall costs are lowered.
3. Profit margins for participants are higher than in traditional approaches.
4. There is less adversariness and legal costs are significantly reduced.
5. Errors and omissions are reduced by improved communication during both the design and construction phases.
6. There are fewer change orders.

Disadvantages:

1. Transactional exchanges between team members are guided by the construction contract while relational exchanges are not legally enforceable.
2. Ongoing team-building sessions are needed to maintain the benefits of partnering.
3. Unless all team members begin participation at the same time (at the preconstruction phase) they may not form a cohesive team.
4. Partnering cannot work effectively if projects are planned with unrealistic schedules, inadequate budgets, or with constructability problems.

Relational Contracting/Lean Design and Construction

Howell (2000) describes lean construction as a new way to design and build capital facilities. Lean theory, principles, and techniques jointly provide the foundation for a new form of project management. It uses production management techniques to make significant improvements particularly on complex, uncertain, and quick projects. Lean methods have reduced office construction costs by 25% within 18 months and schematic design time from 11 to 2 weeks.

Lean construction departs significantly from current project management practice. Processes are actively controlled and metrics are used in planning system performance to assure reliable workflow and predict project outcomes. Performance is optimized at the project level. Whereas current project management approaches reduce total performance by attempting to optimize each activity, lean construction succeeds by optimizing at the project level, as opposed to the less effective current project management approaches, which reduce total performance by attempting to optimize each activity.

Lean applications in design and construction are continually evolving. The most successful applications have been observed with forms of contract that reward cooperation and collaboration between the parties that are actively involved in delivering design and construction. The Integrated Form of Agreement (IFOA) is one form of contract that has been successfully applied to lean construction.

Figure 1.9 represents one interpretation of lean project delivery, namely Integrated Project Delivery (IPD). The team comprises the GC/CM working closely with designers and the primary subcontractors. The team members have relational contracts between them, as delineated by those entities encircled by the dotted line; these are described as "Single pact." The right-hand box (at-risk pool) "rescues" any individual firm that encounters financial difficulty; by the same token, team behavior is likely to leave a surplus in the account. This surplus is shared between the members of the integrated team.

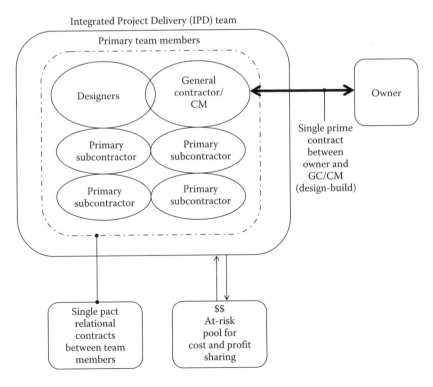

FIGURE 1.9
A representation of Integrated Project Delivery (IPD).

The foregoing concepts have been embodied in variations in delivery methods termed IPD and lean project delivery. This topic is discussed in detail in Chapters 3 through 7.

Forms of Contract

There are several categories of construction contracts that cover the spectrum of available approaches, all of which may be integrated with the foregoing construction delivery methods. Some categories of contract formats may be better suited to particular delivery methods, although the number of contract formats is almost infinite, just as there are no two projects exactly alike. The more frequently used formats are lump-sum contracting, GMP, GMP with cost-savings sharing, and cost-reimbursable contracts (cost-plus). The cost-plus compensation formats may include cost-plus with GMP, cost-plus with guaranteed maximum and incentive, and cost-plus with GMP and provision for escalation.

Fixed Lump-Sum Price

The contractor receives a fixed sum for completion of the contract. This fixed sum may be adjusted for such things as owner changes in the work, differing site conditions, suspensions or delays, defective specifications, and so on.

Guaranteed Maximum Price

The contractor provides cost estimates throughout the design process and, at some point prior to completion of the construction documents, submits a GMP proposal. The GMP proposal is based on a scope of work outlined in the preliminary design documents, and usually proposes that the contractor be reimbursed for the cost of the work, plus a predetermined fee with a maximum fee specified. Because of the incomplete nature of the design, contingencies and allowances are included in the GMP and the contractor is expected to build the job within the GMP limits, unless there is an owner-directed scope change.

Cost Plus a Fee

The contractor receives its actual "cost of the work plus a fee for overhead and profit." Definitions of cost (such as in AIA document A111) include the contractor's direct job-site costs. Agreement on cost definitions is essential to avoid disputes over allocation of items to reimbursable cost or to fee. Since an open "cost-plus" arrangement is in effect a blank check, cost controls, and audits are essential. Fees may be negotiated either as a fixed lump-sum or as a percentage of cost. The latter encourages the contractor to spend more to increase its fee.

Fixed Unit Prices

The contractor receives a fixed sum for each unit of work completed. Because the units are not fixed, contracts may provide for renegotiation of the unit-price for substantial variations in unit quantities from estimated quantities.

Advantages and Disadvantages of Different Forms of Contracts

Lump-sum contracting is often based on definitive specifications and requires complete, detailed design. Construction efficiency and quality are maximized in direct proportion with the availability of design detail. The overall time frames required are longest, as separate design and construction contracts and phases are involved. Lump-sum, based on preliminary specifications and complete general specifications, is often used for "turnkey" projects such as plant construction. Time is saved by concurrent design and construction, and the single party responsibility inherent in this format promotes efficient project execution.

Cost-plus contracts are appropriate where the scope of work does not have to be clearly defined (e.g., in major revamping of existing facilities), where technology is not well-defined or needs to be confidential. Minimum time schedules are generally obtained, but the owner/client needs to provide extensive engineering supervision and cost-control. Cost-plus with GMP usually involves at least preliminary drawings and general specifications. Fast time schedules are possible at the expense of high contract prices, due to the contractor's risk exposure. Cost-plus with guaranteed maximum and incentive leads to fast schedules and higher prices than fixed-price contracts, but encourages the contractor to pursue savings as they are shared between both parties.

Cost-plus with a guaranteed maximum and provision for escalation is generally used for long time schedules, where prices may increase substantially, and project definition is preliminary. Tight owner cost-control is needed. Time and materials contracts that are based on a general scope of work assure the contractor of a reasonable profit, reduce the scope definition/proposal time needed, but require extensive client supervision. Bonus/penalty, time, and completion clauses are used when timely completion is highly critical to the owner. The level of project definition provided greatly affects contract prices. The involved penalties impose high risk on the contractor, and quality is often reduced to meet time schedules. Bonus/penalty, operation, and performance contracts are typically used in process plant construction to guarantee successful plant operation.

Unit-price contracts (flat rate) are best used for repetitious/homogenous tasks such as highway building, or gas transmission piping. Work can proceed effectively even if the parties do not initially know the precise quantities of labor and materials required for the involved work. Unit-price contracts (sliding rate) are appropriate for the foregoing projects, but require extensive client field supervision to ensure that the involved quantities are properly monitored.

Strategies for Improving Construction Performance

Subsequent chapters address lean and productive practices to improve construction performance. These practices include a new paradigm for construction that includes a strategic management approach to focus the parties on optimizing performance at the project level, instead of seeking their own self-interest. These practices include the application of performance improvement and industrial engineering techniques, including the lean methods that have been successfully applied in the manufacturing sector.

The lean methods are based on a participative approach such as relational contracting and lean project delivery to improve the interaction between the members of the project team. Lean is the ultimate in collaboration and cooperation. In relational contracting, companies in the project team put the project first and gains and losses are shared.

Questions for Discussion

1. What are the primary causes of low construction productivity?
2. Discuss the implications for construction quality when comparing the Design-bid-build method of project delivery with DB.
3. What are the factors that constitute construction performance?
4. What advantage does partnering offer in comparison with the traditional DBB method of project delivery?
5. How does optimization of work performance in lean construction differ from optimization in traditional construction?

References

Bureau of Labor Statistics. 2006. *Fatal and non-fatal occupational statistic from 1992–2004*. Online at http://www.bls.gov.

Egan, Sir John. 1998. *Rethinking construction: the report of the Construction Task Force*. [The Egan Report] Department of Environment, Transport and Regions and HMSO, London, UK.

Forbes, L. 1999. *An engineering management-based investigation of owner satisfaction, quality and performance variables in health care facilities construction. Dissertation*. Miami: University of Miami.

Howell, G. A. 2000. White paper for Berkeley/Stanford CE & M Workshop. *Proceedings, construction engineering and management workshop*. Palo Alto, CA: Stanford University.

Kubal, M. T. 1994. *Engineered quality in construction—Partnering and TQM*. New York: McGraw-Hill Inc.

Proverbs, Holt, and Cheok, 2002. Summary of J. Egan, Report, *Rethinking Construction*.

Construction Industry Cost Effectiveness Task Force. 1983. More Construction for the Money. *Summary report of the Construction Industry Cost Effectiveness Project, Business Roundtable*, 11, USA.

2

Productivity and Performance Measurement in Construction

An understanding of construction productivity and performance is essential for deploying improvement strategies successfully. As in any other enterprise, the principle of continuous improvement involves learning from current or past activity and using that knowledge to improve future activities. That knowledge transfer works best when enabled with the PDCA cycle (plan, do, check, and act), which in turn is highly dependent on measurement (i.e., a quantification of accomplishment). Productivity and performance measurement provide a foundation for improving design and construction delivery, regardless of the methods utilized in each respective project. This foundation is especially helpful with lean construction methods, as they are based on a culture of learning and continuous improvement.

Definition of Productivity

Productivity is the measure of how well resources are brought together in organizations and utilized for accomplishing a set of goals. Productivity reaches the highest level of performance with the least expenditure of resources. Productivity is measured as the ratio of outputs to inputs. In the construction environment it may be represented as the constant-in-place value divided by inputs such as the dollar value of material and labor. Productivity measurements may be used to evaluate the effectiveness of using supervision, labor, equipment, materials, and so on to produce a building or structure at the lowest feasible cost.

The Bureau of Labor Statistics (BLS) defines productivity as a ratio relating output (goods and services) to one of the inputs (labor, capital, energy, etc.), that are associated with that output. There are four classical factors: land, materials, capital, and labor.

Overriding these four factors is a fifth factor; namely, technology. In the manufacturing industry productivity is usually described in quantitative terms, but in the construction industry it is usually understood in qualitative terms. Consequently, construction productivity cannot be analyzed scientifically at the working level. For productivity analysis it is essential to identify the significant productivity factors, quantify them, and establish the relationship of these factors with each other. The work situation in construction is project-oriented and much more dynamic than that in the manufacturing industry, introducing many complex factors affecting productivity in construction. These factors involve: managerial, technological, regulatory, labor related, engineering, and craft-related items.

Importance of Productivity

In 2006, new-construction-put-in-place accounted for roughly 8.0% of the gross domestic product (GDP). In addition, over 7.6 million people were employed in the U.S. construction industry in 2007. They were involved in design, new construction, renovation construction, equipment and materials manufacturing, and supply, thus making the design and construction industry the largest manufacturing industry in the United States. Consequently, it has a great impact on the state of the economy.

Productivity is an important component of the overall concept of performance. Although there are differences of opinion on what constitutes performance, several definitions refer to it as a combination of productivity, quality, timeliness, budget adherence, and safety. Productivity is primarily measured by cost. Satisfactory productivity is work accomplished at a fair price to the owner and with reasonable profit for the contractor. The other factors are:

- Quality: conformance to specifications, satisfying owner's needs
- Timeliness: on schedule, everything on hand when needed
- Safety: accident-free projects
- Budget adherence: completion within the contract price

Traditional construction project management tools do not address productivity; they include schedule slippages and cost overruns. The construction industry as a whole measures performance in terms of completion on time, completion within budget, and meeting construction codes. Quality is defined very subjectively and by default is considered to be satisfactory if construction codes are met.

Many contractors assume that it is impossible to meet desirable benchmarks in all four factors simultaneously and that there is a zero-sum relationship between them. In other words, an accelerated schedule results in an increased cost and lower levels of safety and quality. By the same token, higher quality is seen as requiring a higher cost and a longer schedule, while safety levels may remain unchanged.

Thomas and colleagues (1990) define performance in seven dimensions: effectiveness, efficiency, productivity, profitability, innovation, quality of work life, and quality. They emphasize four of the dimensions: effectiveness, efficiency, productivity, and quality. Ballard and Howell (1994) defined performance from the perspective of lean construction as percentage of assignments completed (PAC) for construction work planned for execution in the field.

Construction organizations (designers and constructors) would benefit significantly by establishing formal performance improvement programs that use quantitative approaches to improve productivity and quality. These quantitative approaches are critical—they involve applying measurement techniques to construction processes; these measurements are an essential ingredient for process improvement. As the saying goes "you cannot improve what you cannot measure."

At the macro level, productivity measures provide a consistent reference (outputs divided by inputs) that facilitates comparison between all construction project types, as well as other industries. Contractors need a tool to drive performance improvement through internal or external benchmarking. Measurement may not directly lead to performance improvement; however, performance improvement over time can be achieved through recognizing the need for improvement.

Construction productivity is a major concern, especially when compared to other industries. Nationally, productivity in the construction sector has not kept pace with other industries. At the firm level, it has a direct impact on profitability. As reported by the U.S. Department of Commerce, construction productivity has been rising at a much slower rate than other industries; between 1990 and 2000 it rose by approximately 0.8% compared to more than 2% for all U.S. industries. However, productivity in the construction industry is largely unmeasured, and those measures that do exist are contradictory and conflicting. The BLS does not maintain an official productivity index for the construction industry; it is the only exception to the industries that are tracked.

Productivity Trends in the United States

Construction productivity trends carry immense consequences for the economy as a whole. According to the BLS, during the past 40 years, the productivity of non-farm industries has increased by over 100%, while that of construction industry has been stagnating. Construction costs have been increasing at the same time. Raw materials such as steel, copper, and cement have been rising, especially in the face of escalating global demand. Labor costs are a major component of most construction projects-in the vicinity of 40%, yet on many construction sites a large percentage of the daily labor hours are unproductive. While it is difficult to have absolute measurements of construction productivity, it is still a concern in light of how such measurements are used for other industries.

Two independent methodologies demonstrate that total construction productivity increased between 1966 and 2003, on the order of 33%, or 0.78% per year. Figure 2.1 shows one estimate of productivity growth between 1964 and 2003. This level of productivity

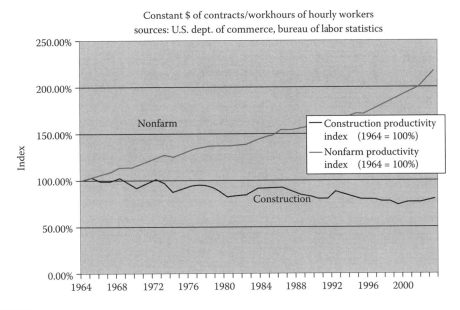

FIGURE 2.1
U.S. productivity gap (1964–2003). (From Paul Teicholz, PhD, Professor (Research) Emeritus, Department of Civil and Environmental Engineering, Stanford University. With permission.)

improvement, however, is less than one-half that of U.S. non-agricultural productivity gains during the same period, which averaged 1.75% annually between 1966 and 2003. Some of this differential in measurement is due to the fact that construction output is affected by the conditions at the job site at that time, and the conditions are continually changing. Construction productivity measurement is complicated due to the heterogeneous conditions in the industry and the varying nature of the output or product, while other industrial products are factory-based and assembly line produced.

Lean Construction: Impact on Productivity

The application of lean thinking to construction has concentrated on workflow throughput in the construction process. Variability in the flow of work often results in longer cycle times and reduces system throughput by increasing the amount of waste in a process. High levels of variability in labor productivity have a negative impact on project performance. Consequently, lean improvement initiatives can best improve project performance by reducing variability in labor productivity instead of output.

Unlike manufacturing, construction is a project-based production process that is conducted mostly outdoors as opposed to the controlled environment of manufacturing facilities. Lean construction attempts to manage and improve construction processes with minimum cost and maximum value by considering customer needs (Koskela and Howell 2002).

Variability is common on construction projects and must be managed effectively; lean production principles suggest that better labor and cost performance can be achieved by reducing output variability. According to Koskela, variability in the flow of work often extends cycle times and reduces system throughput by increasing the amount of waste (i.e., non-value-adding activity in a process).

Studies of a number of projects have identified various causes of below-budget labor productivity as:

- Overstaffing
- Interference with other crews
- Workforce management
- Insufficient work to perform
- Weather
- Equipment
- Design error
- Rework
- Material shortages

Potential for Productivity Improvement

The foregoing causes clearly need to be addressed to improve labor productivity. Activity sampling studies of construction operations in the field have shown that the working portion of activities generally consumes 40–60% of labor hours, and by the same token 40–60% of labor hours are unproductive. There are many reasons for lost time—poor communications, waiting on assignments, waiting on resources, double material handling, rework, accidents, late or inaccurate job status reports, lack of supervision, and so on. One-third of these losses reflect issues that are within management's control. Construction profitability is directly linked to labor productivity. Industry-wide studies suggest that most construction projects yield net profits of 2–3% of the total project cost. The calculations below are broad approximations, but they illustrate the potential impact of performance improvement initiatives.

A hypothetical example:

Contract price	= $10,000,000
Labor cost (40%)	= $4,000,000
Other costs, materials, overheads, etc.	= $5,700,000
Net profit	= $300,000

Assuming a 5% reduction in labor cost due to productivity improvement:

Savings in labor cost = $4,000,000 × 0.05 = $200,000
Revised net profit = $300,000 + $200,000 = $500,000.

Hence, a 5% improvement in labor productivity can improve profitability by 66.7%. Similarly, the value of lost labor hours due to management inefficiencies

= $4,000,000 × 1/3 = $1,333,333.

A 50% reduction in management-based losses would save $1,333,333/2 = $666,667:

Revised net profit = $300,000 + $666,667 = $966,667.

A 50% improvement in labor deployment would improve profits by

967 × 100%/300 = 322%.

Further gains can be derived by addressing other construction processes and supply chain management issues.

Recent surveys of the top 400 U.S. contractors have indicated that cost control, scheduling, design practices, labor training, and quality control are perceived as having considerable room for productivity improvement. Identified as needing more improvement compared with previous surveys were prefabrication, new materials, value engineering, specifications, labor availability, labor training, and quality control, whereas those needing less improvement were field inspection and labor contract agreements. Also, respondents have indicated consistently over the years that they are willing to participate in activities related to improving construction productivity, but they are not interested in funding any such activities.

Unit labor costs in constant dollars and daily output factors were compared over decades for each task. Direct work rate data from 72 projects in Austin, Texas over a 25-year period were also examined. Increasing the direct work rate usually increases construction productivity. Depressed real wages and technological advances appear to be the two greatest reasons for an increase in productivity. The data also indicate that management practices were not a leading contributor to construction productivity changes over time. Subsequent studies are required to add weight to these observations and can be based on the approach presented here. Labor productivity is of central importance to the economic health of the United States economy as well as that of other countries. Due to the size of the construction industry, productivity changes within it have significant direct effects on the national productivity and economic well-being of the United States.

Factors Affecting Construction Productivity

Several factors have traditionally limited productivity growth in the construction industry and they continue to affect it. They have also inhibited the adoption of quality initiatives; several are listed next (Forbes and Golomski 2001).

Ineffective Management Practices

As described in Chapter 1, studies have found that poor management was responsible for over half of the time wasted on a job site. Good management is required for profitability and project success. Ineffective management has been cited as the primary cause of poor productivity rather than an unmotivated and unskilled workforce (Sanvido 1988). Consequently, there has been significant research on making management more effective in supporting workers in the field.

Four primary ways of increasing productivity through management include (1) planning, (2) resource supply and control, (3) supply of information and feedback, and (4) selection of the right people to control certain functions.

Industry attitudes. Owners have not specifically demanded productivity and quality, and contracts have historically implied that code compliance is an adequate standard. There is a general lack of productivity/quality awareness in the industry among all parties, including owners. Owners have come to accept industry pricing, which accommodates the inefficiencies of design and construction providers. This pricing responds to marketplace forces—they have not been able to influence the productivity of the industry—prices have simply become higher on a per unit basis. By contrast, manufacturing activities have become cheaper over time.

Focus on Inspection

Code enforcement representatives of government agencies carry out inspection. Their role is to inspect critical aspects of the construction process by limited inspections on a number of items including reinforcing elements and concrete samples—but not workmanship.

Growth of Specialization

In an earlier twentieth-century era the concept of "Master Builder" placed designers with the responsibility to determine the details and methods used for translating the design to the finished construction. Subsequently, design and construction roles became clearly separated from a legal perspective; designers deferred the responsibility for interpretation to builders. In the present era, contractors prepare shop drawings to represent this interpretation. Consequently, there is not a great emphasis on the constructability of designs. In effect, designers are not usually involved with decisions on construction methods. Except in such situations as design/build projects, designers do not have an opportunity to design the construction process.

Unclear Performance Standards

Architect/engineer contracts are said to be unclear with respect to professional standards of performance, often leading to unmet expectations. Construction owners feel that typical A/E contracts protect designers at the owner's expense. For example, prevailing contract language relieves designers of any liability in the case of a lawsuit or arbitration between an owner and contractor. An outgrowth of this is the practice of "substantial completion" where a job is usable but has 5% of the remaining work in the form of a "punch list." An owner often has a very difficult time in persuading a contractor to finish that work.

Growing Emergence of Subcontracting

The subcontractors are often priced in a manner that does not reflect the contract with the owner—even if the owner pays a high price, the subcontractor may still have to work with inadequate budgets, often compromising productivity and quality as a result. As an increasing percentage of work tasks is subcontracted, the fragmentation of the project team is proportionally increased. In traditional project-delivery methods, the tendency of these subcontractors to seek local optimization increasingly jeopardizes attempts at project-level optimization.

Slow Adoption of Innovation

Small contractors often lack the expertise or financial resources to adopt technological advances—adoption is inhibited further by fear and uncertainty. Roofing contractors, for example, tend to use the same time-honored methods to ensure that supplies and equipment are on site each day. Items that are frequently forgotten are delivered by expediters. Few large companies, and virtually no small companies have implemented the concept of a quality or productivity manager to track performance.

Lack of Benchmarking

There is little, if any, benchmarking—many manufacturers and service organizations have become preeminent by adopting the best practices of benchmarked organizations. Construction has done very little of this due to distrust, fear of losing competitive advantage, but more likely, simply by being anachronistic.

The concept of construction performance does not emphasize productivity and quality initiatives. The work of many researchers has revealed an industry tendency to measure performance in terms of the following: completion on time, completion within budget, and meeting construction codes. Very little attention has been directed to owner satisfaction as a performance measure.

Crisis Orientation

The construction industry has been resistant to change until major events prompt it to change. Significant changes have been sparked primarily by catastrophes of one kind or another. For example, major revisions were made in U.S. engineering codes after the failure of a structure in the Kansas City Hyatt Regency Hotel. Hurricane Hugo caused major damage to buildings in South Carolina in 1989, resulting in changes to building codes and inspection procedures. Hurricane Andrew devastated Dade County, Florida in August 1992, resulting in a major scrutiny of building codes and their enforcement.

The industry is highly fragmented, both vertically and horizontally. Industrial manufacturers have historically been motivated to invest in productivity and quality-related research and development activity as they benefit from it. On the other hand, in the fragmented construction industry, there is comparatively little research because architects and engineers have neither the resources nor incentive to fund research. Constructors have made few attempts to influence innovation in architectural, engineering, or product design.

Labor Shortages

There is a secular downward trend in the availability of skilled labor in the industry. The traditional sources of workers—vocational school programs and apprenticeships—have not been keeping up with the demand. In the area of technicians and equipment operators alone there is a projected shortage. Studies in a number of U.S. states have also pointed to a shortage of skilled labor that is not being met through the public education system or through union-based programs. In the so-called building boom (2000–2006) much of the labor requirements in the United States were met with migrant labor from Central and South America. Although the latter part of the first decade of the millennium is characterized by economic contraction, a shortage of skilled labor is likely to be visible again once market conditions improve.

Project Uniqueness

Projects in construction are never designed or built in exactly the same manner as previous projects. Environmental factors such as landscape, weather, and physical location force every project to be unique from its predecessors. Aesthetic factors create uniqueness from project to project, and have a significant impact on major project characteristics. Whereas most construction personnel find this uniqueness to be an attractive element for a career in construction, it can have an adverse effect on construction productivity. Project uniqueness requires modifications in the construction processes. These modifications require workers to go through a learning curve at the beginning stages of each project activity.

Technology Impacts

Despite a decrease in industry level measures in construction productivity, there has been a steady increase in construction productivity at the activity level. Equipment technology has been one factor that may explain that increase. Studies of improvements in equipment technology and partial factor productivity point to a positive correlation between those factors. In one study five technology factors—energy, control, functional range, information processing, and ergonomics—were examined for 200 activities over a 22-year time period. Through ANOVA and regression analyses, it was found the activities that experienced a significant change in equipment technology also witnessed substantially greater long-term improvements in partial factor productivity than those that did not experience a change.

Technology has had a tremendous effect on overall productivity. All but the most basic of tasks on a site have seen changes due to advances in technology over recent years. Tools and machinery have increased both in power and complexity. These advances in technology can significantly modify skill requirements and make it difficult to isolate the contributions of technology, management, and labor to productivity. Introducing new technology can be more difficult in the construction industry than in other industries. Innovation barriers such as diverse standards, industry fragmentation, business cycles, risk aversion, and so on can create an inhospitable climate for innovations. In many regions of the United States, labor costs for many skills are relatively low. There is less motivation to automate a task when the labor associated with it is not expensive. Due to such impediments, firms are naturally reluctant to try a new technology, particularly if it amounts to putting the entire company on the line. Should the new technology prove effective, the firm gains only a temporary strategic advantage as other firms are likely to follow suit.

Real Wage Trends

In recent years, productivity gains have continued while pay increases have not kept pace. Worker productivity rose 16.6% from 2000 to 2005, while total compensation for the median worker rose 7.2%, according to Labor Department statistics analyzed by the Economic Policy Institute, a liberal research group. Benefits accounted for most of the increase. Real wages have fallen in the construction industry at a more rapid pace over the past 30 years than have wages for most American workers. An increasing percentage of open and merit shop work partially drove this downward trend. It is estimated that union labor declined from approximately 70% of the construction work force in the 1970s to 20% in the 1990s.

Inadequate Construction Training

There is currently a lack of formal training in construction—the lowest of any major sector of the economy. This lack of training is due to practical concerns such as employers completing the increased percentage of non-union work. In general, the work force of contractors is highly mobile. For this reason, contractors are often reticent to invest capital in training those who may soon be someone else's employees. The result may be a decrease in the construction workforce average capability level. It is unclear how this affects productivity. More effective utilization of large narrow skilled and core multiskilled workforces may even result in higher productivity on some projects.

Labor organization through cross training and multiple skills can reduce unit labor costs. Contracts that create flexible work rules on the job site promise productivity benefits as well. Barriers between trades have historically been a source of problems in construction. Reduction in the percentage of the workforce composed of organized labor and improved project agreements with remaining construction labor organizations have reduced this problem.

Productivity Ratios

Productivity measurements may be applied at different levels, ranging from an individual work task to a firm, an industry, or a country. In the construction environment, productivity measurements evaluate the effectiveness with which management skills, workers, materials, equipment, tools, and working space are used to produce a completed building, plant, structure, or other facility at the lowest feasible cost.

Partial productivity is the ratio of output to one class of input. For example, material productivity is the ratio of output to the associated material input.

Similarly, labor productivity is the ratio of output to labor input, and is a partial productivity measure. These ratios can be configured in different ways; in physical terms, labor productivity for roofing may be calculated as the ratio of x square feet of roofing/y labor hours. Productivity may also be measured in financial terms, such as dollar value of roofing/dollar value of associated labor.

Value added productivity is based on output produced that adds value.

Total Productivity

Total productivity is a ratio of output to all inputs. All input resources are factored in this principle. Tracking the productivity changes that occur at different time periods is the most useful application of total productivity. Sumanth (1984) points to the limitations of partial productivity measures, which are measured by the ratio of output to one class of input such as labor productivity. Such measures—if used alone can be misleading—do not have the ability to explain overall cost increases, and tend to shift blame to the wrong areas of management control.

Total productivity (TP) may be defined as:

$$TP = \frac{T(s) \ (\text{total sales or value of work})}{\substack{\text{labor cost } (M_1) + \text{materials cost } (M_2) + \text{machinery cost } (M_3) + \\ \text{money cost } (M_4) + \text{management cost } (M_5) + \text{technology cost } (M_6)}}$$

or

$$TP = T(s)/M_1 + M_2 + M_3 + M_4 + M_5 + M_6. \tag{2.1}$$

Since

$$P_i = T(s)/M_i$$

$$Pt = \frac{1}{1/P_1 + 1/P_2 + 1/P_3 + 1/P_4 + 1/P_5 + 1/P_6}. \tag{2.2}$$

The above-mentioned factors are expressed as constant dollars (or other currency) for a reference period. To increase total productivity, it is necessary to determine what partial productivity factor (P_i) has the greatest short-term and long-term potential effect on total productivity.

The terms productivity, effectiveness, and efficiency may be combined as follows:

$$\frac{\text{Productivity}}{\text{Index}} = \frac{\text{output obtained}}{\text{input supplied}}$$

$$= \frac{\text{performance achieved}}{\text{resources consumed}} \tag{2.3}$$

$$= \frac{\text{effectiveness}}{\text{efficiency}}$$

Therefore, productivity is the combination of effectiveness and efficiency. To increase productivity, the ratio(s) mentioned in Equation 2.1 must increase. This can be achieved by increasing the output, reducing the input, or permitting changes in both such that the rate of increase in output is greater than that for input. An increase in productivity can be achieved in five ways as follows (Gerald 1997):

(1) Reduced costs: $\dfrac{\text{output at same level}}{\text{input decreasing}}$

(2) Managed growth: $\dfrac{\text{output increasing}}{\text{input increasing (slower)}}$

(3) Reengineering $\dfrac{\text{output increasing}}{\text{input constant}}$

(4) Paring-down $\dfrac{\text{output down}}{\text{input down (faster)}}$

(5) Effective working: $\dfrac{\text{output increasing}}{\text{input decreasing}}$

These strategies often have unintended consequences. Cost reduction usually involves decreasing expenses, services, training, and advertisement (Forbes, Ahmed, and Azhar 2001). Management generally views people as a direct expense. Unless the respective costs are prioritized against the organization's objectives, this method may result in the wrong identification of cost-cutting items and can become more defensive and contraction-oriented.

Growth management may involve an investment that yields a greater return than the cost of investment. It may be in the form of capital, technological improvements, system redesigns, training and/or organizational restructuring. Increased productivity may also be achieved by reducing the cost of production inputs through a reengineering of the design and production processes. Paring down means reducing both output and input amounts, with the input diminishing at a higher rate. This "sloughing off" seeks to improve marginally unproductive facilities. If an organization has a number of production facilities that vary in productivity, it would progressively eliminate the use of low-producing facilities.

Working effectively involves a total output increase with a decrease in input. This can be accomplished through a combination of changes in work and management procedures and through proper training of workers. It also requires the motivation of all involved to "produce more with less" and perform to high-quality standards. Simply producing more may be detrimental to an organization if the increased outputs do not meet the quality expectations of its customers. Crosby (1979) points to the high cost of not doing jobs right the first time.

Construction Productivity Measurement

Although several publications exist on the subject of construction productivity, there is neither an agreed upon definition of work activities nor a standard productivity measurement system. Researchers have concluded that it is difficult to obtain a standard method to measure construction labor productivity because of project complexity and the unique characteristics of construction projects. The uniqueness and nonrepetitive operations of construction projects make it difficult to develop a standard productivity definition and measure.

Construction Progress Measurement

Several methods are proposed for measuring construction performance that include a productivity component: units completed, incremental milestone, start/finish, supervisor opinion, cost ratio, weighted unit, and earned value.

Units Completed

This method assumes that construction progresses linearly. If an activity requires 1000 cubic yards of concrete and 300 have been poured, then the percentage of completion is calculated as 300/1000 = 0.3 or 30%. This method falls short as the assumption of linearity does not hold true in many project types.

Incremental Milestone

Percentage of completion is calculated based on a construction organization's experience. A database that catalogs previous projects can provide a reasonable estimate of job progress. For example:

Description	Completion by Activity
Foundation work:	
Excavation	5%
Form work	50%
Place rebars	20%
Pour and finish	10%
Strip forms/finish	10%
Backfill	5%
Total	100%

If work on a building foundation has progressed as far as the placing of rebars, then the percentage in completion of the foundation milestone would be (5% + 50% + 20%) = 75%.

　　Other milestones can be evaluated by the same method (e.g., for piping, structural steel, etc.). Productivity calculations can be developed by comparing the labor hours utilized with the projected/budgeted labor hours.

Start/Finish

The status of activities can be estimated based on their initiation and completion. This method is suitable for activities that have a shorter duration than the reporting period. The incremental milestone method is convenient when activities have clear sequential sub-tasks. However, when an activity does not have clearly defined subtasks or they are not entirely sequential, the start/finish method is an effective way to measure project performance. For example, the work involved in painting a building exterior can be measured in stages using this method:

- Work started: 10% completed
- Building pressure cleaned and prepped: 25% completed
- Main color applied: 65%
- Trim and details applied: 95%
- Clean-up and project completion: 100%

Although similar to the incremental milestone method, the primary difference is that some tasks can be started before other tasks are completed.

Supervisor Opinion

Experienced supervisors have a wealth of knowledge that can provide reasonable estimates for certain categories of jobs that can be readily visualized. The most subjective method of measuring performance is opinion. This method is highly biased by the personality and benefit of the estimator. Some supervisors are overly optimistic and may provide inflated performance measurements while a company engineer may be overly critical of the current progress. Ideally, this method should be used for minor tasks. Examples of these jobs would be: painting, landscaping, dewatering, and installing architectural trim.

Cost Ratio

The ratio of costs or work hours expended to the forecasted values can be used for estimating progress on construction management or administration activities. Examples of appropriate activities are quality assurance, construction administration, and project management. This method is not reliable for work-in-place activities.

$$\% \text{ Complete} = \frac{\% \text{ actual cost of work to date}}{\text{forecasted cost at completion}}.$$

The cost incurred to date can also be used to estimate the work progress. For example, if an activity was budgeted to cost $20,000 and the cost incurred at a particular date was $10,000, then the estimated percentage complete under the cost ratio method would be 10,000/20,000 = 0.5 or 50%. This method provides no independent information on the actual percentage completed or any possible errors in the activity budget: The cost forecast will always be the budgeted amount. Good judgment is required especially when dealing with assemblies that have a large portion of the materials in the warehouse but no assembly has been started.

TABLE 2.1

Earned Value Analysis (Simplified Structure)

Weight	Subtask	Unit	Quantity Total	Equivalent Steel Tons	Quantity To Date	Earned Tons
0.02	Foundation bolts	each	200	1	200	1
0.28	Trusses	each	100	14	50	7
0.30	Columns	each	30	15	20	10
0.20	Beams	each	50	10	30	6
0.20	Roofing-steel deck	square foot	5000	10	1000	2
1.0		steel	ton	50		26

Note: Percent complete = Earned tons to date/equivalent tons = 26/50 = 52%.

Weighted or Equivalent Units

This method calculates percentage of completion based on the concept of "earned value" (Table 2.1). In the case of erecting a steel building, subtasks are defined that are quantified by the unit of measure.

Earned Value

Earned value management (EVM) is a project management technique for measuring progress in an objective manner. EVM has the unique ability to combine measurements of scope, schedule, and cost in a single-integrated system. When properly applied, EVM provides an early warning of performance problems. Additionally, EVM promises to improve the definition of project scope, prevent scope creep, communicate objective progress to management, and keep the project team focused on achieving progress. The earned value concept uses worker hours or dollars to compare a given activity with others (Griffis, Farr, and Morris 2000). The foregoing methods can be used to determine the percentage of completion for each activity. Multiplying estimated completion for each activity by its budgeted values provides earned value.

Percent complete = (Earned work hours or dollars on all accounts)/budgeted work hour or dollars.

Earned Value Management (EVM) Application

Essential features of an EVM application include:

1. A project plan that identifies work to be accomplished.
2. A valuation of planned work, called planned value (**PV**) or budgeted cost of work scheduled (**BCWS**).
3. Predefined "earning rules" (also called metrics) to quantify the accomplishment of work, called earned value (**EV**) or budgeted cost of work performed (**BCWP**).

The EVM application in large or complex projects includes additional features, such as indicators and forecasts of cost performance (over or under budget) and schedule performance (behind or ahead of schedule). The most basic requirement of an EVM system, however, is that it quantifies progress using PV and EV. Definitions of these are:

- BCWS (PV): Budgeted cost of work scheduled (planned value): Sum of the budgeted costs for all planned work scheduled to be completed to-date or on any given date.
- BCWP (EV): Budgeted cost of work performed (earned value): Total of the budgeted costs of the activities (that make up a job or project) completed to-date or on any given date; also called achieved cost or earned value.
- ACWP (AC): Actual cost of work performed: Actual cost incurred for the work performed in a given period, including all labor, materials, other direct costs, overheads, and so on.
- BAC: Budget at completion: The original projected cost of the project.
- CV: Cost variance: Difference between an actual cost and the associated budgeted or estimated cost.

$$CV = BCWP - ACWP$$

- SV: Schedule variance: Any deviation from the baseline plan of a project, measured by comparing budgeted cost of work scheduled with budgeted cost of work performed.

$$SV = BCWP - BCWS$$

- CPI: Cost performance index: A measure of cost efficiency on a project. It is the ratio of earned value (EV) to actual cost of work performed (ACWP). A value of 1.0 means the project is on budget. A value greater than 1.0 means the project is under budget, while a value less than 1.0 means the project is over budget.

$$CPI = BCWP/ACWP$$

- SPI: Schedule performance index: A measure of schedule efficiency on a project. It is the ratio of earned value (EV) to planned value (PV). A value of 1.0 means the project is on schedule. A value greater than 1.0 means the project is ahead of schedule while a value less than 1.0 means the project is behind schedule.

$$SPI = BCWP/BCWS$$

Productivity Estimation Based on Worth

Alfeld (1988) equates performance (worth) to the ratio of accomplishment (value) to methods (cost). Finished work is regarded as an accomplishment and a construction technique is a method. Performance is a combination of methods and accomplishment. Worthy performance occurs when the value of accomplishment is greater than the cost of methods used.

Alfeld developed another measure of performance, i.e., the "performance ability ratio" (PAR) as the ratio of exemplar performance to current performance. Exemplar performance is the historically best instance in which the value of accomplishment exceeds the cost of the methods. Performance ability ratio (PAR) is calculated as exemplar performance/current performance. As the PAR value exceeds the ideal of 1.0, the need for improvement increases. Selecting an exemplar rate the question arises: Whose exemplar should be used?

1. This job's?
2. The company's best?
3. The industry's best?

Any of these measures can be appropriate. In most cases, construction organizations prefer to set the exemplar as their historical best performance. In pricing projects, this approach can provide a competitive advantage; bids that are based on industry norms may fail to distinguish a company from other bidders. This approach is valid only if the company's historical performance is better than the industry's performance.

Computing PAR Values

Tables 2.2 and 2.3 provide examples of calculations to establish PAR values. In Table 2.2 productivity is calculated for installing gypsum board (drywall) of different thicknesses and in different configurations. For reference purposes, this calculation is built on data from the *RS Means Estimating Manual*. It is based on a crew of two carpenters working for an 8-hour day. The productivity values are calculated as the ratio of square feet of gypsum board installed to the number of worker hours required. The productivity values derived are used as a starting point in productivity improvement by using them as exemplar rates; that is, the best rate for performing each work task.

Table 2.3 shows calculations of work productivity for each of the four reference codes in Table 2.2. Productivity values are determined in column C by dividing production by work hours. Those values are compared with the exemplar values in Table 2.3 to calculate the PAR for each work activity.

TABLE 2.2

Productivity Calculations for Gypsum Board Installation

Reference Code	Gypsum Board Installation (RS Means)	Crew Size (Carpenters)	Work Hours Per Day	Production (Square Feet Per Day)	Exemplar Rate Productivity (Square Feet Per Work Hour)
0150	3/8" wall installation (standard) unfinished	2	16	2000	125
0200	3/8" ceiling installation unfinished	2	16	1800	112.5
0350	½" walls (standard)— taped and finished	2	16	965	60.31
1050	½" ceilings—taped and finished	2	16	765	47.81

TABLE 2.3

PAR Value Calculations for Gypsum Board Installation

Reference Code	Work Activity: Gypsum Board Field Installation	A: Production (Square Feet) Observed	B: Work Hours	C = A/B: Productivity Square feet Per Work Hour	D: Exemplar Rate	E = D/C: PAR Values	F: Priority for Action
0150	3/8″ wall installation (standard) unfinished	2105	32	65.78	125	1.9	1
0200	3/8″ ceiling installation unfinished	4153	48	86.52	112.5	1.3	3
0350	½″ walls (standard)—taped and finished	689	20	34.45	60.31	1.75	2
1050	½″ ceilings—taped and finished	637	16	39.81	47.81	1.2	4

Using PAR Values to Prioritize Corrective Action

The PAR values in excess of 1.0 indicate that a task has room for improvement relative to the exemplar values on which it is based. Those tasks that have the largest PAR value are the ones that have the greatest improvement potential and should be addressed first in planning for productivity improvement. In column E the exemplar rate (in column D) is divided by the job productivity (C). In this case, the PAR values range between 1.2 and 1.9; they are prioritized in column F. The PAR value of 1.2 for ½″ ceilings—taped and finished—indicates performance that is slightly worse than the exemplar rate. By contrast, the PAR value for 3/8″ wall installation is 1.9 and indicates great improvement potential.

The ranking in column F shows how these work activities should be prioritized for performance improvement. This exercise should be carried out for all work activities, especially those on the critical path of a project.

Using PAR Values in Project Management

A project management team can use the methodology above to decide how best to increase crew performance to be in line with the initial estimates. To do this, changes in the methods may be needed. Simply adding crews to a job may defeat the purpose by increasing labor costs disproportionately. Whereas the example shown uses industry-wide values from the Means manual, foremen can work with crews to identify best methods that have higher levels of productivity. These best methods can then become the new exemplar rates for a contractor and for the job. These values are critical. Low-productivity levels are likely to result in schedule delays as the rate of work production will fall short of the estimator's projections. On the other hand, the attainment of higher exemplar rates will lead to earlier job completion, at lower labor costs, and very likely higher profits.

In the deployment of The Last Planner® System, weekly work plan (WWP) meetings may be used to continually improve exemplar rates and work performance. Quite often, performance below exemplar rates is due to poor planning, poor coordination, and promises not kept. The LPS process seeks to improve coordination through a detailed, ongoing evaluation of reasons for noncompletion (RNC).

Setting Up a Performance Measurement Program

Productivity measurement is essential for a construction organization's leaders to gauge performance on a number of selected indicators. These indicators provide measurements that serve as a foundation for improvement—as the old adage goes "what gets measured, gets done." Performance measurement systems should track several factors representing components of organizational performance such as: productivity indices, quality measurements, safety measurements, schedule compliance, and budget compliance. Improvement can be achieved by a variety of initiatives that include lean construction, total quality management (TQM), six sigma, and ISO 9001:2000. While most organizations have systems that keep track of financial performance, construction productivity calls for a number of specific measures. They include:

- Labor productivity
- Crew productivity
- Exemplar rates
- PAR
- Nonproductive time
- Safety performance
- Lost time accidents
- Rework labor hours
- Schedule productivity
- Subcontractor productivity
- Where applicable: Lean indicators
  - Percentage of projects complete (PPC)
  - Project waste index (PWI)

Labor productivity is a critical component of construction productivity. Labor costs account for approximately 40% of total costs for a broad range of projects. These costs may account for an even greater percentage in the case of small projects. Work sampling studies of construction operations indicate unproductive labor activity in the range of 40–60%. Figure 2.2 shows typical reasons for non-productive labor hours. In light of the probability of lost labor hours, the performance measurement of labor productivity offers perhaps the highest return on investment (ROI) of all productivity measurements.

A successful program starts with the top management's vision. They must have a commitment to excellence, with an emphasis on a culture of continuous improvement and on management by fact. It should be included in their strategic management program, integral with their vision, mission, and values statement. In that context, productivity measurement provides the organization with the facts that it needs to make the tactical changes that are required to achieve and sustain higher levels of performance. Leaders must communicate to the entire organization its vision, mission, and values so that the continuous improvement philosophy can be incorporated in all activities, especially with construction projects.

At its simplest level, a productivity measurement program may be headed by an analyst that reports to a senior management level, such as a VP. In turn, the VP should maintain an active productivity agenda in executive meetings to keep all decision makers aware of the trends in the organization's productivity through its respective measures. Productivity

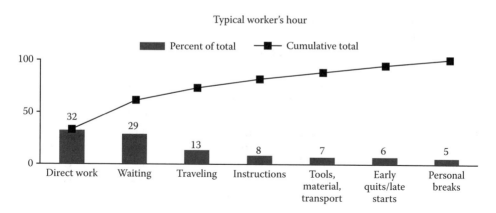

FIGURE 2.2
Workers' time utilization. Typical worker's hours.

measures should be selected based on key performance indicators (KPI). Critical success factors (CSF) should be identified and these should be closely linked with the organization's KPI.

Lean Construction Measurement

If the organization adopts lean construction as its protocol, performance measurement is an absolute requirement in order to determine the extent of accomplishment for a WWP. It is also a necessity to determine the reasons for not completing a task as planned. The analyst role may be assigned to a scheduler or other staff member who has been trained in the lean construction methodology. Other traditional measurements such as labor and crew productivity are still relevant as a means of tracking overall project performance.

Transitioning from Productivity Measurement to Performance Improvement and Lean Construction

The foregoing discussions emphasize the importance of productivity measurement as a component of the greater issue of performance improvement. In this context, the underlying purpose of measurement is to support initiatives to improve construction performance. These initiatives can include lean construction and integrated project delivery, TQM, ISO 9000, or six sigma. Construction organizations have a particular challenge to adopt this paradigm. Traditional construction project management tools do not address productivity; they include schedule slippages and cost overruns. Forbes and Golomski (2001) observe that the construction industry as a whole measures performance in terms of completion on time, within budget, and meeting construction codes.

Code enforcement officials ensure that minimum standards are met for building reinforcement and mechanical and electrical systems that have life-safety implications, but these standards may be met while the quality of finish and workmanship can be below expectations. Owner/client satisfaction is rarely considered under this scenario. Contracts are often awarded on the basis of low bids; most construction activity is subcontracted, and as a whole there has been a tendency to offer minimally acceptable quality in order to be price-competitive.

Construction organizations (designers and constructors) would benefit significantly by establishing formal performance improvement programs that build on the knowledge gained from the measurement approaches that have been discussed above. Lean construction can be deployed through this organizational structure. To be effective in improving the organization, management needs to commit to having a support staff for one or more individuals trained in the use of productivity/quality improvement techniques. Having them trained in lean construction methodologies would be a plus. Industrial engineers could be appropriate candidates for this position, provided they understand the construction environment.

This position should report to top management, and should be versed in the construction process as well as in the use of such tools as Pareto charts, cause and effect diagrams, activity sampling, time studies, histograms, and stratification. This position should also have experience with facilitating and leading team efforts such as assisting workers with identifying process improvement approaches. Top management should empower this position to develop and conduct programs and activities that infuse productivity and quality thinking in the workforce from the lowest to the highest levels. Productivity and quality reporting should become part of the organization's operating procedures, to the same extent as project progress and financial status reporting. These efforts can only succeed if management sets a clear example for the importance of productivity and quality and, very importantly, the use of measurement information to continuously improve the efficiency and effectiveness of procedures and activities. It is good to remember the following words of Albert Einstein:

> Not everything that counts can be counted, and not everything that can be counted counts.

> *(Sign hanging in Einstein's office at Princeton—*
> *documented by K. Harris, 1995)*

Productivity measurement is both quantitatively and qualitatively vital to the profitability of a construction company. The central task is the systematic collection of data based on measured levels of output of different workers performing similar tasks, across sectors, and within particular sectors. This exercise will establish norms or reference output rates. On each specific construction project, measurements should be taken and measured rates should be compared with reference rates. Management must keep up with technological innovations worldwide and seek to adapt as well as motivate employees with improved management systems.

To facilitate performance measurement and improvement, constructors need to develop a culture of "building in" quality, and convey that philosophy to management and workers alike. They should begin to adopt Deming's fourth point "End the practice of awarding business on the basis of price tag alone."

Overall, as has been experienced in the manufacturing and other service industries, top management of all involved organizations—owners, designers, constructors, and construction management companies—must be committed to the concept of performance and quality improvement and must provide necessary funding and staff support to ensure success.

Guidance from the Malcolm Baldrige National Quality Award

A design or construction organization can incorporate performance improvement thinking in its business model by adopting selected portions of the award's seven criteria (2007). The organization does not have to compete for an award, but can benefit from the core beliefs on which it is based. The MBQNA's Criterion No. 4 (Measurement, Analysis, and Knowledge Management) represents the practices of best-in-class organizations that seek to establish leadership in management procedures. These organizations continually measure their performance across a number of key success factors (KSF), and use these metrics for the purpose of continuous improvement. The criteria are described as follows:

4.1 Measurement, Analysis, and Improvement of Organizational Performance

How an organization measures, analyzes, reviews, and then improves its performance through the use of data and information at all levels and in all parts of the organization.

4.2 Management of Information, Knowledge, and Information Technology

How an organization manages information, organizational knowledge, and information technology.

By implementing such measurement systems, construction organizations can maintain the information necessary for high levels of performance, then use it as a foundation for the deployment of lean construction.

Questions for Discussion

1. How do you define productivity in construction?
2. How does lean construction impact construction productivity?
3. What are the methods available for measuring construction progress?
4. What are their advantages and disadvantages?
5. How are exemplar rates determined?
6. Briefly explain how the terms "exemplar performance" and "PAR" values can be applied to crew production rates to identify productivity improvement opportunities. To help your explanation, consider a tile-setting task in which crews produce 700 square yards per week. Exemplar performance is 800 square yards per week for a crew of the same size.

References

Alfeld, L. E. 1988. *Construction productivity*. New York: McGraw-Hill, Inc.

Ballard, G., and G. Howell. 1994. Implementing lean construction—stabilizing work flow. *Conference on Lean Construction*. Santiago, Chile, September 1994.

Crosby, P. B. 1979. *Quality is free: The art of hassle-free management*. New York: McGraw-Hill.

Forbes, L. H., and W. A. Golomski. 2001. A contemporary approach to construction quality improvement in *The Best On Quality*, ed. M. N. Sinha, IAQ Book Series, 12: 185–200. Milwaukee, WI: ASQ Quality Press.

Forbes, L., M. Ahmed, and S. Azhar. 2003. Productivity management, improvement, and cost reduction. In *System-based vision for strategic and creative design*, ed. F. Bontempi. Lisse, The Netherlands: A.A. Balkema Publishers, Swets and Zeitlinger, B.V.

Griffis, F. H., J. V. Farr, and M. D. Morris. 2000. *Construction planning for engineers*. New York: McGraw-Hill.

Koskela, L., and G. Howell. 2002. The underlying theory of project management is obsolete. *Proceedings of PMI research conference 2002*. Seattle WA, eds. D. P. Slevin, D. I. Cleland, and J. K. Pinto. Project Management Institute.

Malcolm Baldrige National Quality Award. 2007. *Criteria for performance excellence*. Milwaukee: American Society for Quality.

Sanvido, V. E. 1988. Conceptual construction process model. *Journal of Construction Engineering and Management* 114(2): 294–311.

Sumanth, D. J. 1984. *Productivity engineering and management*. New York: McGraw Hill, Inc.

Thomas, H. R., W. F. Maloney, M. W. Horner, G. R. Smith, V. K. Handa, and S. R. Sanders. 1990. Modeling construction labor productivity. *Journal of Construction Engineering and Management* 116(4): 705–726.

Bibliography

Adrian, J., and D. Adrian. 1995. *Total productivity and quality management in construction*. Champaign, IL: Stipes Publishing, LLC.

Alarcon, L. F. and A. Serpell. 2004. *Performance measuring, benchmarking, and modeling of construction projects*. Santiago, Chile: Pontificia Universidad Catolica de Chile.

Ashford, J. 1989. *The Management of quality in construction*. London: E & F.N. Spon.

Forbes, L., 1999. An engineering-management-based investigation of owner satisfaction, quality and performance variables in health care facilities construction. Dissertation. Miami: University of Miami.

Gerald, F. 1997. *Building a strong economy: The economics of the construction industry*. Livonia, MI: Sharpe, Inc.

Oberlender, G. D. 2000. Figure 9-15 in *Project management for engineering and construction*, 2nd ed., 228. New York: McGraw-Hill.

Web Sites

Bureau of Labor Statistics (BLS) On line: http://www.bls.gov
http://www.businessdictionary.com/
http://www.bls.gov/iif/osheval.htm
http://www.cit.cornell.edu/computer/robohelp/cpmm/Glossary_Words/CPI.htm
http://en.wikipedia.org/wiki/Earned_value_management
http://pmbook.ce.cmu.edu/12_Cost_Control,_Monitoring,_and_Accounting.html

3

Foundations of Lean Construction

This chapter provides an important foundation with an explanation of how lean concepts apply to the construction environment. The history of lean manufacturing is explained, and how the Toyota Corporation used lean techniques to minimize waste in its processes to accomplish its preeminence in automotive manufacturing. The chapter also describes the work of researchers who sought to adapt Toyota's approach to the construction environment. Lean principles are described in the context of construction projects as a foundation for the application of integrated delivery approaches and The Last Planner® System (LPS).

Defining Lean Construction

Lean construction has been defined in several ways as the concept continues to evolve. The following descriptions are among the most established examples to date. Greg Howell and Glenn Ballard, co-founders of the Lean Construction Institute (LCI), view lean construction as a new way to manage construction. The objective, principles, and techniques of lean construction taken together form the basis for a new project delivery process. Unlike approaches such as design-build and programmatic improvement efforts (partnering and TQM), lean construction provides the foundation for an operations-based project delivery system. From its roots in the Toyota Production System (TPS), this new way of designing and making capital facilities makes possible significant improvements in complex, uncertain, and quick projects.

The Construction Industry Institute (CII) has defined lean construction as " the continuous process of eliminating waste, meeting or exceeding all customer requirements, focusing on the entire value stream, and pursuing perfection in the execution of a constructed project" (CII Lean Principles in Construction Project Team, PT 191).

Lauri Koskela (2002) described lean construction as a way to design production systems to minimize waste of materials, time, and effort in order to generate the maximum possible amount of value.

William Lichtig (2006) observed that lean construction aims to embody the benefits of the Master Builder concept. Essentially, lean construction recognizes that desired ends affect the means to achieve these ends and that available means will affect realized ends.

Lean Theory

In order to appreciate how lean techniques apply to the design and construction processes, we first need to understand how the lean approach improves performance in the manufacturing industry where it originated and subsequently in the service industry.

Lean production is lean because it uses less of everything compared to mass production. Womack, Jones, and Roos (1990), in their book *The Machine That Changed the World*, chronicled the genesis of lean thinking in the automotive industry. They explain how lean production uses half of the resources that are typically consumed: the labor in the factory, the manufacturing space, the investment in tools, materials, and so on. It results in fewer defects and generates an increasing variety of products. Essentially, a lean system provides what is needed, in the amount needed, when it is needed.

Henry Ford pioneered the concept of continuous flow in his manufacturing plants but subsequently placed his company's emphasis on mass production to meet the huge consumer demand that came after World War II. Eiji Toyoda and Taiichi Ohno of the Toyota Motor Company perfected the Just-In-Time (JIT) concept as an alternative to mass production. The term "lean" is attributed to John Krafcik (1988), a researcher with the International Motor Vehicle Program (IMVP) at MIT. The IMVP was created in the 1980s to study the automotive industry internationally and to examine the differences between mass production and the innovative methods used by Japanese manufacturers.

The researchers used the term lean to designate what they saw as a new and superior type of production system for motor vehicles (Womack, Jones, and Roos 1990). The researchers were persuaded by the difference in comparative performance measures between Japanese, American, and European companies. Subsequently, it was found that the Japanese advantage was largely because of Toyota, and many academics and practitioners now believe that the Toyota Product Development System and the TPS are exemplars for a new and superior way of designing and making all kinds of goods and services.

Lean production can best be described by comparing it with craft production and mass production. Craft production was practiced in the preindustrial era—it was based on highly-skilled workers making products for customers, one at a time. Examples of craft production would be custom furniture, art works, and hand-built automobiles, products that are generally too expensive for most consumers. Mass production, by contrast, uses unskilled or semiskilled workers to produce standardized products in large quantities using highly specialized machines, based on processes designed by specialized professionals.

Mass producers are concerned with large production quantities with a narrow range of products and tend to generate a level of defects that they consider to be acceptable. By contrast, lean producers seek perfection, aiming for zero defects, zero inventories, declining costs, and a wide variety of products. Lean producers are focused on value-added process flow; that is, a continuous addition of value to a product or service (one-piece flow) by avoiding or eliminating interruptions or activities that do not add value. Examples of nonvalue-adding activities would be wasted material and wasted time waiting for value to be added. Lean producers address the root causes of wastes, such as defective processes, poor work layouts, lack of standardization, errors by operators, errors in communication, and poor decisions by management.

When lean first became popular and more widely known, there was a tendency to see it as a collection of tools and techniques, but it is now widely recognized as a fundamental business philosophy. "(Our) final conclusion is that lean cannot be reduced to a set of rules or tools. It must be approached as a system of thinking and behavior that is shared throughout the value stream" (Diekmann et. al. 2004).

Henry Ford established many practices that are embodied in lean thinking even today. Ford's "Lean" Business Norms circa 1903–1947 included the following principles:

1. That work environments be spotlessly clean.

2. That business leaders think in terms of serving their communities and society at large.

3. That production techniques not be taken for granted but continuously improved.

4. That primary industries should help their suppliers and the service industries to produce cheaper and better products in less time.

5. That managers should not remain in their offices but should walk around, know their workers, and be capable of doing the work themselves.

6. That workers should be trained and have the opportunity to better themselves and make product improvements.

The Toyota Experience: Building on Ford's Principles

The Toyota Corporation has been highly successful in automotive manufacturing and is recognized as an outstanding example of a lean organization. With regard to Toyota's recalls in 2009 and 2010, it must be noted that these problems were due to failures of management. The Toyota Production System is still a preeminent example of lean manufacturing. Toyota began as a loom company in 1926, founded by Sakichi Toyoda, a loom inventor. They began auto manufacturing in the 1930s, influenced by Henry Ford's continuous flow manufacturing. In the latter part of 1949, Toyota adopted statistical quality-control methods learned through courses provided by Japanese Union of Scientists and Engineers (JUSE). At that time Eiji Toyoda visited Ford Motor in Detroit to gain an understanding of their mass production system. Ford's operation was recognized as the world's most efficient, with a production rate of 7000 cars per day, while Toyota had only produced 2685 cars in 13 years up to that point. In effect, during 1948 a typical American autoworker produced 10 times the output of the Japanese autoworker.

Toyoda studied Ford's operation for 3 months and saw possibilities for improving the auto production plant in Japan. Taichii Ohno, Toyota's chief engineer, had also visited Detroit on a number of occasions. Toyoda and Ohno recognized that there was much waste or *muda* (Japanese term for much waste) everywhere in Ford's operation. For example, there was a high inventory cost to keep parts that were later found to be defective. There was muda of manpower, muda of waiting, muda of transport, and muda of facilities. They noted that only assembly line workers were adding any value to the production process. Specialists were responsible for designing the production process and directing workers to follow it. The foremen served only to ensure that assembly line workers followed orders, and they in turn simply performed repetitive tasks. Shigeo Shingo, an industrial engineer at Toyota identified seven wastes in mass production systems:

1. Overproducing

2. Idle time waste (waiting time/queue time)

3. Transporting/conveyance waste

4. Processing waste: waste in the work itself

5. Inventory waste (having unnecessary stock on hand)

6. Wasted operator motion (using unnecessary motion)
7. Producing defective goods (waste of rejected production)

Ohno realized that all these items of waste represented a financial loss:

- Overproducing led to product quantities greater than the market needed.
- Waiting time between stages in production resulted in lost labor.
- Transporting product from one location to another increased energy costs.
- Processing using inefficient methods required more inputs than necessary to produce needed outputs.
- Unnecessary stock resulted in excessive inventory carrying costs.
- Wasted motion led to nonproductive time (that incurred cost).
- Defective production would have to be corrected before it could be paid for by the customer/client.

These wastes are shown in Figure 3.1. An eighth waste is shown but will be discussed later in the chapter.

Toyota had the challenge of meeting diverse customer needs with a small capital base, as opposed to the Detroit strategy of having a large inventory of cars from which customers could pick and choose. The Detroit strategy was also based on minimizing the cost of each car instead of minimizing wasted inventory. While mass production was effective in minimizing unit costs, any defects that developed in the fast moving process would be replicated in many cars before they could be detected. Toyota's efforts to improve productivity enabled them to produce larger production volumes at low cost, overly optimistic

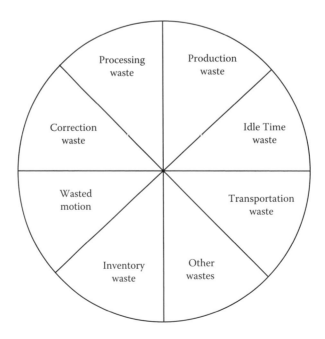

FIGURE 3.1
Production wastes: eight categories.

sales forecasts left them with large inventories of unsold vehicles in the 1960s that led to huge financial losses. Japanese consumers also preferred customized features that would prove to be uneconomic with mass production techniques and the long-setup times they required. Toyoda and Ohno decided that the human and material resources waste they saw in the United States was not appropriate for Toyota, or Japan, for that matter. They decided that a better strategy would be to enable the production system to change rapidly between different models and types of vehicles while producing minimal amounts of waste.

Ohno found a new way to coordinate the flow of parts in the supply system on a day-to-day basis so that parts would only be produced at each step to meet the immediate demand of the succeeding operation. This led to the development of the Toyota Production system (Ohno 1988). At each work station, teams of workers controlled their local operations-based on continuous observation and relationships instead of by specific metrics. The workflow of a Toyota factory regulated itself through "takt time," a beat like a metronome, based on a term used by German musicians. This beat matched the daily production demand.

Engineers were required to sense the needs of the operation and opportunities for improvement by standing in a chalk circle for several hours at a time. Later, a card called a kanban (Japanese for "card") in 1958 was used as a communication tool between different production lines, and became an important part of the JIT system. Ohno had been inspired with this idea by observing the U.S. supermarket system in 1950. He wanted every member of the production system to pay attention to the prevention of potential problems and so reduce waste (muda). This system worked so well that in 1959 Toyota produced 100,000 cars per year for the first time—a major accomplishment.

Toyota was subjected to heavy pressure around that time with the failure of the new Toyota Crown in the United States, and Japan's decision to liberalize trade. This spurred a strategic decision—to compete on the basis of quality. Toyota instituted total quality control (TQC) with the "kanban" system in 1963. Their improved quality and productivity were recognized with the Deming Prize.

Toyota's Production System

From the 1960s to the 1980s, this system evolved into the highly successful TPS that provided a competitive advantage through outstanding levels of production, high quality, and low costs. The TPS has been described as a combination of methods with consistent goals—cost reduction, quality assurance, and respect for humanity—to ensure sustainable growth. It has four main elements: JIT, autonomation, workforce flexibility, and creative thinking.

The Toyota Corporation has since evolved the TPS to become "The Toyota Way."

Toyota was able to produce a car to meet each owner's specifications in real time. Workers performed their own quality control without inspectors. With the JIT system, external suppliers delivered parts directly to work stations without being warehoused. In the mid-1970s, Toyota reduced the time needed to produce a car from 15 days to 1 day, using JIT. It is important to note that Ohno's improvement of Toyota's production process was not necessarily a new technology, but rather the result of involving all participants in a new philosophy of avoiding waste of any type.

It is to Toyota's credit that they shared knowledge with the rest of the world through visitor tours of their facilities. They formed strategic alliances with competitors such as General Motors. They allowed their employees Ohno and Shingo to publish books about Toyota's JIT success. Toyota's success was also due to strategic relationships with suppliers.

Adversarial relationships were considered wasteful; instead, Toyota would pick two or three suppliers for each component and guarantee them a specific volume of business. This trusting relationship reduced waste, cut costs, promoted innovation, and increased profits for all stakeholders.

Just-In-Time (JIT) Concept

Whereas mass production is a so-called push system that dictates production volume based on market forecasts, JIT is a pull system that responds to actual customer demand. In essence, products are pulled from the JIT system. While push systems are susceptible to over or underproducing based on the accuracy of forecasts, JIT only commits the resources needed to meet the actual demands. The customer is the driver of production. Mass production emphasizes the production of the largest possible quantities, as setups and adjustments of machines are time consuming. JIT, on the other hand, seeks to build small lots that are better capable of providing variety for the customer. Companies that have used the Toyota approach have reduced setup times by 90% or more. Omark Industries, one of the earliest U.S. companies to adopt the TPS was able to reduce machine setup time from 8 hours to 64 seconds. Overall, JIT facilitates lower costs, shorten lead and cycle time, and leads to improved quality.

The kanban system minimizes inventories through backward requests. Kanban cards follow a product through various stages of production and provide visual control. They contain information such as the part name, description, process instructions, and quantity required. In keeping with the pull system, kanban cards signal "upstream" when production or the release of inventory is required. In effect, inventory is not created until it is needed. In effect, Toyota's underlying strategy was to produce a car to meet the requirements of a specific customer, instead of maintaining a large inventory of cars from which the customer could choose. The application of JIT results in a lower inventory of raw materials and of parts, less work in process, and shorter lead times. It also leads to a reduction in floor space, less overhead, and lower costs. However, there is a risk that inventories may fall to critically low levels. To be successful, JIT is dependent on very high quality and reduced setup times—the concept calls for raw materials and components to reach a production operation in desired quantities when needed and not before. The supplier and customer have to work closely for this to occur and there is little to no inventory.

Production leveling accommodates fluctuations in demand and the number of setup activities is reduced so that little time is required for changing over to different product mixes. (In modern manufacturing, the kanban card has been replaced by an electronic equivalent.)

Toyota's Way

The Toyota Corporation has continued to evolve beyond the foundation of the TPS. The TPS was represented by two pillars—JIT and autonomation. Liker (2004) describes 14 principles in the contemporary Toyota Way. The Toyota Way represents an important foundation for lean construction. Economics were a factor in the Japanese use of human resources; it proved too expensive to rely on few highly qualified people. Lean thinking as represented in *The Toyota Way*, sees the most cost effective way of improving product quality as empowering all people in organizations (Table 3.1). This varies significantly from the prevailing model in the construction industry, which emphasizes power and control.

TABLE 3.1

A Representation of the Toyota Way

	Toyota Foundations	Principles
1	Problem Solving (Continuous Improvement and Learning	Continual organizational learning View the situation firsthand to thoroughly understand it Make decisions slowly by consensus—consider all options: Implement rapidly
2	People and Partners (Respect, Challenge, and Grow Them)	Grow Leaders who live the philosophy respect, develop, and challenge people and teams Respect, challenge and help suppliers
3	Process (Eliminate Waste)	Create process 'flow' to reveal problems Use pull system to avoid overproduction Level out workload Stop when there is a quality problem Standardize tasks for continuous improvement Use visual control—transparency Use only reliable, tested technology
4	Philosophy (Long-Term Thinking)	Base management decisions on a long-term philosophy, even at the expense of short-term financial goals

Accomplishing a Lean State

Lean can be characterized in terms of objectives, principles, and methods or tools. According to Glenn Ballard, co-founder of the Lean Construction Institute, the lean ideal is to provide a custom product exactly fit for purpose, delivered instantly, and with no waste. If one attempts to conduct business or deliver projects while maximizing value and minimizing waste, then in a minimal sense that represents lean. But it would make little sense to call someone lean who claims to pursue those objectives but disregards fundamental principles. These principles are identified in the Construction Industry Institute's report (CII PT 191), which focus on the process level, or the 14 principles defined in Jeffrey Liker's *The Toyota Way*, as described above that deal with the lean enterprise. Ballard points out that the same holds for those claiming allegiance to these principles but not using available methods. For example, a candidate might proclaim the principle of creating connected process flow, but fail to use pull mechanisms to release work between specialists or fail to reduce batch sizes or fail to right size work-in-progress inventories. Lastly, since lean is a never ending journey in pursuit of perfection, it is more appropriate to ask about the rate of learning rather than the level of conformance to the ideal—remember the fable of the tortoise and the hare.

How can you tell if a project or company is lean? Ballard suggests you assess yourself by answering the following questions:

- Are you pursuing the lean ideal?
- Are you following the appropriate principles when striving for the lean ideal?
- Are you using the best methods for implementing those principles?
- How fast are you learning as a project organization?

Origins of Lean Construction

Early work by Frank Gilbreth in the 1890s indicated the potential for manufacturing approaches to be used in construction to improve speed and labor efficiency. Gilbreth is considered to be the father of industrial engineering as he later built on Frederick Taylor's principles. Gilbreth saw an opportunity in bricklaying—he looked at the way bricklayers worked and felt that they could perform more efficiently. He developed the idea of what he called "speed work." Gilbreth noted that many of the motions made by bricklayers were unnecessary and did not really contribute to the job. They were accustomed to bending down and retrieving each brick from a pile—they would turn it over in their hands before plastering it in place, taking extra time. Frank made several improvements to the job. He had the working stock of bricks located on a platform that made it unnecessary for workers to bend excessively, thereby reducing retrieval time. He also arranged for lower-skilled (and lower-paid) workers to transport the bricks to the platform, allowing the craftsmen to focus on applying their expertise. He developed a "best method" that reduced the number of motions from 18 to 4-½, minimized fatigue and increased productivity.

Gilbreth experimented with system improvements such as finding the optimum load that a worker could manage with a shovel in order to safely carry out the maximum amount of work each day. He developed labor standards that could improve the predictability of work output. Gilbreth had his own contracting team and went on to compete very profitably in the contracting business in the early 1900s. With his wife, Lillian Gilbreth, he developed a body of knowledge that led to the field of Industrial Engineering.

In most of the twentieth century, construction productivity grew but more slowly than manufacturing productivity. As described in Chapter 2, construction productivity was observed to decline at a rate of 0.6% annually over a 40-year period to 2007. Nonfarm productivity grew by 1.8% annually.

Construction practice transitioned from the Master Builder concept, in which one entity was responsible for coordinating both design and construction, to a highly fragmented industry. Designers wished to limit their liability that could result from construction failures, and developed contract language that insulated them from the construction responsibility. Productivity growth was primarily due to the application of technology; better tools and equipment increased labor efficiency. Innovations such as prefabricated building elements and tilt-up construction also improved the speed of the process. Computer applications for planning and scheduling also improved overall efficiency. Yet the industry was fraught with distrust and adversarial relations, resulting in behaviors that were both counterproductive and costly.

It has been known for many years that construction performance has much room for improvement. The U.S. Business Roundtable was formed by a number of industry leaders to identify improved methods for industries including construction. Their findings pointed to major shortcomings in the construction industry, and the Roundtable appointed the Construction Competitiveness Committee to study it in more detail. The Construction Industry Institute (CII) grew out of the Committee, and was created in 1983 and based at the University of Texas–Austin. The CII has been studying industry shortcomings for many years in order to develop system improvements.

A relatively small percentage of total work hours in construction is spent productively. A 1990 CII-sponsored study by Michael Pappas (1990) noted that in steel erection, only 11.4% of recorded work hours represented value-added work. In process piping projects, the percentage of value-added time was even less, at 7.5%. Hammarlund and Ryden 1989 and Nielsen and Kristensen 2001 observed that transformation activities in construction occurred during only 30% of the working time available in projects. Of greater concern than these activities was the time spent in waiting and transporting—as these do not add value. In fact, material handling and transporting greatly increase material costs (Bertelsen 1993).

Lauri Koskela (1992) addressed the application of manufacturing techniques to construction. Koskela spent a year at Stanford University as a visiting scholar and conducted a study titled "Applications of The New Production Philosophy to Construction." He drew parallels between both fields by characterizing construction as a form of production. Koskela looked to the manufacturing industry for a new direction for construction. Specifically, he modeled this new production philosophy after the Toyota's highly successful production system (TPS). Researchers had recommended such solutions for the construction industry's underperformance as industrialization (prefabrication and modularization), automation and robotics, and information technology to reduce fragmentation. Koskela proposed a new approach that was not based on technology, but rather on the principles of a production philosophy. He noted its evolution through three stages:

1. Tools, such as kanban and quality circles
2. A manufacturing method
3. A management philosophy (lean production, JIT/TQC, etc.)

Koskela inferred from a number of productivity-based studies on U.S. and European manufacturing plants that the most successful methods were based on the JIT philosophy. Manufacturing studies by other researchers such as Schonberger (1986) and Harmon and Peterson (1990) reinforced this observation.

In a typical production process, material is processed (converted) in a number of discrete stages. It is also inspected, moved from one operation to another, or made to wait. Inspection and waiting are considered flow activities.

Conversion activities are considered to add value while flow activities do not. Koskela visualized lean construction as a flow process combined with conversion activities, and noted that only the conversion activities add value. This was the transformation-flow-value (TVF) theory of construction. Production improvements can be derived by eliminating or reducing flow activities while making conversion activities more efficient. Koskela proposed that production should be improved by reducing or eliminating flow activities, while conversion activities should be made to be more efficient. He pointed to earlier studies that showed that only 3–20% of the steps in a typical process add value and that their share of the total cycle time is only 0.5–5%.

Koskela attributed the predominance of nonvalue-adding activities to three root causes: design, ignorance, and the inherent nature of production. Design was due to the subdivision of tasks; each added subtask increased the incidence of inspecting, waiting, and moving. A natural tendency for processes to evolve over time without close analysis leads to ignorance of their inherent wastefulness. The inherent nature of production is that events such as defects and accidents add to nonvalue-added (NVA) steps and time between different conversion activities.

Koskela listed the following heuristic principles:

- Reduce the share of nonvalue-adding activities
- Increase output value through systematic consideration of customer requirements

- Reduce variability
- Reduce the cycle time
- Simplify by minimizing the number of steps, parts, and linkages
- Increase output flexibility
- Increase process transparency
- Focus control on the complete process
- Build continuous improvement into the process
- Balance flow improvement with conversion improvement
- Benchmark

Points to note in relation to the heuristic principles are:

- Nonvalue-adding activities can be reduced by identification, measurement, and redesign.
- Output value can be improved by identifying the supplier and customer for each activity, and clarifying customers' value needs.
- Variability of construction activity duration increases the relative volume of NVA activities. The adoption of standard procedures can streamline both conversion and flow processes.
- Process control requires measurement and designated controlling authority—that authority may be cross-functional process owners or self-directed teams. Team building and cooperation with suppliers are proposed for optimizing total flow when multiple organizations are involved.
- Flow improvement and conversion improvement should be balanced—with complex production processes, improvement of flow (NVA) activities has a higher payback than seeking technology-based enhancements to conversion activities.

Other researchers built on this foundation and determined that construction-specific tools would be needed in order to adapt construction to the lean production principle.

Glenn Ballard heard Koskela speak at University of California–Berkeley and they began a collaboration on lean production that led to the staging of a conference in Helsinki, Finland in August 1993. That conference became the first meeting of the International Group for Lean Construction (IGLC). As Ballard reports, it was decided in that meeting to use the expression "Lean Construction." Ballard and Greg Howell subsequently co-founded The LCI in the United States in 1997. Other organizations dedicated to lean construction were formed in Chile, Denmark, the UK, and several other countries.

Ballard pioneered the development of The Last Planner® System (LPS) in 1992. The LPS was based on the concept of reducing the hierarchical layers of construction management and empowering field-based actors in the construction process to optimize the allocation of available resources in the weekly planning, scheduling, and execution of work. Ballard further refined the LPS in 1998, transitioning from weekly work planning to look ahead planning and to phase scheduling. This refinement focused on managing flows in the construction process, reducing flow variation from the plan, and using buffers to limit the impacts of any remaining variability in these flows.

Members of both IGLC and LCI performed studies to further develop the lean construction concept. Ballard set about developing what would later come to be called The Lean

Project Delivery System™. Studies identified that traditional project management methods normally accomplished only approximately 50% of scheduled weekly work plans (Ballard and Howell 1994). Many white papers were generated by the LCI that furthered the development of the LPDS—Ballard's White Paper #8 introduced The Lean Project Delivery System™ in 2000 (2000b).

The LPS was applied to the design process in 1998, representing another milestone. The design process was further improved with the application of target value design (TVD). The first TVD paper was published in 2004.

Adoption of Relational Contracting

Another milestone in the evolution of lean construction was the adoption of relational forms of contracting. At an LCI meeting in 2003 in Las Vegas, Nevada, the attendees discussed the need for a special type of contract for lean; while they determined that such a contract was not necessary, they agreed that prevailing forms of contract could be obstacles to lean. In November 2004, LCI held an international symposium on relational contracting in Atlanta, Georgia. Several types of contracts were discussed that were relational to varying degrees; from the cases presented, there was evidence of a synergy between relational contracts and lean construction. Will Lichtig, an attorney with McDonough, Holland and Allen, PC, and an LCI member, agreed to develop a form of relational contract that later became known as the Integrated Form of Agreement (IFOA). This form of contract was subsequently adopted in several major lean construction projects—notably those established by the Sutter Health System in California.

Bertelsen and Koskela (2004) addressed future directions for lean construction as a product strategy and a process strategy. The product strategy would involve simplifying construction to allow components to be fabricated off-site under relatively predictable conditions; this would facilitate the use of manufacturing techniques, with great increases in productivity. The process strategy would involve improving the on-site construction process.

With regard to the product strategy, Bertelsen noted a trend in construction toward prefabrication and modularization; the use of prefabricated structural steel and concrete slabs increasingly changing on-site work to assembly and observing tolerances. Despite the uniqueness of a given project or facility, the manufacture of modules and subassemblies in controlled manufacturing environments reduced the overall project complexity, placing less emphasis on the management of materials and craft workers on-site. The tradition of building in place had always been subject to high levels of inefficiency. On the other hand, the product strategy seemed to rely on supply-chain management, with various suppliers increasingly manufacturing and supplying components and subassemblies such as windows, beams, roof trusses, and so on.

Bertelsen described open systems and closed systems in the product strategy. Open systems are made to standard specifications and can be combined with systems from other manufacturers, resulting in high levels of architectural design flexibility. Conversely, closed systems are proprietary and offer less flexibility in terms of cross combination. Manufactured housing is a typical example of a closed system. He observed that closed systems seemed to benefit greatly from mass production approaches, but had limitations in the amount of flexibility that they offered. Overall, he saw the product strategy as being more adaptable than the process strategy in addressing special needs such as affordable housing—where uniformity was more tolerable.

The process strategy has become the primary focus of lean construction as it is most applicable to on-site construction. Koskela viewed construction as "unique products of art on a very large scale." Construction complexity originated from this unique nature and from associated undocumented production processes—a temporary production system with shared resources between subcontractors working on different projects simultaneously. It occurs in an ad hoc organization that is a complex social system.

Process improvements include enhancements to the flow of materials—this complements the last planner as described by Bertelsen and Nielsen 1997, and Arbulu and Ballard 2004. Information flow improvements may similarly offer opportunities for improvement. Macomber and Howell (2003) proposed management by conversations as a means of improving the flow of information to improve the construction process.

Bertelsen commented on the development of "construction physics" as analogous to factory physics described by Hopp and Spearman (2000), but noted that the latter was based on production as an ordered system, whereas construction runs counter to a direct application of these principles. Instead, construction physics should be based on maximizing value and minimizing waste. As the lean construction principles evolved, they have been tested and refined through both application and research.

Owen Matthews of Westbrook Air conditioning and Plumbing a mechanical contractor in Orlando, FL, coined the term "Integrated Project Delivery" (IPD) to represent the application of lean principles in a team-based framework. Westbrook Service adapted relational contracting to include lean principles in a chiller plant installation project in Orlando, FL. This was documented by Matthews and Howell (2005).

A major health facility construction program by Sutter Health in California was based on the IFOA to facilitate lean collaboration by construction participants in a legally enforceable framework. Will Lichtig did pioneer work in developing the IFOA and applied it to Sutter projects as their legal counsel.

The IFOA was subsequently adopted by the construction industry in a standard form of contract termed "ConsensusDocs 300." Twenty two leading construction associations united to publish a consensus set of contract documents called ConsensusDOCS in 2007. They succeeded in creating a standard contract for the industry that they felt was fair to all parties. When ConsensusDOCS was released, two large stakeholder organizations, AGC and COAA converted their contract documents to comply with the consensus process. Several other variations on the theme of lean construction have been developed and successfully applied to construction projects:

- Lean project delivery with an integrated agreement
- Responsibility-based project delivery
- Lean production management (LPM) with a major refinery project

Currently, lean construction is still in its early evolution and is slowly but gradually being adopted. Greg Howell, director of the LCI explains the organization's strategy to mainstream the lean philosophy in the U.S. construction industry through the formation of regional chapters. These chapters are associated with university-based research centers that serve as learning laboratories. The original model for this structure was LCI's Northern California chapter in association with University of California–Berkeley's Project

Production Systems Laboratory. Members of the Construction industry are beginning to fund the needed research and is actively engaged in experimentation, the findings from which are shared through LCI and groups termed P2SLs. Chapters dedicated to lean construction have been formed in a number of other countries including: the United Kingdom, Denmark, Finland, Australia, Brazil, Chile, and Peru. Similar chapters are starting up in Singapore, Indonesia, Ecuador, and Colombia.

Lean Design and Construction

Lean design and construction involve the application of lean methods/techniques to the design and construction processes, in order to derive the benefits that have been clearly established in manufacturing operations. Benefits include:

- Lower costs
- Fewer delays
- Less uncertainty
- Less waste
- More efficient buildings/facilities
- Higher user satisfaction

The application of lean principles results in better utilization of resources—especially labor and material. It also results in better construction quality in completed facilities, greater owner/client satisfaction, higher levels of safety, and ultimately greater profitability for clients, builders, and design professionals. Lean construction uses production management techniques to make significant improvements particularly on complex, uncertain, and quick projects Howell (2000). There have been many examples of success with projects based on lean methods: a small office construction project has had building costs reduced by 25% and schematic design time cut from 11 to 2 weeks.

To not use lean construction means major project failures, resulting in both cost and schedule overruns. As shown in Table 3.2, many major projects in the United States have exceeded budgets by 57–460%.

TABLE 3.2

Examples of Problematic Construction Projects

Name of Project	Budget $Millions	Final Cost $Millions	Percent Cost Growth %
The Boston Big Dig (2005)	2600	14,600	460
Kennedy Center Parking Lot	28	88	210
Denver Airport (1995)	1700	4800	180
CapitolHill Visitor Center (2008)	265	621	134
Hanford Nuclear Facility (2001)	715	1600	120
National Ignition Laser Facility	2100	3300	57

Deficiencies in Traditional Construction Methods

Conventional construction is based on craft production methods, hence it is slow and expensive in comparison to mass and lean production methods that have been used successfully by manufacturers of autos and consumer goods. These craft production methods are carried out by many different specialists—shell contractors, masons, roofers, mechanical and electrical subcontractors, and so on. While they may be highly skilled as specialists, they are interdependent, yet they typically have separate contracts with a central entity—a general contractor, construction manager, and so on. These contracts place the parties to construction projects in adversarial roles, with penalties for underperformance. In turn, construction profit margins are subject to many areas of risk such as material price increases or shortages, labor shortages, adverse weather conditions, and so on. Consequently, the parties tend to have little intercommunication and often proceed as quickly as possible, frequently resulting in work that interrupts the sequence of upstream tasks. There have been many examples of plumbers and HVAC subcontractors who competed for the same ceiling space, resulting in poor layouts for both disciplines; or the drywall subcontractor who completed an interior wall before the electrical subcontractor was able to place the required conduit and outlet boxes.

Typically, projects are subdivided into activities; in turn the resource requirements and time frames for each activity are placed on a CPM chart, guided by a master schedule. Constructors are in fact construction managers who subcontract activities, generally by discipline. Once the work starts, the CM attempts to control it by comparing progress with preplanned schedules, usually at weekly meetings.

- Costs are reduced by finding ways to increase productivity—such as use of special equipment or better methods
- Activity durations are adjusted by work force or other adjustments
- Attempts are made to improve quality and safety by inspection and enforcement

Unfortunately, these efforts to improve productivity have limited impact on the overall project. As schedules could quickly go off track, CMs are often in a reactive mode, adjusting staffing or sequencing to maintain schedules. Errors often result from poor communication and unplanned events. Attempts to get back on schedule may lead to cost overruns, quality deficiencies, or lapses in safety practices.

Traditional project management is very limited in its ability to reduce project variability. It uses the critical path method (CPM) to establish overall project schedules and keeps track of whether projects are on time or not. Gantt charts show critical timelines. The master schedule defines construction milestones and tasks but does not give workers incentive to work together or define best methods.

Earned value analysis helps project managers to evaluate performance. However, these tools do not control project variability. Even though these tools are available to project managers, projects are still delayed and have cost overruns. Traditional project management also involves a culture of "pushing" work assignments to subcontractors in order to meet the master schedule, whether or not these producers have all the needed resources to complete those assignments in a given week.

Philosophical Differences between Lean Construction and Traditional Construction

Lean construction departs significantly from traditional project management practice. Processes are actively controlled, and metrics are used in planning system performance to assure reliable workflow and predict project outcomes. With lean methods performance is optimized at the project level, whereas current project management approaches reduce total performance by attempting to optimize each activity. Traditional construction approaches reward individual crew performance—crews may focus on their tasks to the detriment of other crews. In the lean approach, all involved disciplines are rewarded for completing major sections of the project. Lean construction succeeds by optimizing at the project level, as opposed to the local optimization of an individual subcontractor.

Lean construction involves the basic infrastructure that is used in conventional construction. It requires an infrastructure of design professionals, construction organizations (builders/contactors), trades subcontractors, and project managers. These roles are described in the Project Management Body of Knowledge (PMBOK) established by the Project Management Institute (PMI). However, lean involves a philosophy that builds on the basic infrastructure and institutes a different approach to managing projects.

Lean construction involves better short-term planning and control that improves the timely completion of job tasks, and reduces the variability of work output that tends to happen with traditional project management methods. It emphasizes having work flow between crews without interruption. Consequently there is more cooperation between these crews and a joint focus on completing the overall project as opposed to self-interest in their own work task.

Lean construction does not replace CPM or other tools that define the overall work schedule, but it works within them to improve the delivery of short-term assignments. CPM is seen as playing a strategic role, identifying major project milestones and the sequencing of critical activities, while lean planning is seen as tactical in nature, planning the work flow for the next four or five weeks. *It is difficult for CPM to plan and track individual short duration tasks.* One important principle in lean construction is to have the smooth work flow handled by the trades that are in direct contact with it. The "last planner" concept relies on empowering the foreman and crew leaders to decide on the specific tasks to be done in the following weeks, as they are closest to the work and (once attuned to the lean methodology) are more familiar with its relation to each person's capabilities.

The PMBOK views planning as the key to success, and management by control as most important to organize cost and time and achieve a specific scope of work. These requirements seem to be very firm and leave no room for deviation. They view projects as controllable and relatively predictable—based on the critical path analysis—if the long-established planning and control tools are utilized. They also view each project as unique and not very adaptable to the methods used in continuous operations.

Lean construction, on the other hand, recognizes the planning that is so important to the PMBOK is often limited in effectiveness by unplanned events that occur in almost every project. Lean construction promotes the application of cutting edge approaches used in manufacturing. The Last Planner® technique for scheduling is effective because it focuses on a short-time horizon so it becomes a technique for continuous production.

Lean construction focuses on adding value as opposed to controlling cost and schedule. It also focuses on flexibility and learning in order to cope with uncertainty and unplanned events. This is especially important with projects that are not routine and are very complex. The CPM is really an approximation of how work should be done and is not as effective in handling the details of what can actually be done.

Overall, the PMBOK works well with projects that are relatively simple and predictable. When there is a lot of complexity and uncertainty, lean construction is likely to be more effective. Such projects require learning new lessons as the project progresses and using this knowledge in planning successive stages of the work.

Barriers to Applying Manufacturing Methods to Construction

Researchers have described the obstacles that limit the application of manufacturing techniques to the construction industry, and referred to the differences between the respective fields as a major factor. Consequently the lean tools and techniques that have been successfully applied in manufacturing have to be adapted to the construction environment. These differences include

- Construction "products" are made in response to individualized orders. The same design is rarely used more than once, whereas most manufacturing is repetitious.
- Manufacturing activity is usually carried out in a controlled environment—generally indoors, but the location of construction projects is not constant—it differs from one project to the next.
- Manufacturing processes can be designed in detail—this lends to optimization.
- Construction processes are determined by crews in the field where there is the potential for much variability.
- Manufacturing projects tend to be of short duration and are more easily quantified and more easily improved in real time. On the other hand, construction projects are generally of long duration and more complex.
- Owners often commission architectural design and construction separately—working relations between architects/engineers, contractors and subcontractors vary on a project-by-project basis.
- The construction industry has become highly specialized, partly because of liability concerns. Designers document the desired facilities with drawings and specifications that relegate construction details to the contractor.
- Organization of and facilities for production are dispensed with after project completion.
- At completion, each project is appraised differently.

Characteristics of Lean Construction

Lean methods are a departure from conventional methods and their adoption requires changing the behavior of people. Cultural change is the most compelling quest along with the physical transformation of an organization. One cannot force change on people. They have to be engaged so that the intrinsic satisfaction of outstanding performance will motivate them. Chapter 6 describes strategies for engaging construction stakeholders. In a lean culture, people have to be treated as the only appreciating asset. Lean enables organizations to have Responsiveness, Reliability, and Relevance.

Lean Principles

Five lean principles described by Womack and Jones (1996) apply to any organization. The principles are:

1. Value: It is critical to identify the value actually desired by customers and provide it. Lean organizations resist the tendency to persuade customers that they desire what the organization finds easiest to provide.

2. Value stream: Mapping the value stream for each product or service exposes waste and facilitates its removal; establishing cooperation between the participants and stakeholders results in lean processes.

3. Flow: It is necessary to make value creating steps flow. Business, job site, and supply flows depend on an effective value stream with few or no impediments.

4. Pull: Under the lean transformation, the efforts of all participants are to stabilize pulls in keeping with the demands of the customer.

5. Perfection: Strive for perfection, although it may never be achieved. Develop work instructions and procedures and establish quality controls.

Value

Value-adding activities transform materials and information into products and services needed by the customer. Value is not necessarily economic; in the process environment it is obtained when desired products and services are delivered. Wandahl and Bejder (2003) differentiate between product and process values. Product value relates to tangible aspects of a product, such as material composition, price of construction, flexibility, and so on. Process value in the construction environment relates to stages in the building process and interaction between producers such as time, communication, and so forth. The interaction between producers can best be represented by process value. Nonvalue-adding activities use up resources but do not directly contribute to the product or service. One example of value-adding compared with nonvalue-adding activities can be illustrated by considering the installation of an electrical outlet in a preinstalled outlet box on a construction site.

- The box has a length of electrical cable protruding from it.
- The electrician takes hold of the cable with one hand and reaches with the other for a wire stripper/cutter.
- The electrician cuts the wire to the appropriate length, and strips the insulation to expose ¾ inch of bare conductor for each wire. The electrician uses a pair of narrow-pointed pliers to make a loop on the bare conductors, in anticipation of securing them to connecting screws on the outlet. He or she may have misplaced

the pliers from a job done on the previous day. It could take five minutes to track down the tool.

- The electrician may then make a short search to find a new outlet device in the bottom of a tool box. On retrieving the outlet, the electrician may realize that it is not a ground fault interrupter (GFI) outlet specified in the plans, and then extends the search to find the right device. It does not take long—perhaps three minutes.

The steps required to attach the outlet and mount it to the outlet box could take two minutes, representing value-added activities. The time spent locating tools and the outlet took eight minutes—all of it nonvalue adding. In this case, a job duration of 10 minutes included 20% value-added and 80% NVA.

Value Stream

Wherever a process exists, it is accompanied by a value stream that represents the actions needed to create a product or service from its inception as raw material until it reaches a customer in its completed form. Value streams include both value-added and NVA activities. Value stream maps illustrate processes, identify sources of waste, and distinguish them from the value-adding activities that are critical to customers' needs.

Value Stream Mapping

Value stream mapping (VSM) shows material and information flows required to produce outputs. It helps users to understand the process, identify sources of waste by distinguishing between NVA and value-added activities. The VSM requires identifying each step in a process, with an emphasis on where action takes place. In drawing the map it is necessary to collect data on such factors as cycle time, changeover time, working time, scrap rates, and production batch sizes. The completed value stream map of the current state reveals areas that can be improved. Typically, the map is studied by a team that understands the process, and can make changes to the map to reflect recommended improvements. The map is changed to create a future state map with improved processes that maximize value-added time and minimize waste.

Flow

The goal of lean thinking is to have a continuous flow of a product from one activity to the next in a process without delays, stoppages, or storage as *work in process*. This concept is the so-called one-piece flow or single-piece flow. Examples of flow are:

- Business flow: Related with the information of a project (specifications, contracts, plans, etc.).
- Job site flow: Involves the activities and the way they have to be done.
- Supply flow: Refers to materials involved in a project. This is similar to any other supply chain.

Pull

In the JIT philosophy, production is "pulled" from the system by the customer demand. This is the antithesis of mass production that is based on large batch sizes—a "push" philosophy in which production forecasts project customers' needs and produce outputs based on those assumptions.

Perfection

Perfection may not be attainable in the construction environment, but represents a desirable future state—defects would be minimized, as would other categories of waste that delay delivery, incur additional cost, and detract from meeting the customers' needs.

Systems Perspective of Lean

The time a product spends in a production system is an important measure of efficiency (Figure 3.2). Assumptions:

1. The cost of a product is related to the length of time in the system: the shorter the time, the lower the cost to the producer and customer.
2. The shorter the time in the system, the better the producer can meet the customer's delivery requirements.
3. The shorter the time in the system, the smaller the probability of operational problems.

Description of System Components

Move Time

It represents the time required to move a product or service from one work station to another or from a queue to a processing activity. Move time does not add value. Lean methods minimize move time by optimizing the layout of the production facilities.

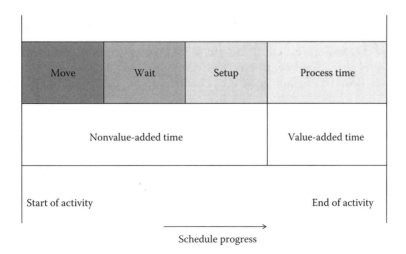

FIGURE 3.2
Impact of value-added vs. non-value added time in a typical construction process.

Wait Time

It is the sum of all phases in a system in which a product or service is waiting to be transformed. Efficiency is negatively impacted the longer the wait time, as it does not add value. This wait time is directly influenced by such issues as equipment downtime, material shortages, and unbalanced lines. Wait time is generally the longest phase in most systems (i.e., the greatest NVA activity).

Setup Time

Preparations are made for a process by adjusting equipment, material, procedures, and so forth in anticipation of processing activities. No value is added in this phase.

Process Time

This represents the only value-added phase; that is, the time a product is actually undergoing transformation by equipment and/or operators. It is the sum total of all processing activities.

Reducing or Eliminating Waste

Lean construction seeks to reduce or eliminate waste in construction activities. Improving the efficiency of individual activities does not necessarily improve the efficiency of an overall process. Waste can occur in the handoffs from one activity to another, or one trade to another, in the form of delays, defects that need to be corrected, and so on. Leanness is increased by reducing or eliminating the waste that occurs between activities in order to increase overall process efficiency.

   Adoption of the TPS applies the principles used in manufacturing to the construction environment. The TPS addresses:

1. Overproducing: When one discipline is so absorbed in self-interest, it may seek to carry out as much work as possible between payment cycles without considering the impact on other disciplines or trades. After all, the standard practice is for crews to request payment for work done within each payment cycle; the more work that has been completed, the greater the dollar value of payment requests. An overzealous drywall contractor may exceed a schedule and install drywall before an electrical subcontractor has completed conduit work for wall outlets or control devices. To get back on schedule, some partial demolition may be needed.

2. Idle time waste (waiting time/queue time): Poor coordination results in significant lost/idle time in projects. An impending concrete pour for a floor slab is planned by scheduling a team of journeymen and laborers who help to direct the pour from a concrete truck over the reinforcing bars and vibrate it to eliminate voids. A motorized float is used to level the surface. If the crew foreman did not have a firm commitment from the supplier, the crew could be left waiting for an hour or more because of a late delivery. Ten workers at an average of $20.00 per hour incur an idle time waste of $200 per hour. If the pumping equipment and the vibrators are rented, the idle time incurs additional cost as well.

3. Transporting/conveyance waste: Excessive transport of materials or equipment does not add value, but consumes resources. A poor site layout can result

in long-travel distances to bring materials to the point of use. For example, if drain pipes for installation underground are stored at a location far from the planned excavation, then the retrieval and transportation could incur significant expense. A front-end loader and its operator could consume additional hours in the process, whereas a more thoughtful location would reduce the transportation costs.

4. Processing waste—waste in the work itself: Studies have pointed to high levels of material waste in construction, depending on the region involved. Bossink and Brouwers (1996) found that 9% by weight of purchased materials are wasted. In the Dutch construction industry, from 1 to 10% by weight of each material item becomes solid waste. In Brazilian construction, 20–30% of purchased materials are not used and become waste. As material costs are 50–60% of project costs, material waste can cause cost overruns. Material waste could be due to:

Design	Poor anticipation of design impacts
Procurement	Wrong material or suboptimal order quantities
Material handling	Damage due to poor handling
Operation	Mistakes, poor use of material sizes
Inefficiencies	Poor waste management

5. Inventory waste (having unnecessary stock on hand): Holding excessive amounts of inventory (materials and supplies) has negative consequences:

   a. The cost of material is significant and involves tying up capital; this could negatively impact cash flow.

   b. Storing excessive inventory involves a carrying cost for storage.

   c. Managing the storage of materials long before they are needed runs the risk of spoilage or "shrinkage" (i.e., unexplained loss of material).

6. Wasted operator motion (using unnecessary motion): Inexperienced trades may not be familiar with best practices. Poor job organization could cause the operator to make additional steps to carry out the job. For example, experienced trades use templates or jigs for positioning bathroom fixtures, and setup kits with the appropriate tools and fasteners so that each installation proceeds quickly without risk of error or delay. Inexperienced workers have difficulty in attaining similar levels of productivity.

7. Producing defective goods (waste of rejected production): Construction that does not meet codes may be rejected by an inspector. It then has to be corrected by extensive, costly rework. Construction errors require correction and inhibit downstream activities because of the time delays that usually result.

Other Categories of Waste

"Making do": This waste is caused when a task is initiated without ensuring that all needed resources are available, such as some materials, or workers that have a skill for the work involved. It also occurs when information or equipment needed by the task are unavailable. If the resources become unavailable after the task has started, then continuing the task is also considered to be making do. It is also seen as equivalent to negative buffering (Koskela, 2004).

"Not speaking and not listening" are two great wastes described by Macomber and Howell (2004). These relate to the waste of human potential observed by Henry Ford in 1926—he thought that nurturing human potential through education would tap workers' creativity with positive results for themselves and for his firm (Ford 1926). Ford thought that failing to nurture this potential would result in degrading performance.

Not listening is caused by the command and control style of management, so prevalent in traditional construction projects; it makes those in power disinclined to listen to those who are closest to the problem. Management structures are established with rigid procedures in the interest of efficiency; these procedures often dictate a chain of command that obstructs communication from those in the trenches to the decision makers. This results in institutionalized not listening.

"Not speaking" comes as a conditioned response to overt or implied signals from management that saying something that runs counter to prevailing beliefs may label one as a complainer or even troublemaker. Children are culturally conditioned to not speak and that habit often continues into adulthood. Most individuals adopt the attitude that not speaking is a reasonable, prudent action. This inhibition denies management critical information about problems that can be averted, or about creative ideas that can benefit all stakeholders.

As Macomber and Howell imply, leaders who wish to be successful in lean construction should cultivate the art of listening as a master skill. They should also create the circumstances for others to speak.

Lean Construction Fundamentals

The Construction Industry Institute CII identified five lean construction principles in their study PT 191:

- Customer focus
- Culture and people
- Workplace organization and standardization
- Elimination of waste
- Continuous improvement and built-in quality

An earlier CII-sponsored study headed by Professor Victor Sanvido, investigated project delivery methods (Sanvido and Konchar 1999). The study findings identified several critical keys to project success:

1. A knowledgeable, trustworthy, and decisive facility owner/developer;
2. A team with relevant experience and chemistry assembled as early as possible, but certainly before 25% of the project design is complete; and
3. A contract that encourages and rewards organizations for behaving as a team.

Relational contracting has emerged as a means of embodying the concepts and success factors listed above; it serves as an excellent vehicle for lean construction.

1. Impeccable coordination

2. Organizing projects as production systems

3. Projects are a collective enterprise—optimize the project, not the pieces

FIGURE 3.3
Three connected opportunities. (Courtesy of Lean Project Consulting).

Three Connected Opportunities

Three connected opportunities in design and construction projects as a foundation for lean construction (Figure 3.3).

These opportunities are described as follows:

1. Impeccable coordination seeks to overcome the unpredictability that is typical of traditional construction projects; lack of coordination results in an average of only 55% or fewer of promised tasks being completed in a specific week as promised. Project success depends on the predictability of workflow that results when commitments are met between various disciplines and trades involved in a project.

2. Organizing projects as production systems align the roles of the parties in the project to maximize overall performance. It emphasizes productions system design to meet the owner's value proposition; conversations between contractors and designers inform the process of translating design to the built environment. Project execution strategies take advantage of technology or best practices such as prefabrication, modularization, and concurrent multitrade coordination.

3. Projects are a collective enterprise. Aligning financial incentives with project-wide optimization motivates project team members to adopt an investment mindset for improving performance. Sharing resources avoids expensive duplication and waste—the savings derived benefit both the team and the owner/client. Team orientation and trust are essential for mobilizing creativity and reducing waste.

Five Big Ideas™

The concept of these ideas proposed by Lean Project Consulting aligns the physics of work (how work gets done), organizations and contracts with lean methodology. It is especially valuable with large, complex projects. The development and application of these ideas is described in Chapter 6 (Figure 3.4). The following description is excerpted from a paper on a Sutter Health construction project (Lichtig 2005):

1. Collaborate, really collaborate

2. Increase relatedness among all project participants

3. Projects are networks of commitments

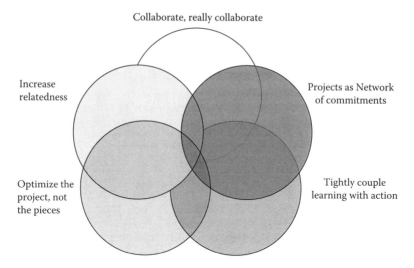

Collaborate, really collaborate

Increase
relatedness

Projects as Network
of commitments

Optimize the
project, not
the pieces

Tightly couple
learning with action

FIGURE 3.4
The Five Big Ideas. (The Five Big Ideas are the intellectual property of Lean Project Consulting.)

    4. Optimize the project, not the pieces
    5. Tightly couple action with learning

1. Collaboration overcomes the fragmentation that has come to characterize the design and building process. Expectation gaps between designers, builders, and owners lead to poor constructability and low quality. Close collaboration between team members secures the best decisions and outcomes in relation to design and construction choices. Design is an iterative conversation; the choice of ends affects means and available means affect ends. Collaborative design and planning maximizes positive iterations and reduces negative iterations.

2. Increase relatedness among all project participants. People come together on AEC projects as strangers. The chief impediment to transforming the design and delivery of capital projects is an insufficient relatedness of project participants. An active pursuit of relationships between team members promotes trust, openness, and learning that are essential for growth, as well as for creating a willingness to innovate.

3. Projects are networks of commitments. Projects are not processes. They are not value streams. Project management's role in the project environment is to activate and articulate unique networks of commitments between the many participants that form a project team. These commitments bind team members to manage and direct projects in real time. This contrasts with the common sense understanding that limits planning as predicting, managing as controlling, and leadership as setting direction.

4. Optimize the project, not the pieces. Conventional practice conditions project managers to push aggressively to maximize speed and minimize cost. Pushing for high productivity at the task level may maximize local performance but reduce the predictability of work released downstream. Lack of work reliability increases project durations, complicates coordination, and reduces trust. Project optimization involves having project teams collaborate to identify and implement solutions that are best

for the overall project, as opposed to meeting only the self-interest of specific team members. This approach avoids conflicts between trades, and increases site safety by as much as 50%.

5. Tightly couple action with learning. This big idea proposes that work should be carried out in a manner that facilitates ready observation of the results of specific actions. In order for this to occur, assessments have to be made at intervals that provide actionable information; discovering deficiencies long after the fact makes corrective action difficult. The lessons so learned should be used for continuous improvement of costs, schedule, and overall project value. Doing work as single-piece flow avoids producing batches that do not meet customer expectations. The traditional separation of planning, execution, and control contributes to poor project performance and to declining expectations of what is possible.

The observation of the foregoing principles facilitates the deployment of lean design and construction. In particular, they focus construction stakeholders on common objectives and commit them to collaborating closely in order to secure project success. Two important systems represent a critical foundation for lean construction deployment: The Lean Project Delivery System™ and The Last Planner® System. These systems are described in detail in subsequent chapters.

Questions for Discussion

1. How would you define lean construction?
2. How does lean construction differ from traditional construction delivery methods?
3. Why has the Toyota Production System been so successful in auto manufacturing?
4. How can the Toyota philosophy apply to construction, which lacks the uniformity and predictability of manufacturing cars in an enclosed building?
5. The "Command and Control" structure has worked well in the military and in many business situations. What are its shortcomings in construction?

   Describe the types of waste found in construction projects. Provide examples from the Design Phase. Provide examples from the Construction/Assembly Phase.

Appendix: ConsensusDocs

ConsensusDocs: (www.consensusdocs.org/) The Web site explains that, before ConsensusDOCS, there were a variety of construction associations that produced standard forms of construction contracts. Acceptance by other industry groups was low; standard contracts published by one association were perceived as ultimately favoring that association's membership. ConsensusDOCS is said to be widely accepted for their new contract documents, because all the parties were invited to the drafting table and had a full vote in deciding final contract terms.

ConsensusDocs Endorsing Organizations

American Subcontractors Association (ASA)
Associated Builders and Contractors (ABC)
Associated General Contractors of America (AGC)
Associated Specialty Contractors (ASC)
Association of the Wall and Ceiling Industry (AWCI)
Construction Financial Management Association (CFMA)
Construction Industry Roundtable (CIRT)
Construction Owners Association of America (COAA)
Construction Users' Roundtable (CURT)
Finishing Contractors Association (FCA)
Lean Construction Institute (LCI)
Mechanical Contractors Association of America (MCAA)
National Association of Electrical Distributors (NAED)
National Association of State Facilities Administrators (NAFSA)
National Association of Surety Bond Producers (NASBP)
National Electrical Contractors Association (NECA)
National Insulation Association (NIA)
National Roofing Contractors Association (NRCA)
National Subcontractors Alliance (NSA)
Painting and Decorating Contractors of America (PDCA)
Plumbing, Heating, Cooling Contractors Association (PHCC)
Sheet Metal and Air Conditioning Contractors' National Association (SMACNA)
Surety & Fidelity Association of America (SFAA)

References

Arabulu, R. and G. Ballard. 2004. Lean Supply Systems in Construction. *Proceedings, 12th International Conference on Lean Construction*, Helsingor, Denmark, 547–59.

Ballard, G. 2000a. *The Last Planner® System of production control*. PhD Thesis. Faculty of Engineering. School of Civil Engineering. Birmingham, AL: The University of Birmingham.

Ballard, G. 2000b. *Lean project delivery system™*. LCI White Paper—8. Lean Construction Institute.

Ballard, G., and G. Howell. 1994. *Implementing lean construction—Stabilizing work flow. Conference on lean construction*. Santiago, Chile, September 1994.

Bertelsen, S. 1993. *Construction logistics I and II, materials management in the construction process*. (in Danish) Denmark: Boligministeriet, Bygge-og, Boligstyrelsen, Kobenhavn.

Bertelsen, S., and J. Nielsen. 1997. Just-In-Time Logistics in the Supply of Building Materials. *1st International Conference on Construction Industry Development* Singapore.

Bertelsen, S., and L. Koskela. 2004. Construction beyond lean: A new understanding of construction management. Proceedings of the 12th annual conference of the international group for lean construction (IGLC-12). Elsinore, Denmark.

Bossink, B. A. G., and H. J. H. Brouwers. 1996. Construction waste: Quantification and source evaluation. *Journal of Construction Engineering and Management* March: 55–60.

Diekmann et al. 2004. *Executive summary, research report RT191*, 4. Austin, Texas: Construction Industry Institute.

Ford, H. 1926. *Today and tomorrow*. New York: Productivity Press.

Hammarlund, Y., and R. Ryden. 1989. Effektivitetet i VVS-branschen, Arbetstidens utnyjande, (Effectivity in the Plumbing Industry – the Use of the Working Hours, in Swedish), Svenska Byggbranschens utvecklingsfond, Sweden.

Harmon, R. D., and L. D. Peterson. 1990. *Reinventing the factory: Productivity breakthroughs in manufacturing today*. New York: The Free Press.

Hopp, W. J., and M. L. Spearman. 2000. *Factory physics: Foundations of manufacturing management*, 2nd ed. Boston: Irwin/McGraw-Hill.

Howell, G. A. 2000. White paper for Berkeley/Stanford CE & M workshop. Proceedings, construction engineering and management workshop. Palo Alto, CA: Stanford University.

Koskela, L. 1992. *Application of the new production philosophy to construction*. CIFE Technical Report #72, pp. 15–17, 24. Stanford, CA: Stanford University.

Koskela, L. 2002. The theory of project management: Explanation to novel methods. Proceedings of the international group for lean construction (IGLC-10), 10th Annual Conference. Gramado, Brazil.

Koskela, L. 2004. Making do- the eighth category of waste. *Proceedings of the Twelfth Annual Conference of the International Group for Lean Construction* (IGLC-12), Copenhagen, Denmark.

Krafcik, J. 1988. Triumph of the lean production system. *Sloan Management Review* 31(1): 41–52.

Lichtig, W. A. 2005. Sutter health: Developing a contracting model to support lean project delivery. *Lean Construction Journal* 2(1): 105–112.

Lichtig, W. A. 2006. The integrated agreement for lean project delivery. *American Bar Association, Construction Lawyer* 26(3).

Liker, J. K. 2004. *The Toyota way*. New York: McGraw-Hill.

Macomber, H., and G. Howell. 2003. Linguistic action: Contributing to the theory of lean construction. *Proc. Eleventh Annual Conference of the International Group for Lean Construction* (IGLC-11), Blacksburg, VA, USA.

Macomber, H., and G. Howell. 2004. Two great wastes in organizations. Proceedings of the 12th annual conference of the international group for lean construction (IGLC-12). Elsinore, Denmark.

Matthews, O., and G. Howell. 2005. Integrated project delivery an example of relational contracting. *Lean Construction Journal* 2(1): 46–61.

Nielsen, A., and E. Kristensen. 2001. Tidsstudie af vaegelementmontagen pa NOVI Park 6 (Time study of the erection of concrete walls on the NOVI Park 6 Project): Part of a not-published master thesis, Aalborg University.

Ohno, T. 1988. *Toyota production system*. Cambridge, MA: Productivity Press.

Pappas, M. 1990. *Evaluating innovative construction management methods through the assessment of intermediate impacts*. Master of Science Thesis, University of Texas at Austin (CII supported).

Sanvido, V., and M. Konchar. 1999. *Selecting project delivery systems: Comparing design build, design bid-build and construction management at risk (CII)*. State College, PA: The Project Delivery Institute.

Schonberger, R. J. 1986. *World class manufacturing: The lessons of simplicity applied*. New York: Free Press.

Wandahl, S., and E. Bejder. 2003. Value-based management in the supply chain of construction projects. *Proceedings of the Eleventh Annual Conference of the International Group for Lean Construction* (IGLC-11), Blacksburg, VA, USA.

Womack, J., and D. Jones. 1996. Lean thinking. New York: Simon & Schuster.

Womack, J., D. Jones, and D. Roos. 1990. *The machine that changed the world*. New York: Harper Collins.

Bibliography

Ballard, G. 1994. *The last planner. Northern California construction institute spring conference.* Monterey, CA, April 1994.

Howell, G. A. 1999. What is lean construction. Proceedings of the 7th annual conference of the international group for lean construction. Berkeley, CA: UCLA, pp. 2–10.

Howell, G. A., A. Laufer, and G. Ballard. 1993a. Interaction between subcycles: One key to improved methods. *Journal of Construction Engineering and Management, ASCE* 119(4).

Howell, G. A., A. Laufer, and G. Ballard. 1993b. Uncertainty and project objectives. *Project Appraisal Journal* 8:37–43. Guildford, England.

Macomber, H., and G. Howell. 2005. *Using study action teams to propel lean implementations.* Louisville, CO: Lean Project Consulting Inc.

Polat, G., and G. Ballard. 2004. Waste in Turkish Construction: Need for Lean Construction Techniques, *Proceedings of the Twelfth Annual Conference of the International Group for Lea Construction* (IGLC-12), Copenhagen, Denmark.

4

Lean Process Management

Operation of The Lean Project Delivery System™

The AEC community has long been aware of deficiencies in the design and construction processes that are evidenced by cost overruns, project delays, and quality and performance shortfalls in the finished construction. Poor design and documentation quality have been identified as a major factor in reducing the overall performance and efficiency of construction projects. Consequently, they have directly caused many projects to run over budget, over time schedules, and to be plagued with rework, change orders (variations), and disputes. Various case studies have identified design and documentation deficiency as the major contributor for construction contract variations.

The Lean Project Delivery System™ (LPDS) provides a means of addressing these shortcomings and improving the entire design and construction process (Ballard 2000; Ballard and Howell 2003). It was developed by Glenn Ballard in 2000 and subsequently refined. Whereas traditional industry practice has separated the roles of designers and constructors, the LPDS sees the activities of these professionals as a continuum for project management to achieve three fundamental goals (Koskela 2000):

- Deliver the product
- Maximize value
- Minimize waste

Fundamental to the LPDS is the deployment of the "Five Big Ideas" as described in Chapter 3:

1. Collaborate, really collaborate
2. Increase relatedness among all project participants
3. Projects as networks of commitments
4. Optimize the project, not the pieces
5. Tightly couple learning with action

The first big idea requires assembling and empowering of all the people and resources necessary to develop appropriate design solutions and explore their impacts on construction delivery. Inherent in the use of the LPDS is also the use of The Last Planner® System (LPS). It comprises

1. The master pull schedule
2. The look-ahead schedule
3. The weekly work plan (WWP)

Structure of The Lean Project Delivery System

The LPDS comprises a number of phases that capture the intent of the traditional project phases, but juxtaposes them in such a manner as to apply production system design principles to enhance the delivery of the entire project from predesign to completion and use. The phases are:

1. Project definition
2. Lean design
3. Lean supply
4. Lean assembly
5. Use/completion

The LPDS model represents these phases as a series of overlapping triangles. As described by Ballard and Howell (2002), these phases influence each other "so a conversation is necessary among the various stakeholders." In the diagram (Figure 4.1), work structuring provides a foundation for the process and sets the stage for production control, which is represented by a horizontal bar. Work structuring is a term developed by the Lean Construction Institute to indicate process design. It is a process of subdividing work such that pieces are different from one production unit to the next to promote flow and throughput, and to have work organized and executed to benefit the project as a whole. Production is redefined from "monitoring results" to "making work flow according to plan, replanning when it cannot." Plans to complete are continually updated in a proactive rather than reactive mode, looking ahead instead of backward.

The LPDS was updated to include additional concepts and methods drawn from the Toyota Product Development System, especially target costing and set-based design. These have been adapted to the construction industry and integrated with computer modeling and relational forms of contract (Ballard 2008).

The LPDS model improves project delivery through the following characteristics:

1. Downstream stakeholders are involved in front-end planning and design through cross-functional teams.
2. Project control has the job of execution as opposed to reliance on after-the-fact variance detection.
3. Pull techniques are used to govern the flow of materials and information through networks of cooperating specialists.

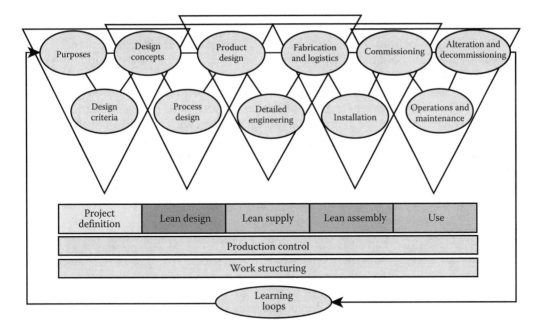

FIGURE 4.1
The Lean Project Delivery System™. (Adapted from Ballard, G. and Howell, G., *Building Research and Information*, 2003a)

4. Capacity and inventory buffers are used to absorb variability.
5. Feedback loops are incorporated at every level, dedicated to rapid system adjustment (i.e., learning).

Project Definition

Project definition typically involves developing project alternatives at a conceptual level, analyzing project risks and the economic payoff, and developing a financial plan. A decision is made to proceed if the owner considers conditions to be favorable; a plan is developed for the organization and control of the project. Effective project scope definition enables all involved parties to understand the owners needs, and to work toward meeting those needs. The project definition phase comprises:

- Needs and values determination
- Design criteria
- Conceptual design

In the needs and values determination, design professionals assist the owner/client in clarifying a value proposition; that is, the purpose of the project and the needs to be served. A design criteria document describes specific needs to be met, such as size, space proximities/adjacencies, and energy efficiency requirements. The conceptual design uses the design criteria and value proposition to define an outline design that serves as a starting point for the design phase.

Establishing Design Criteria

The design criteria defines the owner's basis of design (BOD). It reflects the owner's needs and wants that must be satisfied by the design of the project. This may include the use of spaces, their sizes, finishes, and activities to be performed within them. For example, a room that houses equipment with a high-sensible heat load will need power outlets/connections with adequate power capacity ratings. It will also need cooling designed to offset the respective heat loads. The aspect ratio—length versus width may influence the operations to be housed—a manufacturing assembly operation has different operational flow requirements than an office environment with a high-occupant load. Owners such as institutions—schools, universities, hospitals, and franchise retailers—may have formal design criteria developed through many years of experience. Other owners may require more intensive evaluation and analysis in order to develop project-specific design criteria.

Lean Design Phase

The lean design phase comprises: conceptual design, process design, and product design. It builds on the output from the project definition phase, but with a deviation from traditional design practice. Traditional design is somewhat linear. In architectural projects (as opposed to EPC projects) an architectural firm serves as the principal designer, and is supported by design engineers such as civil/structural, mechanical, and electrical. The design team often starts by designing the design process; they use Post-It notes on a wall and apply the Activity Definition Model (ADM) to ensure that design assignments have necessary prerequisite work completed and that no constraints would delay the process.

In conventional projects, the architect provides preliminary drawings to the other disciplines at various stages in the design process and the engineers apply their respective design parameters to the architectural framework. Although this work may benefit from the cost efficiencies of 2-D or 3-D CAD, the process is not generally interactive—it is linear and additive. It is often punctuated by design changes originating from scope changes or value engineering exercises necessitated when someone determines that the budget may not accommodate the design features that have been developed. There is very little involvement, if any, of contractors/builders in the design process in design-bid-build projects.

In design-build projects, a single entity contracts with the owner/client; there is an interplay between designers and builders, but the owner/client is not generally included in this process and does not have a fiduciary relationship with the designer of record. In CM@ Risk the builder is brought on board with the primary motive of becoming familiar with the project and its environment; the risk of unforeseen conditions is reduced.

The foregoing traditional methods separate design from construction—designers focus on product design and not process design in order not to assume the constructor's risk. This risk avoidance has other negative consequences. Research findings suggest that designers can in fact have a strong influence on construction safety. In 1985 the International Labor Office recommended that designers give consideration to the safety of workers who are involved in erecting buildings. In 1991, the European Foundation for the Improvement of Living and Working Conditions concluded that about 60% of fatal accidents in construction are the result of decisions made before the site work begins. A 1994 study of the United Kingdom's construction industry found a causal link between design decisions and safe construction.

As shown in the second triangle, lean design is a significant departure from the foregoing scenarios. Design is done with both the construction product and the process in mind. Constructability reviews and value engineering are not seen as tools to apply in a problem-solving mode, but rather, are continually integrated with decision making in the design process. This is accomplished with cross-functional teams that include architects/engineers, contractors, subcontractors, and various specialists who collaborate with each other and interactively make decisions that are optimal for the product and the process.

Lean Supply

The third triangle, lean supply, comprises product design, detailed engineering, fabrication, (Ballard and Howell 2003a); it requires up-front product and process design to define what is needed and when it should be delivered. This is especially important with engineered to order components as utilized in EPC projects. Lean supply also includes reducing the lead time for project information requirements.

Traditional projects depend on procurement specialists or buyers to ensure that materials are available for installation; they operate as a functional silo that is decoupled from project workflow. Flaws in this process result in material shortages or incorrect materials on site when they are needed that often cause serious project delays and losses in construction quality. Alternatively, material inventories accumulate, creating an unsafe condition on-site, while tying up scarce capital. These situations represent examples of construction waste. Lean supply addresses these problems through three main approaches (Arbulu and Koerckel, 2005): (1) improving workflow reliability—maintain constraint identification and removal, (2) using web-based project management software to increase transparency across value streams, and (3) Linking production workflow with material supply.

Lean Assembly

Lean assembly is practiced in the actual construction of a project, putting materials, systems, and components in place to create a completed facility. As described later in the chapter, in The Last Planner® System of production control (LPS), work structuring culminates in the form of schedules that represent specific project goals. Schedules are created for each phase of the project, beginning at the design phase and ending at project completion. The production control provided by the LPS deploys the activities necessary to accomplish those schedules. As shown in the diagram (Figure 4.1) production control and work structuring refer to the management of production throughout the project.

The lean assembly triangle contains fabrication and logistics, installation, and commissioning. Prefabrication is a production technique that can enable a contractor to operate in a lean manner by reducing the many nonvalue-added steps that are required in field fabrication. Whereas field work contends with uncertain conditions such as weather, and limitations in the availability of skilled labor, material, and equipment, shop fabrication benefits from a predictable, controlled environment. Tweet/Garot, a specialty contractor in Wisconsin has improved efficiency by prefabricating many of its products for quick and efficient site installation. This includes plumbing work for bathrooms in a given building in congruence with the construction schedule. Over a period of time, Tweet/Garot has progressed to 70% installation in the field and 30% fabrication in the shop. With that new approach, workers spend a relatively short period of time doing installation work. Prefabricated piping is interconnected within minutes because of the precision of design.

The commissioning process provides quality assurance before a facility is completed and accepted by the owner by ensuring that all systems have been installed as promised in keeping with the designer's plans and specifications. It improves the probability that the facility will meet the owner's project requirements (OPR) and performs as expected, to provide user satisfaction.

"Use" refers to a completed facility. Following successful commissioning, the facility should undergo a protracted operations and maintenance phase as it is used. "Alteration and Decommissioning" refers to a future activity when the facility may be repaired, renovated, or taken out of service. "Learning loops" refer to the application of root cause analysis to the LPS on a weekly basis to review PPC values and commitment reliability. Learning is also accomplished through the process of post occupancy evaluation in which a facility is surveyed after occupancy/acceptance to review the consequences of decisions made during the execution of the project. The POE enables project participants to learn from the past.

Production Control and Work Structuring

Production control is represented as a horizontal bar that extends from the very inception of the project to its conclusion (Figure 4.1). It consists of work flow control and production unit control. Work structuring and production control are used throughout a project to manage production. The term work structuring was developed by the Lean Construction Institute (Ballard 1999) to describe construction-related process design. It is a process of subdividing work so pieces are different from one production unit to the next to promote flow and throughput, and to have work organized and executed to benefit the project as a whole. Work structuring is described in greater detail as a part of the LPS.

Lean Design Details

Building design is a complex undertaking. It involves developing a value proposition that represents the owner/client's needs, balancing those needs with available budgets that are frequently underfunded, and meeting codes that are increasingly restrictive. Design professionals have to work with complex technologies and pay close attention to environmental standards in order to configure high-performing buildings that meet twenty-first century standards.

Lean construction advocates simultaneous product design and process design. It contrasts with traditional construction that is based on specialization of the parties—architects/engineers design a facility, convert the design to plans, specifications, and bid documents. These instruments are used as source information for builders/contractors to compete with each other for the award of a contract—and subsequently become contract documents.

Contractors, subcontractors, and suppliers then determine how the design is to be converted into a working facility that serves the needs of the owner. This approach severely limits the builder in exploring options for the production of the facility. It is prone to error and often results in costly rework.

Crosby (1979) considered that the impact of rework at different stages of a project could be defined with the "1/10/100" rule. This rule in construction would mean that changes made during the predesign phase would have an impact of $1 to the project, $10 during the design phase, and up to $100 after the beginning of construction. If rework is deferred until after project completion, the cost to implement changes not picked up previously could be as high as $1000!

Opportunities to capture the expertise of the construction team and develop cost-effective solutions are more difficult after the design phase because design review becomes more costly (complex) and resistance to change becomes greater (Figure 4.2).

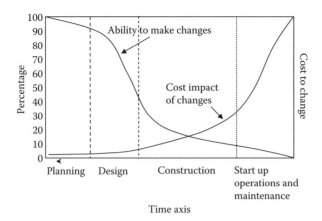

FIGURE 4.2
Relationship between time and impact of design changes.

With the design process generally being managed by traditional project management methods, there is an obvious link between the way design and documentation are managed and the poor level of performance being achieved. For that reason, research is currently being carried out in relation to lean design management, a methodology that promotes the implementation of lean production philosophies within the design and construction process.

Building information modeling (BIM) is a highly productive tool for carrying out lean design. It generally uses 3-D, real time, dynamic building modeling software to increase productivity. A building information model is created for each facility, representing it in terms of building geometry, spatial relationships, geographic information, and quantities and properties of building components.

Lean design with BIM was successfully used on the El Camino Medical Group campus in Mountain View, California. A 250,000 square foot medical office building was included in the project, as well as a 420,000 square foot parking structure, at a cost of $94.5 million. The project schedule was highly accelerated for early occupancy, hence it was necessary to utilize concurrent engineering approaches to start construction activity while the design was still in progress. The general contractor and a number of mechanical subcontractors were engaged during the design phases at approximately the same time as the architect and the structural engineers. The contractors and designers collaborated on the design to derive maximum constructability, lower cost, and an aggressive construction schedule. The team collaborated with a virtual model of the project; design focused not only on the product but also on the construction schedule, including material supply and prefabrication activities. The results at the end of the project were:

- The project was completed six months earlier than would have been achieved using the traditional design-bid-build project delivery method without BIM and lean techniques.
- It was completed under budget.
- Labor productivity was 15–30% better than the industry standard.
- There were no change orders related to field conflict issues.
- There were no field conflicts between the systems coordinated using BIM.

As was experienced in this project, 3-D modeling is one of the tools for improving both the design and construction processes. It not only helps designers to visualize and avoid potential design conflicts between different trades; it simultaneously generates bills of materials. Three dimensional modeling can also be used to simulate the facility design and the construction and fabrication process. The 3-D Modeling is described in greater detail in Chapter 8.

Lean Design Management

The British Institute of Architectural Technologists (BIAT) defines the purpose of lean design: "to improve manufacturability through attention to information coordination and flows at the outset of the project." The BIAT suggests that lean design management should follow the lean thinking principles as follows:

1. Understanding how value is delivered
2. Identifying the customer's point of view of value
3. Achieving flow within the process as waste is removed
4. Achieving pull so information is delivered when needed
5. Constant pursuit of improvement (perfection)

Some tools and techniques suggested for lean design are:

1. Organize cross-functional teams, involve downstream players in upstream decisions.
2. Pursue a set-based design strategy; carry out a simultaneous design of alternatives from which to make the best choice.
3. Minimize negative iteration: Use pull scheduling and a design structure matrix.
4. Apply production control: utilize a small design batch approach, evaluating against project requirements purpose.
5. Use technologies that facilitate lean design such as a web-based interface, BIM.

Designing for Lean Operations

As a design is developed, careful consideration should be given to facilitating lean operations in the completed facility. A poorly laid out facility could be relatively expensive to operate—excessive travel from one point to another is one of the seven wastes documented in the Toyota method. In the case of manufacturing facilities, it would be a good idea to include an industrial engineer with experience in facilities layout. In some cases the size of the expansion may be reduced—the expansion may not even be necessary.

By way of an example, the president of a window manufacturing operation felt a need for more square footage to house the equipment needed for increased production. The factory was located in a warehouse district with adjacent buildings. He acquired additional space from one of these buildings for expansion. Shortly after that he became interested in lean production, and contacted a consultant who specialized in Kaizen for help. Unexpectedly the consultant asked over the phone if he had recently bought additional space. The president was taken by surprise—why would this stranger care if he had bought more space? The consultant explained that by the time clients contacted him they were already having production bottlenecks and felt they needed to expand their facilities. After implementing Kaizen or other lean approaches they often discovered that the original square footage was more than enough!

Once it is determined that increased building size is necessary, then all the stakeholders should be included to the greatest extent possible. They may include:

- Building users
- Designers: architects and/or engineers
- GC/CMs, subcontractors
- Suppliers
- Specialty consultants: sustainability, commissioning

Lean operations and target value design (TVD): TVD involves designing to a specific estimate rather than reducing a design to meet a stated budget. It supports lean by avoiding the wastes incurred in scaling back a completed design. The TVD process was used in one of the projects for Sutter Health System in California. It was decided to apply TVD such that 90% of the program for the facility should be accommodated in 70% of the initially programmed space. Discussions with users (nurses, doctors, and other medical personnel) identified a more efficient layout that met operational needs than was originally thought. Through the TVD process, the project cost was reduced to 13% below the market cost average. In effect, the space requirement of the new facility was reduced by approximately 30% below the originally anticipated square footage. The savings derived from that exercise allowed the owners to include additional features that were excluded due to budget limitations.

Sustainability Issues*

Contemporary design criteria should include responsible sustainability practices. This design must be based on reducing the carbon footprint, including lower energy usage and less utilization of potable water. Buildings account for one-sixth of the world's fresh water usage, one-quarter of its wood harvest, and two-fifths of its material and energy flows. In the United States, buildings and their landscapes are a major source of pollution, using one-third of all energy consumed, producing more than one-third of the carbon dioxide emissions and almost one-half the sulfur dioxide emissions. A typical construction project generates 2-½ pounds of waste per square foot of completed floor space; construction waste accounts for 40% of the total U.S. waste stream.

So-called green buildings use energy, potable water, and other resources more efficiently. They also yield improved indoor air quality and higher worker productivity. The U.S. Green Building Council's leadership in energy and environmental design (LEED) certification system rewards sustainable practices in design and construction with a ranking system that

---

* This subject is covered in detail in Chapter 10.

reflects each facility's compliance with "best" environmental practices. Research findings suggest that LEED-certified buildings have a 3.8% higher occupancy than non-LEED-certified buildings. Buildings that meet the EPA's Energy Star standards are said to have a 3.6% higher occupancy than unrated buildings. Buildings that meet either LEED or Energy Star standards have been shown to have lower energy usage and command higher rental rates.

The adoption of Green Building standards may significantly influence a facility's design criteria and the OPR (see also Chapter 9). It may also increase the number of decision makers involved in the design and construction process, for beneficial reasons. In order to fully apply sustainability principles to project design, the design team should include a professional with recognized credentials in that field. The designation of a LEED accredited professional (LEED® AP) is one of the best established.

LEED projects generally require an accredited professional to serve as a resource in a project from the definition phase through to the occupancy phase. They also generally require a commissioning agent (CxA) for either fundamental commissioning or enhanced commissioning, to serve as an owner's representative and participate in policy-level project decisions.

Sustainability standards are an important component of a facility's design criteria and the owner's project requirements (OPR). These standards require decisions and support structures as early as the project definition and preliminary design phases. The LEED standards, for example, are subdivided into six categories:

Sustainable sites	SS
Water efficiency	WE
Energy and atmosphere	EA
Materials and resources	MR
Indoor environmental quality	EQ
Innovation in design	ID

In order for a project to comply with LEED standards, many early decisions have to be made to set the stage for the conduct of the project. For example, the EA category has the potential for a high-point score because it includes HVAC equipment. Enhanced commissioning contributes to this score; it involves hiring a CxA during the design phase to oversee all commissioning activities. The CxA reviews the OPR, BOD, and design documents prior to the mid-construction documents phase.

Under the MR category, designers have to lay the groundwork for additional points through the choices of materials. LEED encourages the use of existing sites and buildings, including the reuse of building elements such as walls, floors and roofs. The use of rapidly renewable materials is also rewarded—these include bamboo, cotton insulation, agrifiber, linoleum, wheatboard, strawboard, and cork.

Set-Based Design

Set-based design is an approach that defers design decisions to the "last responsible moment" to allow for the evaluation of alternatives that improve constructability. Toyota has used this technique to design new models in a much shorter time than the industry standard. They have learned that the best solutions are hybrids of original design options.

Set-based design involves deep thinking at each level of design for systems, subsystems, assemblies, and components. By having decisions made at the last responsible moment, set-based design avoids loop backs that may need to be made later because of hasty decisions, as past issues have to be revisited. Value stream mapping helps to identify the last responsible moment. While this approach consumes much time and effort from knowledgeable professionals, it derives the best collective decisions.

Target-Value Design

Target value design (TVD) involves designing to a specific estimate instead of estimating based on a detailed design. It seeks to address the problem that affects many projects—various design disciplines work from a common schematic design to do design development in their areas of expertise. Working in their respective offices they are subject to "project creep" and "project growth"; with little cross-functional collaboration, project designs often become overpriced, unconstructable, and behind schedule. Corrective action may include a misuse of value engineering to radically cut the scope of a project, or to suppress certain features that are desirable but unaffordable. Furthermore, the lack of collaboration often results in early design decisions that are later found to be suboptimal, but may be difficult to change. Ultimately, much time and effort are wasted, and the design cycle is longer than it should be. These wastes run counter to the lean philosophy.

- TVD proposes that designers should "do it right the first time" and build constructability into their designs, as opposed to designing first and then evaluating constructability later.
- TVD recommends concurrent design, with various disciplines in ongoing contact, as opposed to periodic reviews.
- Solution sets should be carried forward in the design process to ensure that good alternatives can be available later.

Macomber, Howell and Barberio (2005) propose a number of foundational practices for target value design. These practices promote design conversation, as they see design as a social activity that involves several professionals focusing on meeting the needs of the client. This approach is especially effective in light of the fact that the client's needs can change over time and value assessments need to be repeatedly made to ensure that design decisions meet these needs.

1. Conduct design activities in a Big Room (*Obeya* in Japanese). Toyota has used it successfully, especially in product development, to enhance effective and timely communication. The Obeya is similar in concept to traditional war rooms, and will contain charts, pictures, and graphs on display boards that visually represent program timing, milestones, and progress to date. The boards also display actions or recommendations to resolve delays or technical problems. The Obeya houses project leaders and key staff in close but comfortable proximity to shorten the communication cycle and promote an effective plan, do, check, and act (PDCA) cycle. Spontaneity comes easily as specialists can collaborate readily in key design or construction decisions.

2. Work closely with the client to establish the target value. Designers should guide clients to establish what represents value and how that value is produced. They should ensure that clients are active participants in the process, not passive customers.

3. Once the target value is established, use it to work with a detailed estimate. Have the design team develop a method for estimating the cost of design alternatives as they are developed. Deviations should not continue unchecked; if a particular design feature exceeds the budget allocated for it, then that design should be adjusted promptly in order not to abort further design work that cannot be accepted.

4. Apply concurrent design principles to design both the product and the process that will produce it. This work should be done as a collaboration between architects/engineers, specialty designers, contractors/subcontractors, and the owner. Be flexible to include innovation in this process. Practice reviewing and approving design work as it progresses.

5. Working in small, diverse groups is best. Groups of eight or fewer people establish better group dynamics; it is easier to create a spirit of collegiality and trust that lead in turn to more innovation and learning. Design with the customer in mind. Focus on designing in the sequence of the discipline that will use it. Use the "pull" approach with each design assignment to serve the next discipline. Lean is obtained by meeting downstream needs as opposed to producing what is convenient. Overproduction increases the possibility that the work so produced may not be what is needed for the next discipline to maintain its schedule, and it may lack the collaboration necessary for constructability.

6. Collaboratively plan and replan the project. Planning should involve all stakeholders to continually maintain an actionable schedule. Joint planning will refine practices of coordinating action. This will avoid delay, rework, and out-of-sequence design.

7. Lead the design effort for learning and innovation. Expect the team to learn and produce something surprising. Also expect surprise events to upset the current plan and require more replanning.

8. Learn by carrying out conversations on the results of each design cycle. Include all project participants in order to capture knowledge on success factors. Use this information as a part of the PDCA cycle, and use formal measurement systems, if possible.

A major effort has to be directed at the front end of the process by working with the client and determining what represents value and what it might cost. Subsequently the budget is allocated to various facility systems and cross functional teams collaborate to work within those limits. A critical requirement is for cost estimating to keep up with the design work as it progresses—with very frequent real-time updates. This approach varies substantially from traditional practice, which is highly linear. The budget becomes an influence on the design, rather than an outcome of design.

In general, for this process to work the construction staff must include members who can develop estimates from preliminary drawings instead of contract drawings and specifications. Designers need to be receptive to critical reviews of their designs with regard to both pricing and constructability. The owner (or representatives) must be accessible and be equipped to provide feedback on balancing project costs with specific building features. This may involve value engineering compromises of life-cycle costs versus form, function, and time. Target value design brings a standardized process to the team (Figure 4.3).

Target value design may be integrated with quality function deployment (QFD). QFD is relatively complex and requires prior experience in order to be effective, but it can pay

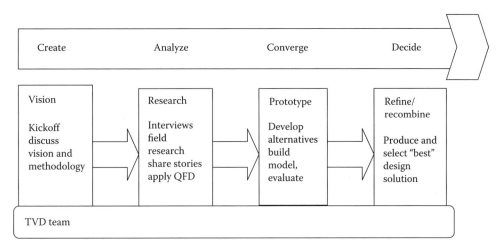

FIGURE 4.3
The target value design process.

significant dividends in capturing the OPR and quantifying the importance of desired features in order to balance their inclusion against budgetary and other limitations. The following outline is explained in greater detail in Chapter 11:

1. QFD serves as a road map for navigating the planning process and always keeping track of customer requirements and satisfaction. This actually helps eliminate human inefficiency.

2. The process of building a QFD matrix can be a good communication facilitator that helps break through barriers between the client and the designer and among members of the design team.

3. QFD can be an excellent tool for evaluating project alternatives, balancing conflicting project requirements, and establishing measurable project performance targets.

4. QFD can be used as a quick sensitivity test when project requirements change.

Ballard and Reiser (2004), documented one of the first projects in recent history to successfully apply TVD. The Saint Olaf College Fieldhouse project involved an athletic facility that was built between 2001 and 2002. The project was delivered on time, within budget, and with more value provided to the owner than comparable projects; the contractor (Boldt) also made a reasonable profit. A similar facility that was built in the region between 1998 and 2000, took 10 months longer to build, at a cost 18% higher.

Target value design was adopted as an essential part of the integrated form of agreement for lean project delivery used in projects for the Sutter Health System. The TVD process is intended to have a project designed within a desired budget in accordance with a detailed estimate, and establishes value, cost, schedule, and constructability as basic components of the design criteria. With the Sutter projects, there was a provision in the contract documents that a project's target cost should never be exceeded without the express approval of the owner. Design activity included a core group of contractors and subcontractors that blended considerations of constructability and the construction process.

Last Planner® System

This system is an important subset of The Lean Project Delivery System™ (LPDS) and is critical to its effective deployment. It accommodates project variability and smooths workflow so that labor and material resources can be maximally productive. In conventional construction management, there is a tendency for project managers to assign work schedules beyond crews' ability in the hope they can complete it. Research has shown that in traditional construction projects only approximately one-half of the tasks assigned for a given week are likely to be completed in that week. Ballard and Howell (1998) published measurement data on work flow variability; their studies of a wide range of projects identified an average plan failure rate of 54%.

Flow variability greatly influences lean practices, as a delay in work completion by one trade directly affects the downstream activities of the next trade. The LPS uses lean methods to provide improved project control. In essence, the last planner is in the best position to match labor and material resources to accomplish assignments in response to downstream demand. Work planning may establish time frames but is not very effective in establishing that the tasks assigned are capable of completion. Stops and starts typically occur or making do (i.e., keeping occupied with noncritical tasks). The LPS technique as refined by Ballard (2000) decentralizes decisions and empowers the crews that are in direct contact with the work to plan and schedule detailed tasks; in effect, they become the last planner. Foremen are empowered; they have a duty to say no to an assignment that fails to meet agreed criteria (Howell 1999).

Principles:

- Assumption: Construction project work plans are really forecasts, on the other hand forecasts are often wrong.
- Planning is more detailed the closer one gets to doing the work.
- Plans are made in collaboration with those who will do the work.
- Constraints should be identified and removed by a team effort.
- Construction team members must make reliable promises.
- Plan failures/breakdowns should be treated as an opportunity for learning, not for negative actions.

The LPS is based on three or four levels of schedules and planning tools. Certain projects such as fast-paced EPC projects and/or projects involving engineered-to-order components may require a daily work plan (Figure 4.4):

1. The master pull schedule
2. The look-ahead schedule
3. The WWP
4. A daily plan (when required)

Master Schedule

The master schedule represents an overall view that identifies major project phases and documents milestones for these phases. In traditional construction, the master schedule

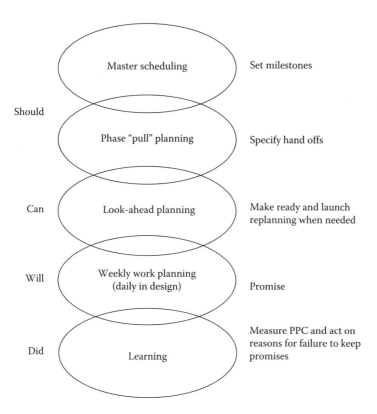

FIGURE 4.4
Interconnected Conversations in The Last Planner® System. (With permission from the Lean Construction Institute. www.leanconstruction.org)

is typically included in the contractor's bid. Hence it is an optimistic forecast of the major stages in carrying out a project, based on the early information provided in the contract documents.

Look-Ahead Schedule

The look-ahead schedule is derived from the phase plan, based on a time frame of 6–8 weeks for construction tasks, and is used for work flow control. It uses items pulled from the master pull schedule.

Weekly Work Plan

The WWP is derived from the look-ahead plan. It provides a detailed work plan specifying hand offs at each phase between trades. It is updated weekly, based on discussion between the last planners for various trades/disciplines; they may be design team leaders, trade foremen, or supervisors, depending on the complexity of the project.

Daily Work Plans

Some projects such as engineer, procure, and construct (EPC) construction that comprise engineered-to-order components need more precise control than is offered by the WWP.

Such engineered components are designated for installation by specific trades in specific locations. The emergence of constraints can quickly disrupt the construction schedule; in a fast moving project, a week's delay may result in significant schedule deviation. Establishing a daily commitment enhances coordination with other trades.

The so-called last planner is the foreman or other professional who prepares the weekly (or daily) planner schedule. This schedule also includes a buffer of work activities based on future work. The weekly accomplishment is measured as percent planned, complete (PPC). The PPC is based on 100% completion within a given week; tasks are considered to be either complete or not complete—there is no partial credit. The last planner also determines physical, specific work *assignments* for the next day. Work selection rules shield production from upstream uncertainty and variation. Production control is aimed at the avoidance of waste, such as waiting for someone to complete a task after the scheduled completion date or time.

Weekly meetings are used to evaluate the team's performance during the preceding week (i.e., what the team DID). Root cause analysis is carried out to determine the reasons for underperformance, and the knowledge gained is used to improve the process used in assigning work in subsequent weeks.

Communication between trades is an essential requirement for the LPS; many assignments require a close collaboration between various trades. For example, erecting a structural wall that contains electrical wiring and plumbing requires having electricians and plumbers work closely with form work carpenters and masons to ensure that conduit and piping are positioned and secured in time for a concrete pour. If either the electrical or mechanical trade fails to carry out their prescribed assignment on schedule, the concrete pour may be delayed—at great cost. Completing the pour without one of those trades may require time consuming corrective work, such as cutting a chase in the wall to retrofit the necessary pipes and/or conduits.

As has been noted previously, improving the efficiency of individual activities does not necessarily improve the efficiency of an overall process. Waste can occur in the hand offs from one activity to another in the form of delays, defects that need to be corrected, and so on.

Requirements for Successful Application

The lean construction philosophy views a project as a promise delivered by people working in a network of commitments supported by linguistic action (Macomber and Howell 2003). Smooth work flow is dependent on having the parties make and keep promises to carry out assignments. Reliable promises downstream enable others (especially upstream) to make reliable promises as well, in anticipation of predictable workflow. When all the parties keep their promises, waste is reduced, productivity is increased, and projects can be completed more rapidly. Promise-keeping by the parties is based on their ability to meet their promises. At the level of WWP, the trades in question must be aware of the scope of upcoming assignments. They must use this knowledge to determine the resource requirements—labor, equipment, materials, information, and so on. Above all, they should ensure they can deliver on their promises.

The reliability of the LPS and lean construction in general, hinges on informed commitments to maintain the trust that is essential for avoiding waste. If subcontractor (Sub) B depends on Sub A to have a work area available by a given day, then Sub B should be able to mobilize crews for that day without the reservation that Sub A might not keep its promises. Should that occur, Sub B could incur significant waste by having workers report for duty without being able to carry out the scheduled tasks. This waste would not only penalize Sub B, but would contribute to an ongoing spiral of distrust that characterizes traditional construction. Each trade schedules hesitantly, cautiously optimistic that other supporting trades may keep to the schedule, but anticipating more often than not that they will have to fall back on other tasks in order to keep the work force busy. This is the waste of making do.

As mentioned in Chapter 3, the LPS of production control sets up and manages schedules in order that:

- The flow of work is structured for the project.
- Workflow is planned within the structure.
- Work is controlled to accomplish the plan.
- Weekly measurement evaluates accomplishment and permits fine tuning for improvement.

Creating a Support System for Managing the Lean Process

A support system with assigned responsibilities is necessary for managing and sustaining the lean process; its size would depend on the size and complexity of the construction project. On a medium to large-sized project, the PM, field engineers, schedulers, and so on serve in a staff role and become facilitators and guardians of the lean process. They remove constraints and focus the team on continuous performance improvement. They maintain the master schedule and look-ahead schedules. They track the attainment of the master schedule by removing constraints and supporting the foremen/last planners, and they enable the foremen's daily work plans to have a higher level of reliability. They track major measures such as earned value. Foremen commit to performing work based on confirmation that resources or materials are available. They also need to confirm that prerequisite work has been done.

In the case of projects that involve a high percentage of engineered-to-order materials or systems, assembly is a very carefully ordered process that may require a fourth level of scheduling (i.e., a daily schedule; P. Gwynn, personal communication, September 2009). Materials and/or systems are designated for specific locations, with very little interchangeability. Examples of this type of project include processing plants and petroleum refineries, which involves large quantities of pipe sections that are engineered for a specific purpose and location. Items such as process piping and ductwork are configured to fit with other building elements. It is critical to have the appropriate systems available for installation in their respective locations. For this reason, daily planning is important for that type of project. Daily planning can also benefit AEC projects.

Information technology support would be needed for contemporary projects requiring the management of large quantities of data. Close collaboration is needed between multiple trades to ensure that hand offs from one work activity to another occurs without extensive delays. Premature starts on a successor activity also represent waste; hence the information on all is voluminous and requires computer support. Several software systems can be customized to the lean environment. Strategic Project Solutions is one of the systems that have been used successfully.

Lean construction applies the principle of autonomation that is an important ingredient of the Toyota Production System; the people closest to the work are empowered to stop production if they determine that the upstream production is defective. The last planners and their crew fulfill this role.

Work Structuring

Work structuring and production control are used throughout a project to manage production. It is a process of subdividing work so that the pieces are different from one production unit to the next to promote flow and throughput, and to have work organized and executed to benefit the project as a whole. Work structuring may be applied to both design processes and the construction processes.

In The Last Planner® System of production control, work structuring culminates in the form of schedules that represent specific project goals. Schedules are created for each phase of the project, beginning at the design phase, and ending at project completion. The production control provided by the LPS deploys the activities necessary to accomplish those schedules. In work structuring, a project is viewed as being comprised of production units (i.e., an individual or team performing construction tasks), and units of work that are capable of being handed from one production unit to another. Production units add value to each unit of work until it is completed (Tsao et al. 2000; Al Sudairi 2004). Work structuring requires determining:

- How work will be subdivided into units and assigned to specialists
- The sequence of these units
- The rules for releasing work from one production unit to the next
- Whether work will be performed in continuous flow or decoupled
- The locations and sizes of decoupling buffers
- The timing of various work units

Work structuring influences the size of work packages; that is, the quantity of work tasks planned for execution and this in turn affects the amount of work in progress (WIP). Koskela (1999) used derived the formula:

$$\text{Cycle time} = \text{WIP/Throughput.}$$

If the throughput is maintained constant, then reductions in WIP reduce cycle time.

This relationship was studied with a construction project at King Faisal University, Saudi Arabia. It was observed that reduction of WIP resulted in reduced cycle time; however, diminishing returns could result by having too many tasks. As the number of tasks increases, crews may need to spend a disproportionate amount of waiting time between tasks or in moving from one task to another. In a project involving cast-in-place concrete processes, for example, the curing time caused excessive delays between steps (Al Sudairi 2004).

Workable Backlog

Workable backlog is used to describe assignments that have met all quality criteria, but may need other prerequisite work to be done before they can be started. For example, the paving of the tennis court for a new university sports center is scheduled for 3 months time. It can be done now, as it is on a remote part of the site that has already been compacted and graded. It is not on the critical path. This assignment will be maintained as workable backlog.

Workable backlog enables crews to continue working productively if there is a constraint that prevents the completion of an item on the WWP. As a rule of thumb it is recommended in committing crews to no more than 75% of their capacity when developing the WWP. The crew can maintain its reliability in passing work on to the downstream crew, when called to do so as they "pull" activities, even if the weekly (or daily) schedule is interrupted.

Process Steps in The Last Planner System of Production Control

The LPS as depicted in Figure 4.5 consists of a number of activities that are noted in the respective symbols. These symbols represent three components that work together to provide production control: (1) look-ahead planning, (2) commitment planning, and (3) learning.

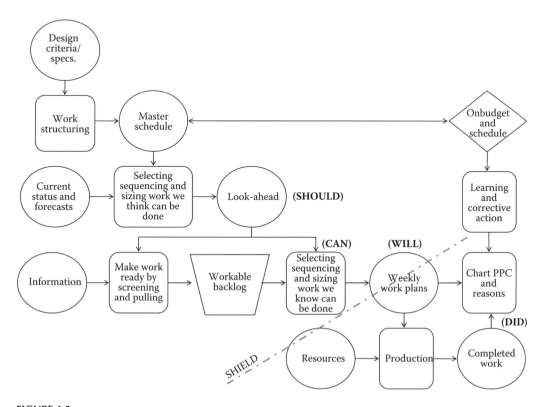

FIGURE 4.5
The Last Planner® System: process flow.

Comparison with Traditional Planning and Scheduling

Planning and scheduling are a standard requirement for construction projects in general. As construction activity is comprised of many activities and tasks, planning is a formal process that identifies what needs to be done, what resources are needed (such as labor, materials, and equipment), and when they are required. A schedule represents the plan in graphical form; typically a bar chart or critical path method (CPM) chart documents each activity on a horizontal time scale. This scale is graduated to represent days, weeks, months, or even years depending on the size and complexity of the project.

Procedurally, planning and scheduling to prepare a master schedule may involve the following steps:

1. The project is subdivided into major work activities.
2. The duration and work force requirements are determined for each activity.
3. The activities are placed in a logical sequence to form the master schedule. For example, an excavation has to be made and reinforcement placed before a concrete foundation can be poured.
4. Project duration is determined from the overall sequence of activities and their respective durations.
5. Overall duration is reconciled with the contractual requirements of each project. Activity durations are adjusted through resource utilization. For example, the time required for placing reinforcement can be accelerated by increasing the number of crews carrying out that activity.
6. The master schedule is further refined, using the CPM to (a) identify the critical path (i.e., the sequence of activities that directly affect overall project duration) and (b) identify "float" or "slack time" that determines how long an activity can be delayed without jeopardizing the schedule. This float enables construction managers to accommodate delays or unforeseen conditions such as bad weather without affecting the completion date.

CPM scheduling involves conducting calculations through the CPM diagram as forward or backward passes. A forward pass determines the earliest start time for each activity (EST). Adding the expected duration of the activity yields the earliest finish time (EFT). A backward pass through the CPM diagram yields the latest start time (LST). Adding the expected activity duration yields the latest finish time (LFT). Each activity is examined to determine the amount of float available; that is, the difference between the earliest start and latest start, or earliest finish and latest finish. In practice, some activities are interdependent and influence the overall schedule. For example, placing concrete blocks for the walls of a one-story structure has to be preceded by pouring the foundation, columns, and the floor slab. Placing the blocks cannot begin until those respective activities are completed, hence the EST for wall construction would be the EFT for the preceding activities. The overall schedule may leave some activities with a float of zero days (i.e., EST–LST = 0). Such activities cannot be delayed without delaying the overall project; they represent the critical path. The critical path is that sequence of activities throughout the entire project that have the minimum amount of total float available.

In traditional construction management production, control relies on the experience of construction managers. Master schedules are often generated preliminarily in the process of bid preparation, using estimators' takeoffs in conjunction with past experience. These schedules have significant limitations, yet they provide the basis of many projects. The available project management tools address what SHOULD be done to meet a master schedule or CPM schedule, as opposed to verifying what CAN be done. Decision makers at a level that is detached from the day-to-day activities in the field often lack the ability to ensure that scheduled work is within a crew's capabilities. They often present crews with optimistic "stretch" goals, which are clearly beyond their capabilities, in the hope that this will motivate them to above average performance. Workflow reliability on these projects often range between 30 and 60% and is linked to low industry performance.

In contrast, the LPS uses process-driven approaches for project control that improve workflow reliability. Improving workflow reliability is synonymous with increasing the accuracy of site demand, and enables planners to better match the supply of resources to that demand. This results in the accomplishment of a higher percentage of planned tasks. Figure 4.5 shows the process starting with work structuring leading to the master schedule. As indicated above, work structuring subdivides the project into manageable production units (Ballard and Howell 1994).

Look-Ahead Planning

Look-ahead planning involves taking activities from the master schedule and placing them in the Look-ahead schedule. This schedule represents a time frame that is generally 6–8 weeks in length for construction tasks. Look-ahead planning:

- Shapes work flow sequence and rate
- Matches work flow to system capacity
- Creates workable backlog: a backlog of work that is ready to be scheduled
- Prepares detailed work plans

Commitment Planning

This identifies work that CAN and WILL be done. It enables quality assignments. The criteria for these assignments are:

- Definition: The work should be specific such that its requirements and completeness are clear.
- Soundness: Prerequisites are identified and constraints are removed.
- Sequence: The sequence of activities should be best for constructability.
- Size: The assignments should be matched to the capacity of the crew.

Constraint-free assignments are kept in the block marked "workable backlog" and sequenced for the WWP. "Shield" indicates how this process shields downstream activities from upstream uncertainty and variation. Learning is based on using performance feedback to improve the system.

From the diagram in Figure 4.5, after WWP have been implemented and completed through the "production" block, the completed work is analyzed in the "Chart PPC and Reasons." Where assignments have not been completed, they are examined and studied through root cause analysis in order to improve future assignments.

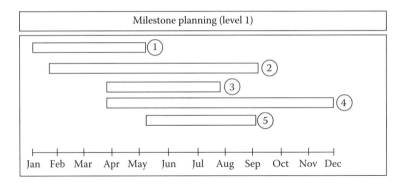

FIGURE 4.6
The master pull schedule. (From Gwynn, P., *ULSD Lean Production Management (LPM)*, Lean Implementation Services, Schereville, IN, 2008. With permission.)

The LPS is based on three levels of schedules and planning tools (Figure 4.6):

1. The master pull schedule: Level 1
2. The look-ahead schedule: Level 2
3. The WWP: Level 3
4. Certain projects may require a daily work plan: Level 4

Master Schedule

The master schedule may be used to establish project feasibility and likely duration. It identifies long lead-time items. If it is prepared before design work is completed, it establishes packages for action by designers and plans their sequencing and timing. The master schedule also includes infrastructure items based on:

- Estimated quantities
- Estimated craft density
- Standard rates for labor
- Labor distribution curves (industry based, or company based) and the appropriate craft density to complete the work safely and cost effectively.

The example illustrated above refers to an oil-refinery project. Much of the work involved installing prefabricated process piping that was stored in spools, each designated for a specific location.

There are significant limitations in the master schedule. It is generally prepared by people who are not going to do the work. It is tantamount to preplanning. While the master schedule is followed closely in traditional projects, it is further modified in lean construction projects. Phase scheduling, on the other hand, gets people involved who are actually going to do the work. For that reason it is generally more realistic to break down a master schedule into a look-ahead schedule and subsequently into a weekly or daily schedule.

Master and Phase Planning

This planning occurs in a meeting of project participants and stakeholders. The steps in the process are:

1. Have an agenda ready for the meeting
2. Introduce the entire team so that stakeholders will know each other
3. Ensure that all required project data are available
4. Use "Post It" notes on a wall-mounted board
5. Use a backward (Pull) approach for assignments
6. Promote creativity (with controlled chaos) to generate a broad cross section of ideas
7. Identify float and verify with stakeholders
8. Document the plan (SPS or MS project are examples of software that can be used for this purpose)
9. Review and fine tune the plan

Reverse Phase Scheduling

Reverse phase scheduling (RPS) is used as a starting point for the LPS. It is a detailed work plan that specifies the hand offs between trades for each project phase (Ballard and Howell 2003). In practice, subcontractors participate extensively in RPS as projects are generally comprised of activities performed by a number of subcontractors. They typically use Post-It notes on a wall-mounted board to plan their schedules in concert with the schedule for each project phase.

Traditional construction emphasizes pushing tasks to meet an optimistic master schedule, with little thought to what CAN and WILL be done in a given time frame. In LPS, the master schedule is refined as a reverse phase schedule by using a team to work backward from the expected completion date. This team comprises "last planners" from different trades/subcontractors. The master schedule and reverse phase schedule indicate what work SHOULD be done in order to meet the schedule.

The critical path is identified and an acceptable amount of float can be introduced to accommodate risk and uncertainty. This approach is based on the pull technique. The resulting phase plan provides a framework for dovetailing the work of various specialists that are needed to carry out their respective work plans. It is more realistic than the master schedule, but does not make allowances for conditions in the field that may fluctuate from week to week.

Comparison between Push Systems and Pull Systems

Production control systems are typically divided into two categories: push and pull systems. Push systems may be described as having production activities that are scheduled. Pull systems are those that have the start of one activity triggered by the completion of another (Spearman, Woodruff, and Hopp 1990). In the manufacturing environment, push systems work best with predictable rates of production (or cycle time), and with material supply systems that can keep pace. In practice, cycle times are often random while supply systems are designed with fixed lead times. With push systems the variability in cycle time often leaves the supply system incapable of keeping pace. Alternatively, it may result in large quantities of WIP in order to compensate.

Pull systems do not schedule the start of jobs, but rather authorize production. The kanban system is a pull system; as parts or components are used in downstream work stations, the kanban card is passed upstream to authorize the production of more parts or components. Studies have shown that there is a less variance in cycle time for a pull system than in an equivalent push system.

The disadvantage of the push philosophy is that most construction tasks are interdependent—a completed project has components that are interwoven to create a functioning system; many disciplines have to have an ongoing interaction with each other, and cannot work uninterrupted at full capacity throughout the project. While gains can be derived by improving the performance of individual trade work assignments, this improvement may not be reflected in the overall completion of a project; hence, lean construction seeks to optimize the entire project.

The flow of work between producers is a hallmark of lean manufacturing and is a major factor in its success. In the Just-In-Time (JIT) concept, tasks are released from each work station when they are "pulled" by a downstream station and not before; this maintains a one-directional flow of work assignments that takes into account the variability that inevitably occurs in matching workloads and capacities. The progress of cars on an expressway is an example of the principles of dependence and variation. There are gaps between the cars and if their speeds are similar these spaces become small. A slight hesitation or delay results in an immediate effect on the flow of cars. An individual car cannot progress until the car in front yields roadway. Lean construction seeks to adapt these principles to the construction environment.

Look-Ahead Schedule

The look-ahead schedule is derived from the phase plan, based on a time frame of 6–8 weeks for construction tasks and is used for work flow control (Figure 4.7). That duration has been identified by experience as the most flexible. Peter Gwynn of Lean Implementation Services points out that longer durations have been found to be unwieldy and subject to

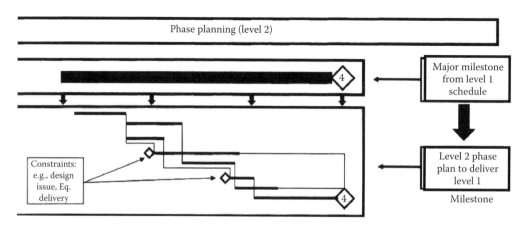

FIGURE 4.7
Level 2 phase schedule and six to eight week look-ahead. (From Gwynn, P., *ULSD Lean Production Management (LPM)*, Lean Implementation Services, Schereville, IN, 2008. With permission.)

unexpected events. It uses items "pulled" from the master pull schedule. It reflects major work items that need to be completed for the milestone dates in the master pull schedule. It identifies the work activities that CAN be done, within the constraints that have been indicated. In effect, it reduces uncertainty by developing assignments that have a high likelihood of successful completion. The trade foremen are specifically asked what CAN be done in light of issues such as weather conditions, availability of crews, availability of materials, and completion of prerequisite work. In the case of design work, the look-ahead schedule could be from 3 to 12 weeks duration. Constraints are identified so they can be planned for ahead of time. The purpose of look-ahead planning is (Ballard and Howell 2003):

- To shape the sequence and rate of work flow
- To match work flow and capacity
- To maintain a workable backlog
- To develop detailed plans for how work is to be performed

Look-ahead planning includes:

- Constraints analysis
- The ADM
- First run studies

Constraints analysis involves observing the rule that no activity can be kept to its schedule date unless the planners are sure there are no outstanding constraints. This approach ensures that only "sound" assignments are made that can be depended upon, be free of limitations, and will not cause problems downstream. Constraints analysis applies to the production aspects of projects, in design, fabrication, or construction.

In the case of construction, the supervisors who are tasked with executing the schedule are assigned to developing and maintaining the look-ahead schedule. They obtain input from the owner, designers, and material and equipment suppliers. They reconcile the work requirements with the contractor's labor estimates for the project. If there is a discrepancy between the look-ahead schedule and the Level 1 schedule, then replanning is done to bring them in line with each other.

Each week, the project team reviews the upcoming items to ensure that they are not obstructed by constraints and can be carried out predictably. If there is evidence of incomplete preparatory work, insufficient labor and/or materials, or unsatisfied requests for information (RFIs), such work assignments are considered defective and should be rejected. This review also includes: current progress, remaining duration estimates, and forecasted deliveries of equipment and materials. It results in a continual update on a workflow plan that is realistic and credible to all project stakeholders.

Look-ahead schedule items should meet the following criteria:

- Manageable size: Schedule items are small enough so they can be detailed to show downstream tasks that prompt work release.
- Readily measurable: Progress and remaining durations should be measurable.
- Free of constraints: It should be made clear where those constraints that have not been resolved are so as not to obstruct the flow of work.

Weekly Work Plan

This project included the installation of metallic piping in a refinery to convey liquids and gases from one process location to another. The piping is supplied to the site wound on large-diameter spools; it is unrolled and secured in place with hardware that is cut and welded in place.

The WWP is derived from the look-ahead plan. The weekly planner schedule delineates the work activities or assignments pulled from the look-ahead schedule that must be initiated to meet the completion dates in that schedule. Eligible activities or assignments are those that have no current constraints, and that have resources available and assigned. For example, responsible supervisors enhance the Level 2 activity into a detailed workflow or plan. The plan is laid out by geographic area and the materials, sequence, and tasks are listed for each area. The pipe spools are identified within each work package and sequenced for optimal installation. Required tasks for each pipe spool are based on experience and documented best practices for installation. This may involve using a library of standard processes, developed and improved over time.

Level 3 activity contains work packages for physical/geographic boundaries; packages are small enough to be

- Managed by one foreman,
- Rapidly replanned when change occurs, and
- Updated quickly and accurately by the foreman for completion.

In Figure 4.8, four chunks (areas) and spools are shown. They are listed on the sheet in a sequence (1–4) and the tasks involve welding, trim and support, and finally QA/QC inspection. In the WWP, constraints identified in the look-ahead schedule are an important consideration. In some projects, the format for updating may include the project manager and the superintendent, or combinations thereof. Each trade foreman develops a weekly work schedule; based on the input from its last planner (the trade foreman may be the last planner).

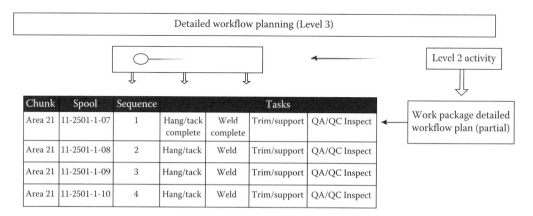

FIGURE 4.8
Level 3 weekly work plan. (From Gwynn, P., *ULSD Lean Production Management (LPM)*, Lean Implementation Services, Schereville, IN, 2008. With permission.)

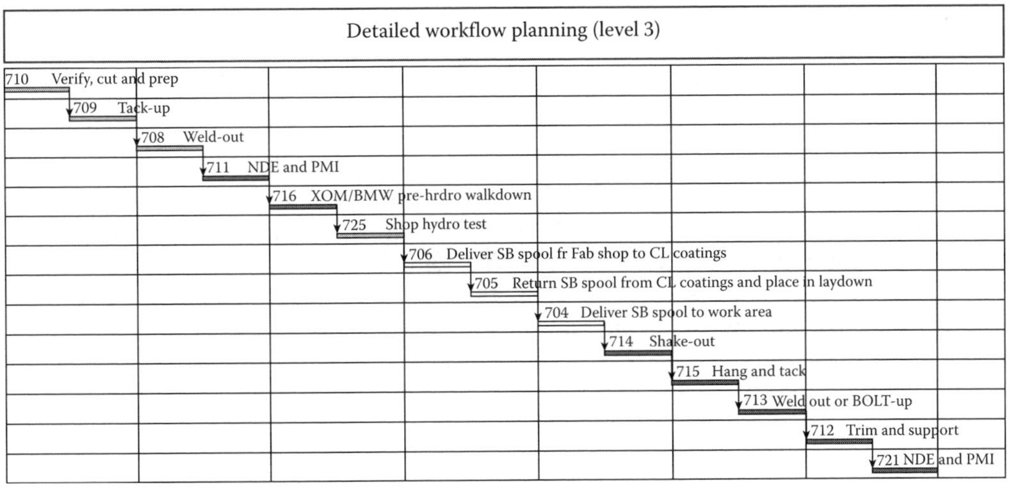

Detailed workflow planning (level 3)						
710 Verify, cut and prep						
↓709 Tack-up						
↓708 Weld-out					
↓711 NDE and PMI					
↓716 XOM/BMW pre-hrdro walkdown				
↓725 Shop hydro test				
↓706 Deliver SB spool fr Fab shop to CL coatings			
↓705 Return SB spool from CL coatings and place in laydown			
↓704 Deliver SB spool to work area		
↓714 Shake-out		
↓715 Hang and tack	
↓713 Weld out or BOLT-up	
↓712   Trim and support
↓721 NDE and PMI

FIGURE 4.9

Detailed workflow planning. Detailed workflow planning using standard processes on a complex project involving ~5000 engineered-to-order pipe spools. Standard workflow plan (Level 3) incorporating each of the ~5000 pipe spools within appropriately defined work packages. In this standard process example, a pipe spool's workflow requires: 14 tasks, five teams, six hand offs between teams, and transport between four site locations. (From Gwynn, P., *ULSD Lean Production Management (LPM)*, Lean Implementation Services, Schereville, IN, 2008. With permission from Lean Implementation Services.)

The detailed workflow plan represents work performed to install pipe spools for a complex project involving approximately 5000 spools (Figure 4.9). The work on these spools is organized into work packages.

Commitment planning is an essential component of weekly or daily planning. It is a method for defining criteria that lead to the selection of quality assignments (Ballard and Howell 1994). It results in commitments to deliver that other actors in the production system can rely on, by following a prerequisite that only sound assignments should be made or accepted. The foremen are empowered to reject any assignments that do not appear to be sound, as they are likely not to be completed and will reduce the PPC accomplishments. Despite this, foremen should be asked to confirm that each assignment that is scheduled WILL be completed in the coming week, based on preparations that are needed for this to occur.

Weekly Work Plan (WWP) meetings are held to review the work accomplished for the past week and plan the work for the following week. It is important to note that planning is really a *conversation*, not just a written document. Each WWP meeting is an essential part of the process—it depends on open and frank communication between all trades and their last planners. In some projects, it is appropriate to include suppliers such as rebar and truss and concrete suppliers. The issues discussed include the actual schedule, site conditions, safety issues, construction methods, manpower needs, and capabilities.

The WWP emphasizes what WILL be done in the upcoming week—from those tasks that CAN be done, based on the informed commitment by the involved trades. It considers "workable backlog" if it is needed as a buffer. The LPS is based on promises—it is imperative that the last planners promise what they WILL do only when they have verified beyond a doubt that they can carry them out.

Criteria for weekly planner schedule items:

- Detailed information: Level 2 tasks are subdivided into tasks that can be managed by one foreman.
- Soundness: Are the assignments workable? Do you understand what is required? Do you have what you need from others?
- Best sequence: Are the assignments selected from those that are sound in priority order and in order of workability? Will doing these assignments release work needed by someone else? Are additional lower priority assignments identified as workable backlog whenever possible?
- Size: Are the assignments sized to the productive capability of each individual or group while still being achievable?
- Learning: Are incomplete assignments tracked, reasons for noncompletion identified, and action taken?

Commitment Reliability

Completed work (DID) is compared with planned work (WILL). The PPC values are calculated. The reasons for noncompletion are recorded and analyzed in order to improve the process. Learning should be a positive experience. It should not be an occasion for casting blame!

Daily Work Plan

The daily workflow plan (Level 4) is appropriate for projects where there is extensive prefabrication and little or no interchangeability of materials/equipment. The Level 4 diagram is similar to the Level 3 weekly diagram, but the foremen commit daily to completing tasks based on the crew's capability and capacity. (See Chapter 7, Case C for a large EPC project.)

Constraint Analysis

A constraint is an obstacle that inhibits the execution of a task that is required in the lookahead plan. It should also be beyond the control of the last planner in learn the forepersons project managers, schedulers, and other staff guide the scheduling process and remove constraints so work tasks can be assigned. They have to maintain communication with the foremen, who as last planners, coordinate the work with their counterparts in other trades. For construction processes to be lean:

- Weekly work plans must be accurate.
- Completed work must be correctly reported.
- Constraints must be indicated early.

Examples of constraints are:

- Adverse weather conditions
- Unexpected events

- Open RFIs
- Design modifications

Constraints should be formally recorded in a log to indicate

- Date of record
- Requestor
- Responsible party—who should take action on the constraint
- Description of the constraint
- Tasks/activities affected by the constraint
- The date required for constraint removal (based on the look-ahead plan)
- Commitment date by the promissory
- The cost impact of the constraint
- Prerequisite work completed
- Resources that are unavailable, if any

Benefits of Constraint Analysis and Management Process

- If constraints are identified early, the chances of removal or mitigation are significantly improved (Table 4.1).
- It clarifies the corrective actions that are needed and is linked to the look-ahead plan.
- It serves as a positive course of action for problem resolution.
- It facilitates open, honest discussions by all parties on obstacles for schedule compliance.
- It documents events/actions for future reference and learning.
- It maintains a record that can be used for performance analysis and tracking.
- A formal process of constraint analysis and management improves the levels of PPC accomplished by work crews.
- It is also an important tool for continuous improvement by maintaining information on actions taken.

Activity Definition Model

The ADM is an important component of The Last Planner® System; it governs assignments that are moved from the phase schedule to the lookahead plan. It represents design tasks or construction processes and enables planners to enhance scheduled tasks to a level of detail where their readiness for execution can be determined. This evaluation requires a determination of constraints by examining directives to carry out tasks and comparing them with available resources as well as the status of prerequisite work. If the prerequisite work for a task has not been carried out, then the task cannot be carried through to completion even if the resources are available.

TABLE 4.1

Constraint Analysis Worksheet, Benefits of the Constraint Analysis and Management Process.

Item No.	Date	Requestor	Responsible Party	Description of Constraint	Date Removal Required	Commit Date	Cost Impact?	Prerequisite Work Complete?	Resources Available
1	3/11	JB	DB	Begin process of determining when control room floor plans will be available for planning electrical work	4/7	4/3	None	Yes	Yes

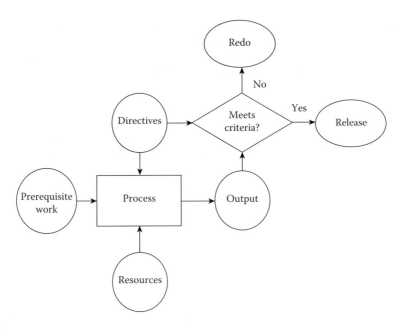

FIGURE 4.10
The Activity Definition Model. (From Ballard, G. and Howell, G., *Building Research and Information*, 2003a. With permission.)

As shown in the diagram of Figure 4.10, directives, prerequisite work, and resources are the three main sources of constraints. Directives may be assignments, design criteria, or specifications. Resources may be labor, tools, equipment, or space. Materials and information are not considered to be resources. These directives advise workers what tasks to carry out and how to execute them. The process ensures that prerequisite work has been carried out before passing transactions through to the output where it is examined in the decision node. Ready assignments are released for scheduling. Assignments that do not meet the criteria are sent to be redone.

Use of Buffers

Ballard and Howell (1995) propose the use of buffers to adapt the JIT concept to construction. Two types of buffers may be used to shield downstream construction processes from flow variation. Plan buffers are inventories of workable assignments. Schedule buffers are materials, tools, equipment, manpower, time, and so on.

Alves et al. (2003) describe two types of buffers: Passive buffers are linked to process flow, such as materials, work in process, documents, time, and space. Active buffers relate to operations where work is performed on a product, such as excess capacity in equipment or labor. There should be a buffer of tasks that are appropriate (sound) for each crew as part of the last planner (Koskela 1999).

In the application of the Last Planner® technique, buffers are built into work plans to maintain the flow of work and to avoid interruptions. These buffers include a "workable backlog" of tasks that are available to be done but do not need to be done; that is, they are not on the critical path. This backlog allows crews to do productive work in case there are interruptions to the WWP. Plan buffers are backlogs of workable assignments. They are based on what CAN be done as opposed to what SHOULD be done. The work that CAN be done is used as a source of the work that crews commit to doing—WILL be done in the next week. This varies from traditional project management that insists on what SHOULD be done—without fully considering whether it CAN be done.

It is recommended to base WWPs on only 75% of the capabilities of the crew, in order to maintain the flow of work to the next crew. Workable backlog provides a source of work assignments that can maintain a smooth flow. As work is pulled from each crew by the downstream crew that leaves spare capacity and increases the probability of completing required tasks. On the other hand, traditional scheduling tends to assign as much work as possible to each crew in the hope that it can be completed resulting in a lower level of performance. An early start on a task that is not critical to the schedule does little to accelerate the project completion date. It simply builds inventory, which is a form of waste.

Visual Management

In lean manufacturing, resources that do not contribute to the desired production are considered to be waste and are excluded from the system (Table 4.2). Visual controls provide transparency; as construction has mobile workstations, the placement in plain view of all indicators and activities allows everyone to understand the status of the system at a glance. The work flow and the respective action plans can be seen on the job site (Moser and dos Santos 2003). Commitment charts are typically used to secure everyone's agreement to actions that support lean endeavors. They have been used to identify best practices for job site safety. Having a document signed by all responsible parties in plain sight provides high visualization; this improves safety compliance. Safety signs that are designed/selected through a team effort also seem to obtain higher compliance.

Project milestones are typically maintained in a trailer under the watchful eyes of project staff. A visualization approach involves posting milestones in the work area for everyone to see. It leads to higher levels of involvement among crews. Similarly, PPC charts in plain view make everyone accountable. One nationally known contractor maintains an electronic display board in the main site trailer, near an entrance way that everyone uses. It is approximately 4×3, and displays an automated countdown of the number of days remaining on the project. It is very effective in keeping people aware of time frames and the overall commitments of the project.

TABLE 4.2

Examples of Mechanisms for Increased Visualization

Requirements for Increased Visualization	Criteria/Change
Commitment charts	Visualization
Safety Signs	Team effort
Mobile signs	Knowledge
Project milestones	Communication
PPC charts	Relation with other tools

Source: Adapted from Salem, O., Solomon, J., Genaidy, A., and Minkarah, I., *Journal of Management in Engineering, ASCE*, October, 2006. With permission.

Questions for Discussion

1. Describe the components of and the process flow for The Lean Project Delivery System™.

2. Describe a staff support structure that can be used for managing lean construction projects.

3. What are the steps in Target Value Design? What are its advantages?

4. Who is the so-called Last Planner? What value does that individual bring beyond the traditional method of project planning and scheduling?

5. What is work structuring? What steps does it involve?

6. What is the purpose of Reverse Phase Scheduling?

7. How can one ensure that the weekly work plan represents the most appropriate work to be performed in a given week?

8. What role does constraint analysis play in The Last Planner® System?

References

Al-Sudairi, A. 2004. Simulation as an aid tool to the best utilization of lean principles. 12th annual conference on lean construction. August 3–4. Helsingor, Denmark.

Alves, T. C. L., and I. D. Tommelein. 2003. Buffering and batching practices in the HVAC industry. *Proceedings 11th Annual Conference of the International Group for Lean Construction IGLC-11*, 22–24 July, Virginia tech., Blacksburg, Virginia, USA.

Arbulu, R., and A. Koerckel. 2005. Linking production-level workflow with materials supply. *Proceedings of the 13th annual conference of the international group for lean construction (IGLC-13)*, July 2005, Sydney, Australia.

Ballard, G. 1999. *Work structuring white—Paper #4*. Las Vegas, Nevada, Lean Construction Institute.

Ballard, G. 2000. *The Last Planner® System of production control*. PhD Thesis. Faculty of Engineering. School of Civil Engineering. Birmingham, AL: The University of Birmingham.

Ballard, G. 2008. The Lean Project Delivery System™: An Update. Lean Construction Journal 2008, pp. 1–19

Ballard, G., and P. Reiser. 2004. The St Olaf fieldhouse project: A casestudy in designing to target cost. *Proceedings 12th Annual Conference of the International Group for Lean Construction.*, IGLC-12, Elsinore, Denmark.

Ballard, G., and G. Howell. 1994. Implementing lean construction—Stabilizing work flow. Conference on lean construction. September 1994, Santiago, Chile.

Ballard, G., and G. Howell. 1995. Toward construction JIT. Proceedings of the 1995 ARMCOM conference, association of researchers in construction management. Sheffield, England.

Ballard, G., and G. Howell. 2003a. Lean Project Management. Building Research and Information, 119–133.

Ballard, G., and G. Howell. 2003b. An update on last planner. Proceedings of the 11th annual conference of the international group for lean construction (IGLC-11), July 2003. Blacksburg, VA.

Bureau of Labor Statistics. 2006. *Fatal and non-fatal occupational statistic from 1992–2004*. Online at http://www.bls.gov.

Crosby, P. 1979. *Quality is free: The art of making quality certain*. New York: Mentor.

Gwynn, P. 2008. *ULSD lean production management (LPM).* Lean Implementation Services, Schereville, IN: author.

Howell, G. A. 1999. What is lean construction. Proceedings of the 7th annual conference of the international group for lean construction. Berkeley, CA: UCLA.

Koskela, L. 1999. Management of production in construction: A theoretical view. *Proceedings 7th Annual Conference International Group for Lean Construction.* IGLC-7, ed. Iris Tommelein, 26–27 July, Berkeley, USA, University of California.

Koskela, L. 2000. *An exploration towards a production theory and its application to construction.* VVT Publications; 408. Online at http://www.inf.vtt.fi/pdf/publications/2000/P408.pdf

Macomber, H., and G. Howell. 2003. Linguistic action: Contributing to the theory of Lean Construction, *Proceedings 11th Annual Conference of the International Group for Lean Construction.*, IGLC-11, 22–24 July, Virginia tech., Blacksburg, Virginia, USA.

Macomber, H., and G. Howell. 2005. *Using study action teams to propel lean implementations.* Louisville, CO: Lean Project Consulting Inc.

Macomber, H., and G. Howell. 2005. 5 Necessary actions for change, 1–2. Lean Project Consulting, Louisville, CO 80027.

Macomber, H., G. Howell, and J. Barberio. 2005. Target value design: Seven foundational practices for delivering surprising client value, 1–2. Lean Project Consulting, Louisville, CO 80027.

Moser, L., and dos Santos. A. 2003. Exploring the role of visual controls on mobile cell manufacturing: a case study on drywall technology. *Proceedings of the Annual Conference (IGLC-11).* International Group for Lean Construction, Blacksburg, VA. http://strobos.cce.vt.edu/IGLC11

Salem, O., J. Solomon, A. Genaidy, and I. Minkarah. 2006. Lean construction: From theory to implementation. *Journal of Management in Engineering, ASCE* 168–175.

Spearman, M. L., D. L. Woodruff, and W. J. Hopp. 1990. CONWIP: A pull alternative to Kanban. *International Journal of Production Research* 28: 879–894.

Tsao, C. C. Y., I. D. Tommelein, E. Swanlund, and G. A. Howell. 2000. Case study for work structuring: Installation of metal door frames. Proceedings of the 8th annual conference of the international group for lean construction (IGLC 8), July 17–19, Brighton, UK.

Bibliography

Alarcon, L. F., A. Grillo, J. Freire, and S. Diethelm. 2001. Learning from collaborative benchmarking in the construction industry. Proceedings of the 9th annual conference of the international group for lean construction (IGLC-9), August 6–8, Singapore, University of Singapore.

Ballard, G. 1994. The last planner. *Northern California construction institute spring conference.* April, 1994. Monterey, CA.

Bertelsen, S. 1993. Construction logistics I and II, materials management in the construction process (in Danish). Kobenhavn, Denmark: Boligministeriet, Bygge-og, Boligstyrelsen.

Fiallo, M., and V. Revelo. 2002. Applying the last planner control system to a construction project: A case study in Quito, Ecuador. *Proceedings, IGLC-10,* August, 2002.

Howell, G. A. 2000. White paper for Berkeley/Stanford CE & M workshop. Proceedings, construction engineering and management workshop. Palo Alto, CA: Stanford University.

Howell, G. A., A. Laufer, and G. Ballard. 1993a. Interaction between subcycles: One key to improved methods. *Journal of Construction Engineering and Management, ASCE* 119(4).

Koskela, L. 1992. *Application of the new production philosophy to construction.* 87. Technical Report #72. Stanford University: CIFE.

Lichtig, W. A. 2005. Sutter Health: Developing a Contracting Model to Support Lean Project Delivery. *Lean Construction Journal,* 2(1): 105–112.

Matthews, O., and G. Howell. 2005. Integrated project delivery an example of relational contracting. *Lean Construction Journal* 2(1): 46–61.

Tsao, C., I. Tommelein, E. Swanlund, and G. Howell. 2004. Work structuring to archive integrated product, process design. *Journal of Construction Engineering and Management* 130(6): 780–89.

Vrijhoef, R., and L. Koskela. 1999. *Role of supply chain management in construction (IGLC 7)*. Berkeley, CA.

Womack, J., and D. Jones. 1996. *Lean thinking*. New York: Simon & Schuster.

Womack, J., D. Jones, and D. Roos. 1990. *The machine that changed the world*. New York: Harper Collins.

5

Lean Process Measurement and Lean Tools/Techniques

Measuring Lean Construction Performance

As in any process improvement endeavor, measurement is all important. In the words of one quality guru "you cannot improve what you cannot measure." Research studies have indicated that approximately 54% of project commitments are completed in the typical weekly schedule, using traditional project delivery. Lean project delivery can raise the reliability of work plans to 85% or more.

Work accomplishment is recorded as a graphical plot of percent plans complete (PPC); it shows the percentage of the assigned plans; that is, commitments that were completed (fractional completion is not considered). Some lean construction practitioners refer to PPC as commitment reliability. In The Last Planner® System, teams commit to completing designated tasks in a given week; failure to complete all of those tasks represents less than 100% reliability. It must be noted that a PPC value does not give a true indication of how efficiently assignments have been carried out (Mohammed and Abdelhamid 2005). It does not measure the level of utilization of a work crew. Instead, it measures production planning effectiveness and workflow reliability.

Observations on Commitment Reliability

Figure 5.1 relates to an EPC project that is described in Chapter 7 (Case C). The figure shows a highly fluctuating level of commitment reliability (as represented by the PPC values). The mean value of reliability ranges between 44 and 48%, indicating that, on the average, fewer than half of the planned assignments are completed in a given week. This value is at or near zero on three occasions and between 80 and 82% on two other occasions in a 6-week period. In the context of lean construction, the desired outcome is to have minimal variability in the hand offs of assignments to downstream crews. Wide fluctuations in the completion of tasks upstream subjects downstream crews to the risk of scheduling labor and materials for tasks that are not in a full state of readiness. Having crews available for delayed assignments represents time consuming and costly waste.

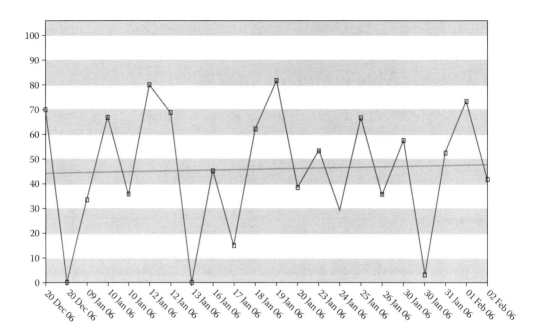

FIGURE 5.1
Commitment reliability. (From Gwynn, P., *ULSD Lean Production Management (LPM)*, Lean Implementation Services, Schereville, IN, self-published, 2008. With permission.)

In this case, the reasons for high variability were studied through a review of reasons for noncompletion (RNC). The most frequently occurring RNCs were found to be short-comings in commitment/planning and inadequate prerequisite work. By addressing these issues, significant improvements were made, variability was reduced, and PPC/reliability was increased from the 45% range to over 80%. These changes accelerated the project from being significantly delayed to being on time with virtually the same resources.

Use of Statistical Process Control

Statistical process control (SPC) and PPC values:

SPC involves the use of control charts to track the performance of a process and diagnose the nature of the variation it exhibits. The control chart is an important statistical tool in Six Sigma management that enables process owners to minimize the variation in processes, and the resulting defects. The chart distinguishes between the common causes of variation and the special causes. In lean-based construction projects, The Last Planner® System requires the tracking of PPC in order to calibrate the reliability of actors in the construction supply chain. PPC charts are essentially "Run Charts" that simply reflect basic job accomplishment.

The SPC concept takes the run chart to another level; it introduces control limits that reflect the range of normal behavior for the process. Typically, the limits are placed three standard deviations above and below the process mean. A process that is stable and operating normally will experience a degree of fluctuation around that mean, but will keep its values inside the control limits.

Types of variation: Gitlow and Levine (2006) describe two types of variation. Chance or common causes of variation create fluctuations or patterns in data that are due to the system itself; examples are hiring, training and supervisory policies and practices. Common cause variation explains fluctuations within the control limits. They are the responsibility of management, who establish policies and procedures.

Special or assignable causes of variation create fluctuations or patterns in data that are not inherent in the process, and are detected when control limits are breached. They occur because of abnormal events that are often avoidable, for example, a missed concrete pour could be caused by a late order from a site office resulting from an oversight.

Companies that have successfully run lean-based projects may investigate refining PPC tracking with the application of SPC; the technique is practical only with established, stable processes. Because of the relatively small number of assignments in construction, as compared with manufacturing, a variables chart ($\bar{X}$, R) would be more appropriate than an attribute chart such as a "p"chart which works best with large sample sizes. (Several texts are available on the subject of SPC).

Learning: Reasons Analysis and Action

At each weekly meeting, time is devoted to learning why certain tasks were not accomplished in the previous week before creating a weekly plan for the work to be executed in the following week. Incomplete plans are studied to determine the root causes (reasons for noncompletion) and why they were not completed in order to improve the effectiveness of future work plans. A Pareto chart reflects the most frequently occurring reasons. Typically, bad planning and poor coordination are frequently occurring causes.

This exercise serves as an all-important learning tool. Without learning, crews are likely to simply repeat the mistakes of the past.

1. Percent plan complete (PPC) measures the degree to which work is completed as planned. It indicates the likelihood of future work accomplishment; however, the value changes as teams learn from past performance and improve future work plans. A CII-sponsored study by Ballard and Kim (2007) of pipe fitters on a major project, identified a positive correlation between PPC and productivity at the 95% confidence level. PPC is calculated by determining how many individual work tasks were initially planned, then computing a ratio of completed plans to planned tasks. There is no credit for plans that are incomplete. This measurement is also used for commitment reliability.

2. Reasons for noncompletion: These reasons are tabulated by studying incomplete plans to determine the root causes for lack of completion. These causes should be categorized in meaningful groups to facilitate identifying system-related causes. A Pareto chart is a convenient tool for ranking the reasons for noncompletion. The knowledge from the root-cause analysis enables crews to avoid obstacles in future work cycles.

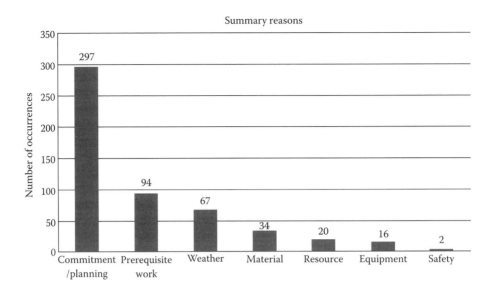

FIGURE 5.2
Tracking of RNC (reasons for noncompletion).

Comments on the RNC

Tracking of RNC points to specific problem categories that are observed to prevent timely completion of scheduled assignments (Figure 5.2). In the example provided, commitment planning and prerequisite work were the most frequently occurring reasons for noncompletion. These categories were subjected to improved methods in the actual project with positive results. The RNC can be further investigated with five-Why analysis.

Five-Why Analysis

The five-Why analysis procedure is a simple, yet effective procedure for finding the root cause of a problem. It is a part of the Toyota problem-solving process, and focuses on involved processes rather than on people. The steps in the process are:

- Ask why a particular process deviates from expectations.
- Ask why the answer is as stated.
- Repeat the questions: asking why until a root cause is found. (If needed, this may extend beyond five questions.)
- Other quality tools such as a Pareto analysis and Fishbone diagrams may be used instead.

Example:

1. Why wasn't the drywall installed in room 302 last week as scheduled?
   a. Answer: The conduit work for electrical and communications was not complete.
2. Why wasn't the conduit work completed?
   a. The electrical subcontractor did not have the material to complete the work.

3. Why wasn't the material available?
   a. The materials were ordered late.
4. Why was the material ordered late?
   a. There was a discrepancy between the plans and specs with regard to the configuration. There was an open request for information (RFI).
5. Why did this RFI remain open?
   a. The communications specialist did not spot the discrepancy until the middle of the week.

The root cause of this problem should prompt the electrical foreman and general foreman to improve the constraints analysis process. Shop drawings should be prepared on a timely basis, and approved through the project manager and designer to avoid last minute surprises. Reliable, sound commitments can only be made when all constraints have been removed.

Rolled PPC

Rolled PPC is analogous to rolled throughput yield (YRT). Mohammed and Abdelhamid (2005) have proposed using rolled PPC to expose the "hidden factory" in a construction operation. This is the rework to correct defects that are caused in subprocesses. It is said to be a more realistic performance measurement tool that reveals the deficiencies in the planning process. The weekly rolled PPC is a value obtained by multiplying the daily PPCs for the week (Table 5.1). It gives an accurate value for measuring the performance of the process without hiding the rework because of incomplete tasks on the previous day(s).

$$Rolled\ PPC = \sum_{i=1}^{i=n} PPC_i$$

Where PPC is the PPC for day i (calculated as assignments completed/assignments made).

The product of the PPCs is $0.084 = 8.4\%$ (rolling PPC).

In general, measured or collected data can be analyzed statistically in various ways to determine areas of improvement. This is necessary for management to identify tasks that need immediate attention and assign priority levels that need to improve the overall production planning process and thus improve crew performance and crew-to-crew hand offs (workflow). Determination of high-priority actions can be achieved by performing statistical analysis on the data.

TABLE 5.1

Tabulation of Rolled PPC Values

Day	March 1	March 2	March 3	March 4	March 5	Cumulative Product
Tasks planned	3	4	4	3	4	
Tasks completed	2	2	2	2	3	
PPC	0.67	0.5	0.5	0.67	0.75	8.4%

Plus-Delta Analysis

Plus-delta analysis is an important component of the learning process. What produced value? Items in this category are listed in the "Plus" column. What might produce more value? This points to items in the delta column (i.e., those that could be improved).

Plus-Delta analysis can be used for a wide range of topics, including construction meetings related to The Last Planner® System, such as the Weekly Work Plan meeting. Plus-Delta enables continuous improvement with the way the meetings are conducted by having meeting participants ask and answer the foregoing questions.

Jim Shug, a consultant, cites the extensive use of Plus/Deltas in the military after tactical/training/combat operations as after action reviews (AAR) and lessons learned meetings. He expresses the opinion that feedback is a gift and great leaders want to know what others think and how to improve. Their intentions may be great, but the perceptions of their actions are reality to those around them.

Shug views Plus/Deltas as a significant step toward developing a learning organization. He recommends that there should be not only be a discussion, but some concrete facts as well to show "reality". "In military training events, they could play back radio transmissions from the battle and snapshots of vehicle/squad positions. This really becomes a wake up call to leaders on the ground and takes shape in a similar way for projects."

With regard to the construction process, Plus-Delta may be applied to performance measures such as those listed in Chapter 2. Ed Anderson, a consultant, emphasizes the BIG FIVE: cost, schedule, quality (scope/spec), safety, and reliability. He states that there can be many subsets of these, but lean advocates simplicity (KISS) first, then when and where needed, to add more detail. Anderson cautions that while measurement is important, proactive lean behavior involves having systems in place that concentrate on eliminating non-conformances, as opposed to measuring and tracking everything.

Some possible measures are listed here for convenience:

- Schedule variance
- Work in process.

See also Chapter 2 for other performance measures; some are listed here for convenience:

- Schedule variance: Percentage difference (positive or negative)
- Work in process: The number of work packages in process
- Number of planned versus unplanned work packages in process
- Yield: Hours worked on planned work/available hours
- Available hours = total shift hours minus breaks and travel time
- Performance factor (PF): measures productivity
- Earned labor hours/actual labor hours expended
- The labor utilization factor for each trade
- The field rating
- Efficiency

Designer performance, valuable measures are:

- Number of design errors
- Processing time for shop drawings
- Number of RFIs submitted
- Response time for RFIs
- Customer satisfaction ratings

Contractor performance, during construction:

- Accidents, days without an accident
- Number of punch list items
- Number of failed inspections
- Estimated cost versus actual cost—by trade

After turnover:

- Number of warranty calls
- Customer satisfaction ratings

These measures are based on observation of the work force. They indicate the portion of available labor hours that are productively utilized. By the same token they provide an approximation of the amount of waste incurred.

Lean Performance Measures

The following factors are appropriate for establishing performance measures for lean processes (the value of these measures varies, depending on the maturity of the respective organization in implementing lean):

- Cycle time: Time it takes from start to finish to complete a task, create a product, or provide a service.
- Value creating time: Process time spent to transform raw materials, parts or components, and information into a usable product or service desired by a customer.
- Lead time: Time taken to move one transaction through the entire process—before another transaction can begin.
- Takt time: Represents the rate at which a customer wants/consumes a product. Ideally, the process should be based on Takt time in order to produce or be in step with (downstream) customer demand. Takt time = available working time per day/customer demand per day.
- Production efficiency: Units produced/Takt time.

Lean Tools and Techniques

Additional lean approaches provide a means of augmenting The Last Planner® System. Some of these approaches are:

First Run Studies	Trying New Construction Ideas on a Pilot Basis
Value stream mapping: (VSM)	The documentation of specific actions to create finished products materials to meet customer demand.
Kaizen:	Continuous improvement philosophy
5S:	It is a system for workplace organization—it promotes lean thinking and application.
Visual management, visual controls	The placement in plain view of all indicators and activities allowing all to understand the status of the system at a glance.
Kanban:	Pull system central to the Just-In-Time (JIT) philosophy.
Poka-yoke:	Error proofing
Preventive and predictive maintenance	Keeping all equipment functional and maintained so it can operate reliably.

How Do Lean Tools/Techniques Work?

First-Run Studies

This involves trying new construction ideas on a pilot basis and applying the PDCA (Plan-Do-Check-Act) methodology. To determine its success and identify the best means, methods, and sequencing for a specific activity. These studies are carried out a few weeks in advance of their actual use in a project, with enough time to obtain needed materials.

A study documented by Salem et al. (2006), reviewed the installation of bumper walls and construction joint installation. Selection occurred in the "PLAN" phase. The test installation was observed carefully in the "DO" phase and videotaped In the "CHECK" phase. The finished work was examined carefully by the project manager, foreman, and crew to identify possible improvements. In the "ACT" phase various recommendations were tested, resulting in a 38% reduction in the cost of bumper walls, and 73% reduction in the cost of construction joints.

Value Stream Mapping

A value stream is all the actions (both value added and nonvalue added) currently required to bring a product through the main flows essential to every product: (1) the production flow from raw material into the arms of the customer, and (2) the design flow from concept to launch. Value stream mapping (VSM) is specific actions to create finished products from raw materials to meet customer demand. VSM focuses on information management and transformation tasks. The VSM process generates (1) a current-state map, (2) a future-state map, and (3) an implementation plan (Rother and Shook 2003). It distinguishes between value-adding and non-value-adding activities. It sets the vision for a what can be accomplished in the future with needed changes.

At Toyota, value stream mapping is referred to as "Material and Information Flow Mapping." In the Toyota production system, mapping is used to develop implementation plans for lean systems; it focuses on establishing flow, eliminating waste, and adding value.

Womack and Jones (1996) recommend that processes be transformed through the following approach:

1. Find a change agent
2. Find a sensei (a teacher whose learning curve can be borrowed)
3. Seize (or create) a crisis to motivate corrective action
4. Map the value stream for all product families
5. Pick something important and get started removing waste quickly

They point out that overzealous people who seek change often bypass mapping the value stream and go directly to step 5 with self-defeating results. Instead, they recommend that Kaizen efforts are most effective when integrated with efforts to create a lean value stream. In the construction environment VSM would be used to examine the process of delivering construction or design services in order to reduce or eliminate non-value-adding steps.

The current-state map represents the existing process. It requires studying operations very carefully in order to fully understand the flow of materials, information, and labor. It is recommended to work backwards from the delivery of the completed product or process to visualize what actually occurs from start to finish. Important statistics are collected and represented on the map, such as: work hours, work time, lead times, unproductive time, value creation time, setup time, and defect rates.

An evaluation of the current-state map shows opportunities for simplifying processes and eliminating or reducing unnecessary steps. This information is used to draw a future-state map with desired process improvements. For example, in a chocolate manufacturing operation the process was found to have 25% scrap rate. Value stream mapping revealed that the rejects were primarily due to improper wrapping. A process redesign led to a reduction in rejects from 25 to 0.05%. Significant reductions were made with in-process inspections, rework, lost time, overtime, and so on. Staffing reductions also yielded saving in direct and indirect costs.

Value Stream Mapping at Tweet/Garot Mechanical Inc.

Tweet/Garot Mechanical, Inc. has used the Kaizen approach in conjunction with value stream mapping to improve their fabrication processes. The attached value stream maps VSM#1 (Figure 5.3) and VSM#2 reflect improvements made to the fabrication of Tee Dampers for a primary customer. Note: On the map, the process is shown in two rows. Activities are labeled as follows: VA = value-added activity and NVA = non-value-added activity. Numerical values are shown on the symbols; they represent the minutes elapsed in carrying out an activity. The symbols are also labeled to indicate if the time elapsed is value added or not. Figure 5.3 shows a customer generating an order (P.O.) for a tee damper, starting with the customer

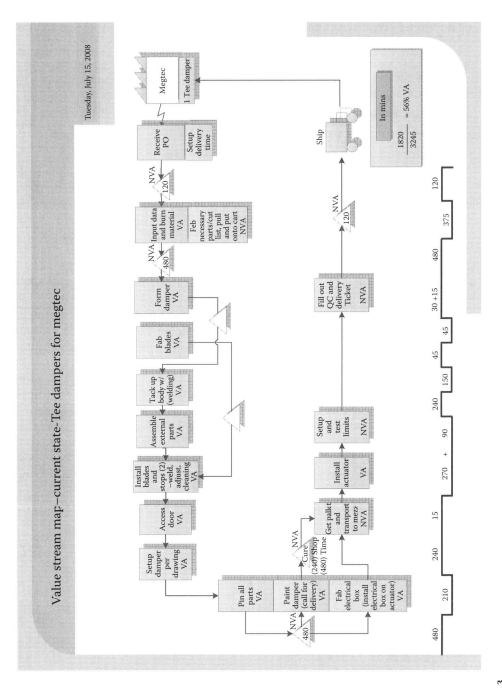

FIGURE 5.3
Diagram of the current-state value stream map 1.

symbol at the upper right hand side of the page. Tweet/Garot receives the P.O. and sets up a delivery time. Two periods of NVA follow in quick succession—120 minutes and 480 minutes. Other symbols with NVA time are visible in the map.

Summarizing these activities yields: total time = 3245 minutes; value-added time = 1820 minutes. The percentage of value-added time is the ratio 1820/3245 = 56% VA.

A Kaizen event was conducted to have stakeholders brainstorm on how to reduce or eliminate NVA activities. The process was streamlined in order to reduce waste and to shorten the steps needed to manufacture the dampers. The new process is shown in the next section.

Future-State Map

The customer originates a purchase order (PO) for a damper. A project manager fast tracks it to a superintendent for scheduling. Three activities immediately follow:

- Data input
- Development of a "cut" list for fabrication
- An order to pull premade parts and hardware from inventory—these items are delivered to the fabrication area in quantities that meet downstream demand.

Fabrication begins and consists of a series of processes (see Figure 5.4).

Summarizing these activities yields: total time = 1170 minutes; value-added time = 732 minutes. The percentage of value-added time is the ratio 732/1170 = 61.5% VA. The percentage improvement is $(61.5 - 56) \times 100/56 = 9.8\%$.

While this improvement seems modest, it has added greatly to Tweet/Garot's performance and profitability by reducing waste. Fabrication errors were significantly reduced.

Kaizen Methodology

Kaizen is the Japanese concept of continual incremental improvement. Literally, Kai means change, and zen means good; hence, Kaizen stands for ongoing changes for beneficial reasons on a never-ending basis. It involves all members of an organization actively participating in making improvements on an ongoing basis. Masaki Imai (1986) of Japan credits Kaizen as the "single most important concept in Japanese management, and Japan's competitive success." Kaizen is rooted in Japanese philosophy—they see improvement as being closely related with the concept of change. According to Imai in the Japanese way of thinking "change is a basic condition of life and should be incremental in order to be healthy, whereas sudden change is seen as unnatural."

The Kaizen approach encourages employees/stakeholders to develop and implement ongoing improvements to the systems that they are most involved with, thereby improving their job performance. It seeks to standardize processes and eliminate or reduce waste.

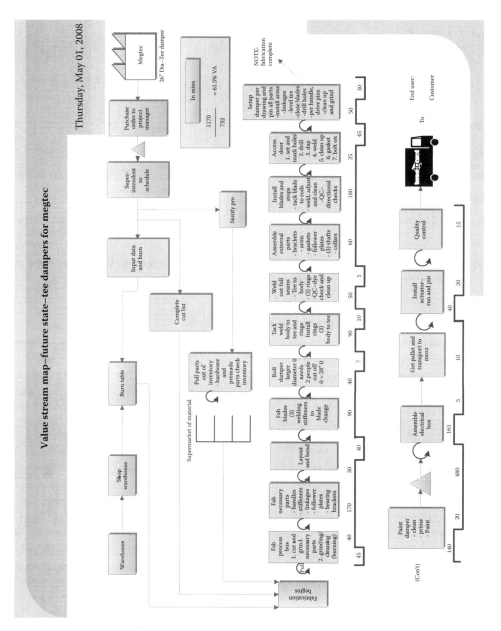

FIGURE 5.4

The future-state value stream map.

It starts with recognizing a problem, and subsequently a need for improvement. It has been said that complacency is the archenemy of Kaizen.

One important consideration for the adaptation of Kaizen to construction-related activities is that improvements are expected to be followed by a period of time for new methods/processes to be standardized. As described, the Kaizen methodology is appropriate for a specialty contractor or supplier. On a construction site the time frames may need to be reduced from 1 week to 2 or 3 days.

Roles in the Kaizen Process

Senior executives/managers: They must first embrace Kaizen as an organizational strategy and communicate this policy to all stakeholders. As with other performance improvement initiatives, top management's commitment, leadership, and active support are critical to the success of Kaizen.

Middle managers: They deploy the policies established by senior executives. They ensure that staff members have the appropriate training and preparation. They establish implementation milestones for supervisory staff.

Supervisors: They work closely with line workers to implement Kaizen at the functional level. They are actively involved in collecting Kaizen suggestions for improvement from employees and coach the respective teams to promote success.

Employees: They are expected to actively pursue self-development through education and training in order to support Kaizen team activities.

Kaizen practitioners study a process firsthand by visiting the involved work area (*gemba*) and observing the activities involved. This approach clarifies the difference between value-added and non-value-added process steps. Kaizen teams reduce non-value-added steps through a prescribed approach: to combine, simplify, or eliminate in order to reduce waste.

There are two primary approaches to Kaizen improvement activities: Flow Kaizen and Process Kaizen. Flow Kaizen emphasizes value stream improvement while process Kaizen focuses on the elimination of waste. While Kaizen involves ongoing, long-term dedication to improvement of the organization and its people, the process periodically conducts short-term activities as "Kaizen events" or "Kaizen breakthrough" methodology. These events are a cross-functional, team-based process for rapid improvement, and are often a week long in duration. They involve a bias for action that harnesses creativity to obtain results. The areas selected for Kaizen events have direct impact on organizational performance

Employees' must contribute and focus 100% to the Kaizen philosophy for the needed organizational transformation. Kaizen events are selected for maximum visibility and impact. They have several advantages:

- Benefits are obtained very rapidly.
- Quick implementation sends a positive message to all stakeholders.
- They provide hands-on training for everyone.
- They allow management to gauge workers' resistance to change.

Typical Kaizen Structure

Preliminary activities:

>Select area/work unit for Kaizen
>>Based on business requirements
>>Based on potential for improvement
>Establish project objectives
>>Document baseline data
>>Set metrics for improvement—numerical goals
>Appoint Kaizen team members
>>Team leader and subteam leader
>Publicize the event—meet with the involved work unit team
>>Explain roles and expectations
>>Engage hearts and minds for the event

Day 1:
Teams meet for conceptual training and problem identification. Training topics include:

- Lean theory and lean production systems
- Kaizen breakthrough methodology
- Team dynamics—multipurpose
- The lean transformation

Day 2:
Evaluate/analyze current processes:

- Document existing processes
- Identify value-added versus non-value-added activities
- Measure cycle times
- Discuss improvement options
- Formulate process improvement—rearrangement of layouts
- Initiate process improvement

Day 3:

- Continue process improvement
- Apply lean tools
- Focus on delivering value to the customer
- Recheck cycle times

Day 4:

- Refine process/system improvements
- Establish new work/process standards

- Document new standard operating procedures
- Recalibrate new cycle times

Day 5:

- Present recommendations and results to management
- Celebrate Kaizen event accomplishments

Kaizen events are short-term in duration but have proven to be very effective. They focus everyone's effort for a few days, but the benefits in improved performance far outweigh the cost of lost production.

Five-Step Plan (5S)

The Five-Step plan is a Japanese approach to improving operations. It helps to create an organized environment in a facility that promotes the application of lean thinking. It involves a sequential process based on the five Ss.

1. *Seiri*: (Straighten up) Segregate and Discard. This involves getting rid of unnecessary items in order to be more organized. This step is applied to such areas as tools, work in process, products, documents, and so on.
2. *Seiton*: (Put things in order) Neatness is obtained by proper rearrangement of the work area and identification of proper locations. The object of the exercise is not just to maintain a neat appearance. Workers can reliably find what they need to do a job without wasting time by looking. They should place tools and equipment in a manner that will improve the flow of work. Everything should be kept in their designated locations—equipment should be returned to its proper place at the end of each job.
3. *Seiso*: (Clean up) Cleaning and daily inspection to avoid the confusion of an untidy work area. Unnecessary items should be disposed of properly in order not to create clutter that inhibits the effective execution of work tasks. After each job, tools and equipments should be cleaned and restored to their proper locations. The root causes of waste and uncleanliness are investigated to ensure that they do not reemerge.
4. *Seiketsu*: (Personal neatness) This refers to standardizing locations for tools, files, equipment, materials, and so on. Color coding and labeling can help to standardize locations. Employees' creativity is actively involved in the development of standardized systems.
5. *Shitsuke*: (Discipline) It provides motivation for employees to sustain the other four Ss. Management can use recognition programs to provide motivation.

Some organizations have taken 5S to a further level. Tweet/Garot added a sixth S (Safety). Chapter 6 has an example of Tweet/Garot's use of an A3 report to support a 6S of their shop welding machines (see Figure 6.3). Their welding machines were equipped with many electrical cords that were not labeled and created confusion and delays on their projects. The report explains their analysis and the savings in time and space that were derived. The development of the Toyota A3 report is described next.

A3 Report

The A3 report is a system for implementing PDCA management, which is simple yet disciplined and rigorous (Sobek and Smalley 2008). The report used by Toyota fits on a single sheet of A3 paper, measuring approximately 11×17 and documents the results of the PDCA cycle in summary form. It has a template that guides users in the steps needed to determine the root causes of an organization's problems and to address them systematically. The A3 report concisely describes problems and the available information so as to understand their ramifications quickly, focusing on improving processes and solving the problems (Figure 5.5).

The development of the A3 report is thought to have been influenced by Taichi Ohno's dislike of reports that were longer than one page. It must be understood that the A3 report is not by itself a complete system for problem solving. In Toyota's case their success in process improvement comes from their underlying philosophy about collaboration and teamwork. The A3 report serves as an information and facilitation tool to support a very practical problem-solving process. A3 thinking as practiced by Toyota is based on seven elements (Sobek and Smalley 2008).

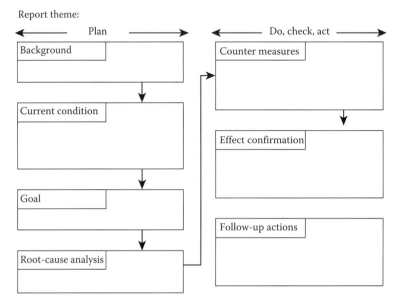

FIGURE 5.5
Typical flow of the problem-solving A3 report.

Logical thinking process: Logical thought processes are employed to recognize the most important problems, using the 80/20 rule, and recognizing the ramifications of the possible courses of remedial action.

Objectivity: Participants are encouraged to express their viewpoints; no single viewpoint drives the process, and consequently there is little opportunity for biases and misconceptions. A3 thinking considers multiple perspectives to promote objectivity and find the solution that benefits all stakeholders.

Results and process: Both results and the process used to obtain them are considered to be equally important. The involvement of people in the process leads to personal development and to more lasting outcomes. The process also provides a mechanism for future refinement, as opposed to a spontaneous idea.

Synthesis, distillation, and visualization: The A3 format is concise and compresses large quantities of information in a graphical format; the visualization it provides aids the synthesis and distillation process.

Alignment: A3 reports enable a practically oriented consensus—three dimensional communication flows between the problem-solving group and others. Horizontal communication considers others at the same level that may be affected by proposed changes; vertical communication examines broader organizational issues. These approaches are inclusive and lead to better alignment between people system wide

Coherence within and consistency across: A3 reports have a logical flow that focuses on root causes of identified problems and provides a coherent approach. This increases the effectiveness of problem resolution.

Systems viewpoint: People are guided (and mentored) to develop a deep understanding of: (1) the purpose of the course of action, (2) how the course of action furthers the organization's goals, needs, and priorities, (3) how it fits into the larger picture and impacts other organizational units.

Overall, the A3 report is not considered a documentation format, but really is an important adjunct to the informed use of the PDCA cycle.

The three most frequently used examples of reports are: (1) the problem-solving report, (2) the proposal report, and (3) the status report. The problem-solving A3 sets the stage for analyzing and improving operations. It consists of seven sections; that is, rectangular boxes that are used for recording information. The sections are:

I. Background
II. Current condition and problem statement
III. Goal statement
IV. Root-cause analysis
V. Countermeasures
VI. Check/confirmation of effect
VII. Follow-up actions

Sections I through IV are usually stacked vertically on the left side of the page; Sections V through VII are similarly stacked on the right side.

Recommendations for report content are as follows:

Background: Provide a clear theme that is relevant to the objectives of the project or project team.

Current condition and problem statement: Describe the problem clearly and visually. Distinguish between facts, observations, opinions.

Goal statement: Describe the goal or target clearly and state what is to be accomplished. State how the goal will be measured or evaluated. State what you expect will improve, by how much, and when.

Root-cause analysis: Show that the analysis was thorough and that it examined the appropriate issues deeply. Use the cause and effect diagram, based on all relevant factors. These may include people, machines/equipment, materials, measurement, and the environment. Use five-Whys/one-How analysis.

Countermeasures: Show that the countermeasures relate specifically to the root causes identified. Show clear steps for these actions. Explain how the effects will be verified, and how problem reoccurrence will be prevented. Provide an implementation plan—who does what and when.

Confirmation of effect: Explain how effectiveness of the countermeasure will be measured. Show that the PDCA "check" item aligns with the previous goal statement. Provide a "plan B" if performance is not improved, and show what may have been missed.

Follow-up actions: Describe what is left to be accomplished Show how problem reoccurrence will be avoided. Explain how the improvement will be standardized and communicated. Show how other organizational units will be advised.

The Proposal A3 report begins with a thematic title and presents a plan for consideration by decision makers. The Status A3 report presents a logical snapshot of a project or problem-solving effort and how it is progressing. The report describes the situation with a theme and explains accomplishments as well as any work that remains to be done. A recommended format is as follows:

Background Issues: Provide a clear theme that is relevant to the objectives of the project or project team.

Current condition and problem statement: Describe the problem clearly, visually. Distinguish between facts, observations, and opinions. Quantify the situation as much as possible. Show the progress made and the action steps taken to date.

Results: Indicate the results obtained to date. Confirm that improvement has occurred and quantify it as much as possible. Show that the metrics used are appropriate. State whether any negative side effects have occurred. If results fall short of expectations, explain possible causes.

Unresolved issues/follow-up action items: Describe any problems that remain. Explain actions needed to meet planned progress. Explain how gains may be sustained.

Kanban and Lean Construction

Pull/Kanban is a system of cascading production and delivery instructions from downstream to upstream activities in which the upstream supplier does not produce until the downstream customer indicates a need. It is an essential part of the Just-In-Time (JIT) process. In JIT, production is driven by customer demand; that is, the needs of the next trade downstream as opposed to "pushing " to meet schedules. Essentially, nothing is built until there is an order for it. JIT avoids wastes such as inventory of raw materials, work in process, or finished assemblies that are not yet needed. JIT is producing only what is needed, when it is needed, and in the quantity that is needed. Therefore, no goods are produced that cannot be sold at a price that is cost effective for the producer.

In its original form, Kanban was a card-based system that improved process management through visual control. Kanban cards follow a product through various production stages, maintaining information on the product name, number, due dates, and so on. Kanban is a visual system, providing status information at all times. In contemporary applications, the card has been replaced by a computer-based system, but the purpose is the same. The kanban tells the supplying process exactly what it wants and how many. The supplying process is not authorized to make more product until the kanban advises it to do so.

The Toyota system utilizes two different types of kanban—the withdrawal kanban and the production kanban. The withdrawal or "move" kanban authorizes the movement of materials or work in process from one process to another. The production kanban authorizes a process to produce another lot of the required size. It also specifies the description of the items, the materials required, the production work center, and so on. The following rules enable the kanban system to work successfully:

1. Do not send defective product to the subsequent process. Stop the process and determine the reason for the defects.
2. The subsequent process withdraws only the specific items needed, when they are needed.
3. Produce only the exact quantity required by the following process. Do not produce more than is authorized
4. Smooth the production load.
5. Adhere to kanban instructions while fine tuning.
6. Stabilize and rationalize the process.

Supply Chain Management and Lean Construction

Supply chain management (SCM) has saved hundreds of millions of dollars in manufacturing while improving customer service through taking a systems view of production activities of autonomous units. It is proposed that similar savings could be derived in construction, especially since subcontractor and supplier production account for the largest percentage of project costs. Studies have indicated project cost increases of up to 10% because of poor supply chain practices.

The term supply chain encompasses all the activities that lead to having an end user provided with a product or service—the chain is comparable to a network that provides a conduit for flows in both directions, such as materials, information, funds, paper, and people. The main elements of SCM are information flow, order fulfillment, and product development with faster response times, less waste, more effective information flow, and smaller amounts of inventory. Studies by Bertelsen (1993) indicated project cost increases of up to 10% because of poor supply chain design. Comparisons with previous research (Jarnbring 1994) on SCM of construction projects in Sweden indicated a potential for cost savings of 10–17% due to inefficiencies caused by lack of coordination between contractors and suppliers.

Supply Chain Management analyzes the impact of facility design on the construction process and enables superior project planning and management, avoiding the fragmented approach of other methods. Through SCM, all parties are kept aware of commitments, schedules, and expedites all work as a virtual corporation that can source, produce, and deliver products with minimal lead time and expense. The SCM application needs to be tailored to the conditions in the geographic area and environment in which projects are executed. Construction supply chains are subject to inefficiencies caused by interdependency with causes in various stages of the chain, primarily due to self-serving actions by individual companies and organizations, as well as divisions of the same organization.

The implementation of SCM in construction is challenging. The industry trend toward subcontracting has resulted in specialization and fragmentation; each subcontractor tends to act in its own interest and the relationships between the various parties often becomes adversarial. The sharing of information is critical for both SCM and lean construction. Central to their business model is the importance of making and keeping commitments. A general contractor (GC), for example, should be able to extract a reliable commitment from an air conditioning subcontractor that a chiller has been ordered from a specified supplier. Using an online information system, it should be possible for the GC and the owner's project manager to verify the supplier's delivery schedule, although pricing information could be kept confidential. The dimensions and capacities of the chiller could be verified long before delivery to ensure that it will be delivered and installed in seamless agreement with the construction schedule. This system allows a JIT approach to be used, so that the chiller is not delivered too far in advance, when on-site storage requirements may become costly and inconvenient. By the same token late deliveries would be avoided. Unavoidable deviations from the construction schedule would be shared with all stakeholders and needed adjustments made.

Supply chain systems facilitate the sharing of resources, staff and expertise, problem solving, improved economic performance, and increased innovative capacity. Supply chain systems also facilitate the tracking of work performance, resource utilization and provide feedback on workforce productivity. This feedback is indispensable for enabling continuous improvement efforts and corrective actions to be taken on a timely basis.

Three case studies were examined in the Netherlands and Finland, to apply different methods of analysis (Vrijhoef and Koskela 1999). The breakdown of the cases was as follows:

Case 1: The installation of concrete wall elements in residential construction. Time buffers in the supply chain were quantitatively assessed.

Case 2: The façade elements in residential construction. The control systems in the supply chain were qualitatively evaluated.

Case 3: A study of the cost impact of the contractor's method of materials acquisition.

The results were informative. Large time buffers developed that were 70–80% of the total lead time. Controllability problems were caused by the parties near the beginning of the chain separate from the activities being studied. This excessive value was caused by adversary relations between the parties and lack of goal congruence. These results pointed to self-interested actions by the parties.

Vrijhoef and Koskela (1999) propose a four-stage approach similar to the Deming cycle to improve construction supply chain performance: Supply chain assessment, supply chain redesign, supply chain control, and continuous supply chain improvement. This involves first evaluating a supply chain to identify waste and problems and determine root causes. The supply chain is then redesigned to address shortcomings, using analytical and feedback systems to monitor system performance, measure waste, and apply corrective measures. This is followed by continuous improvement.

One important function of the supply chain is to ensure the appropriate timing of material acquisitions to avoid excess material inventories; ideally, forecasts of materials requirements should be accurate enough to facilitate JIT in construction (Arbulu and Koerckel 2005). Material costs may exceed 50% of project costs, hence supply chain improvements that benefit material management greatly improve project performance.

Materials fall within three categories that must each be handled differently:

1. Made to stock items (MTS). These are commercial items that are made to meet market forecasts. Examples are consumables such as nails and small tools.

2. Made to order items (MTO). These are standard items such as rebar and concrete that are easily produced to meet project needs.

3. Engineered to order (ETO). Items are customized to the project; they include steel trusses, HVAC ductwork, systems piping for chillers, and plumbing.

A study by Arbulu and Koerckel (2005) on the Heathrow Terminal 5 (T5) project in the UK identified several systems improvements in material supply practices.

A concrete supply operation experienced untimely deliveries throughout the day because of a lack of transparency between site operations and the concrete batch plant. A coordinator was responsible for placing orders, but based his requests on guesstimates; reliability worsened during his absences. The ordering process was unreliable (50–80%) Overlapping orders from different project members at the same time led to overtime operations for both the plant and the project. A computerized system was implemented to integrate concrete orders across the project, significantly reducing the need for ad-hoc orders. Productivity improvement during the first 12 weeks of implementation saved 5000 man hours or 100,000 British pounds.

The T5 project used expanded polystyrene panels to minimize movement in floors installed over soil with high clay content. These panels varied in thickness and were made-to-order materials. Lack of transparency between site operations and the fabrication operation found production levels greatly exceeded requirements and resulted in excessive inventory of a fragile product that could not be double handled. The excessive storage was maintained in trailers or returned to the factory that also held excessive quantities.

Value stream mapping identified the deficiencies in the supply chain; system improvements included: storage trailers used as a kanban system—empty units generated orders, full trailers indicated no need to produce.

- Order sizes were reduced to trailer capacity
- A rolling 2-week look-ahead of orders was maintained
- Shorter lead production times were reduced to 5 days
- Trailers were reallocated weekly between project teams based on forecasts of demand.

The kanban system yielded significant benefits. Inventory capacity was reduced to one day's usage, down from several weeks. This resulted in overall cost reductions of hundreds of thousands of British pounds for the duration of the project.

Questions for Discussion

1. Describe three types of lean-related measurements that are used for lean construction projects.
2. Describe the Kaizen event process. How can it benefit a design or construction organization?
3. What benefits can be derived from the 5S system in construction: (a) Short term? (b) Long term?
4. How can the A3 process improve a construction organization's effectiveness?
5. How can the Plus/Delta technique improve a) the design processes; and b) the construction processes?

References

Arbulu, R., and A. Koerckel. 2005. Linking production-level workflow with materials supply. Proceedings of the 13th annual conference of the international group for lean construction (IGLC-13), July 2005. Sydney, Australia.

Ballard, G., and Y. W. Kim. 2007. Roadmap for lean Implementation at the project level. Research Report 234-11, Construction Industry Institute, 426.

Bertelsen, S. 1993. Construction logistics I and II, materials management in the construction process (in Danish). Kobenhavn, Denmark: Boligministeriet, Bygge-og, Boligstyrelsen.

Gitlow, H., and D. Levine. 2006. *Design for six sigma for green belts and champion*. Prentice Hall, USA.

Gwynn, P. 2008. *ULSD Lean Production Management (LPM)*. Lean Implementation Services. Schereville, IN: author.

Imai, M. 1986. *Kaizen*. New York: McGraw Hill.

Jarnbring, J. 1994. Translated—*Material flow costs on the building site*. Rappot 94:01. Lund, Sweden: Lunds Teknisa Hogskola, Institutionen for Teknisk Logistik.

Mohammed, T., and T. Abdelhamid. 2005. Understanding percent plan complete data using statistical quality control charts. International workshop on innovations in materials and design of civil infrastructure, December 28–29, Cairo, Egypt.

Rother, M., and J. Shook. 2003. *Learning to see*. Lean Enterprise Institute, Inc.

Salem, O., J. Solomon, A. Genaidy, and I. Minkarah. 2006. Lean construction: From theory to implementation. *Journal of Management in Engineering, ASCE*, October.

Sobek, II, D., and A. Smalley. 2008. *Understanding A3 thinking: A critical component of Toyota's PDCA management system*. Boca Raton, FL: CRC Press.

Vrijhoef, R., and L. Koskela. 1999. *Role of supply chain management in construction (IGLC 7)*. Berkeley, CA.

Womack, J., and D. Jones. 1996. *Lean thinking*. New York: Simon & Schuster.

Bibliography

Alarcon, L. F., A. Grillo, J. Freire, and S. Diethelm. 2001. Learning from collaborative benchmarking in the construction industry. Proceedings of the 9th annual conference of the international group for lean construction (IGLC-9), August 6–8. Singapore, University of Singapore.

Ballard, G. 2000. *The Last Planner® System of production control*. PhD Thesis. Faculty of Engineering. School of Civil Engineering. Birmingham, AL: The University of Birmingham.

Ballard, G. and G. Howell. 2003. An update on last planner. Proceedings of the 11th annual conference of the international group for lean construction (IGLC-11). July 2003. Blacksburg, VA.

Fiallo, M., and V. Revelo. 2002. Applying the last planner control system to a construction project: A case study in Quito, Ecuador. Proceedings, IGLC-10. August, 2002.

Howell, G. A. 1999. What is lean construction. Proceedings of the 7th annual conference of the international group for lean construction. UCLA, Berkeley, CA.

Howell, G. A. 2000. White paper for Berkeley/Stanford CE & M workshop. Proceedings, construction engineering and management workshop. Palo Alto, CA: Stanford University.

Koskela, L. 1992. *Application of the new production philosophy to construction*. Technical Report #72. Stanford University: CIFE.Lichtig, W. A. 2006. The integrated agreement for lean project delivery. *American Bar Association, Construction Lawyer* 26(3).

Macomber, H., and G. Howell. 2005. *Using study action teams to propel lean implementations*. Louisville, CO: Lean Project Consulting Inc.

Matthews, O., and G. Howell. 2005. Integrated project delivery an example of relational contracting. *Lean Construction Journal* 2(1): 46–61.

Tsao, C., I. Tommelein, E. Swanlund, and G. Howell. 2004. Work structuring to archive integrated product, process design. *Journal of Construction Engineering and Management* 130(6): 780–89.

6

Lean Construction Applications

This chapter covers two important topics:

- It describes the prerequisites for initiating a lean philosophy within organizations involved in design and/or construction, including "success strategies" for lean coaching. It presents commentary on these topics from recognized members of the lean community—Greg Howell, Robert Blakey and Matt Horvat.

- It shares with readers the journeys of four visionary subcontractor organizations that implemented lean in their internal operations, even when working with projects that did not represent a lean environment.

These organizations are Tweet/Garot Mechanical, Inc., Belair Contracting, The Grunau Company, Inc., and Superior Window Corporation.

Prerequisites for Lean Design and Construction

There are several prerequisites that have to be met before a design or construction organization can successfully implement lean principles.

a. A willingness to change is essential. Lean methods are a departure from conventional methods and their adoption requires changing the behavior of people. Cultural change is the most compelling quest along with the physical transformation of an organization. One cannot force change on people—they have to be engaged so that the intrinsic satisfaction of outstanding performance will motivate them. In a lean culture, people have to be treated as the only appreciating asset. Lean enables organizations to have Responsiveness, Reliability, and Relevance.

b. A commitment to training and learning. Stakeholders at all levels need to be trained in lean techniques in order to become successful participants in lean projects. Lean implementation also requires that completed assignments have to be continually examined as a source of learning for future improvements instead of serving as sources of blame.

c. A quality-oriented culture is needed for successful application of such techniques as Just-In Time (JIT), Lean Construction, and Supply Chain Management. JIT in particular, demands discipline, as there is no room for unreliable suppliers. JIT does not work in an atmosphere of suspicion, distrust, and internal competition.

d. A "Shared Vision" is essential to have all stakeholders on the same page (Macomber and Howell 2005). Promote the importance of a shared vision in which a workforce aligns itself with the direction set by a leader. This alignment is far different from carrying out orders. It is based on a sharing of beliefs and a common view of a future state that benefits everyone and makes them receptive to the changes necessary to reach that future state. In the case of lean construction, the shared vision would lead to a lean mind. A study-Action Team™ (SAT™)* is recommended to develop the shared vision. Members start as a reading group and focus on learning as much as possible in order to bring about change. The SAT members should be volunteers from various groups involved in construction.

e. A commitment to reducing or eliminating waste is a fundamental principle of lean construction (Polat and Ballard 2004). A commitment to improving safety is critical to lean implementation as construction accidents are rivaled only by mining. In 2006, construction accounted for 21% of all deaths and 11% of all disabling injuries/illnesses in private industry in the United States (Bureau of Labor Statistics 2006).

f. A commitment to cost and performance measures. These measures are important indicators of the impacts of lean in construction projects. Benchmarking is based on comparing performance with other organizations (Alarcon et al. 2001). Measurements of percentage projects completed (PPC) and commitment reliability are essential as a foundation for continuous improvement.

g. A willingness to implement lean during the design stages. Tsao and others (2000, 2001) identified that committing more project resources to the earlier design phases improved the impact of lean techniques.

h. Collaborative relationships. Lean requires close collaboration between the parties; the standard forms of contract are adversarial in nature. Relational contracting is a transaction or contracting mechanism that apportions responsibilities and benefits of the contract fairly and transparently based on trust and partnership between the parties. It provides a more efficient and effective system for construction delivery in projects that require close collaboration for execution. Working relations between the stakeholders are improved, and improved efficiency and reduction in conflict lead to better financial returns. The most significant factors that underlie relational contracting are cooperation and dependency between the parties.

i. Information technology makes it possible to effectively manage the construction process to transform physical resources such as money, materials, manpower, and equipment. Integration is greatly improved by the sharing of information between the parties. Building Information Modeling (BIM) is a technology that effects many economies in construction projects (See Chapter 8). It makes a reliable digital representation of the building available for design decision making, high-quality construction document production, construction planning and performance prediction, and cost estimates. Having the ability to keep information up-to-date and accessible in an integrated digital environment gives architects, engineers, builders, and owners a clear, overall vision of all their projects, as well as the ability to make informed decisions quickly.

Organizing Lean Construction

The adoption of the lean construction methodology represents a paradigm shift for many owners, designers, contractors, and suppliers who have been accustomed to traditional

---

* Study-Action Team™ (SAT™) is the intellectual property of Lean Project Consulting, Inc.

construction practices. The success of The Last Planner® System hinges on counterintuitive actions by members of the construction supply chain to optimize the entire project instead of maximizing the efficiency and profitability of their respective portions of the work.

Greg Howell, co-founder of the Lean Construction Institute (LCI) references the opinion of Dean Reed from DPR Construction "that The Last Planner® System (LPS) is a disruptive technology." In effect it reveals issues and opportunities invisible to those operating in current practice. The LPS changes the way work is planned and managed on projects and in construction organizations. Companies implementing LPS always face these contradictions and eventually must decide to transform their company or somehow wall off those projects operating on LPS. The first projects managed on LPS can show companies what will need to change to support wider implementation.

> Greg Howell suggests that at this point the senior management of the respective company (in the project delivery team) must decide their future. Then they must tell this to the people in the company. People need a compelling reason to change; they must be able to understand their role in the change and where it will lead. It is senior management's job to say, "We are going Lean" and then tell why, what it will mean, what it will require, and what it will produce.

Howell recommends that this declaration must be followed with a steady effort to shift the organization to the adoption of a lean culture. It isn't easy to put a company on a learning track, and people will follow their leaders as they explore the ideas and plan for implementation. Regardless of whatever structure is adopted, the focus needs to be on learning.

Some consultants recommend that one should avoid creating a complex structure for lean implementation. Robert Blakey*, founding principal of Strategic Equity Associates, LLC suggests tailoring the structure to the size of the project or implementation effort; an overburden of administration would bog down attempts at cultural change. Another approach would be more along the lines of a "quality circle" where one individual serves as a champion and works with the other members already normally assigned to the project/program team to raise their awareness and to build them into a lean team. Based on that success, a larger effort could be mounted in future projects/programs that might involve the creation of a "council" approach as is applied with total quality management (TQM).

Glenn Ballard (1999) in "The Challenge to Change" suggests that companies should first understand themselves before rushing out to hire a consultant to make a lean transformation. He cites Peter Block (Flawless Consulting) on the three possible roles for a consultant: (1) as a technical expert who tells people what to do to accomplish a specific goal; (2) as a temporary member of the organization, often for unpleasant jobs such as layoffs and reorganization; and (3) as a facilitative consultant that teaches an organization new skills. However, in the case of lean construction the people in the organization need to learn to think, see, and act in a lean mode.

Companies that understand and that have practiced self-transformation can benefit from the fresh ideas brought by a consultant. Those that have not tried to change from within may delude themselves by expecting consultants to bring about change extrinsically, when it needs to occur intrinsically. A lean consultant would benefit a company most by helping it to learn how to change. Ballard views a consultant as "always an outsider." If

---

* Comments by Greg Howell, Robert Blakey and Matt Horvat on organizing lean construction are based on e-mails with L. Forbes between December 2008 and January, 2009.

he or she makes the mistake of acting like management, then real change does not occur as an outsider cannot force anyone to learn new behaviors.

Ballard recommends that companies start with the LPS as it identifies improvement opportunities while freeing up resources to address them. A consultant can prepare project staff (with studied diplomacy) to face unpleasant truths about their need to change their beliefs and actions in order to benefit from improved project outcomes.

Actions for Change

Leading change successfully requires several necessary actions (Macomber and Howell 2005); construction project teams need to be guided in adopting new attitudes and behaviors in order to deploy lean construction. Five specific actions* have been identified as a minimal requirement; while adopting all five does not guarantee success, excluding any one of them will derail lean efforts.

1. Communicate the need for change in a manner that resonates with the people involved. Maintaining the status quo makes everyone vulnerable in a competitive environment. Provide specifics of how proposed changes can have concrete benefits to all involved.

2. Present stakeholders with reasonable performance standards they are expected to meet and secure their agreement. Maintain transparency with setting standards in order to project fairness, and establish on-going measurement in order to hold people accountable to their "customers." These customers may be downstream actors. Reinforce positive behaviors with sincere, public appreciation.

3. Demonstrate clearly the new behaviors instead of merely talking about them, and engage people in emulating them. Tools such as video are excellent for conveying knowledge and new information.

4. Measure, acknowledge, and reward new behavior. Work with people hands-on and try to see them doing something right. Point out the negative consequences of old behaviors and help to correct mistakes.

5. Help to make the changes positive for the participants. That reduces frustration and engages the workforce in responding to benefit all stakeholders, including themselves. A change agent must positively display passion in order to convey a sense of commitment to the work force. To do less is to diminish the value of lean endeavors in the eyes of those who need to be an integral part of the change.

Leadership's Role

The primary responsibility of lean leaders is:

- Ensure processes are running as designed
- Improve processes
- Empower and develop people

---

* Macomber and Howell's five Necessary Actions for Change, Lean Project Consulting, pp. 1–2. With permission.

Leaders must visibly lead the change. The lean transformation is organizational, not just a project commitment. It needs to be continuous and not sporadic. Leaders would do well to heed the words of Aristotle:

> We are what we repeatedly do. Excellence, then, is not an act, but a habit. Aristotle

Lean requires time from people for training, conference space, equipment and supplies ranging from computers to sticky notes. The organization must provide these resources. Mentors are essential; they help people to understand change and their role in it as a positive experience. Training should be Just-In-Time; when people start work on their projects they should already have the necessary lean training, but not too long before.

Desirable leaders' attributes:

- Practice servant leadership
- Teach—be the sensei
- Be a humble learner
- Enable and celebrate the successes of others

Leading Lean with Passion

Jay Berkowitz, President of Superior Window Corporation cites the importance of having a leader with passion, and credits his success with applying lean to his manufacturing and installation operations to that rare attribute. In his words "Lean without passion is like a love lost relationship", i.e., one that has no spark and no energy to sustain itself. He sees passion as the invisible ingredient that drives the lean culture, analogous to a wave of sustained enthusiasm that spreads throughout an organization, influencing everyone to carry out work without waste, overcome resistance to change and adopt new approaches to supplant old and inefficient habits. Berkowitz links his success with lean to a management style that overturns the traditional pyramid to make management accountable and place workers in a position where empowerment, recognition, and performance feedback bring out the passion for self—improvement that resides in everyone. While his company's lean process needs the continuing maintenance of a "Kaizen Professional Officer", the passion of the leader is needed to make the lean initiative thrive. (See Case# 4 on page 161)

Training the Work Force

Construction involves the collaboration of a number of permanent organizations (contractors, subcontractors, suppliers, etc.) to form a temporary organization in order to execute a project. If lean construction is to be adopted, then some orientation and preparation need to be provided to the members of those organizations. Blakey suggests that it takes a "champion" and a firm (general contractors (GC), CM, owner) interested in investing in the process. A model project of limited risk and cost may provide a good foundation for learning; it may involve implementation of only one element of lean construction (such as The Last Planner® System). Thus, the involvement of the firm's top executive may be very limited beyond approval of a model project. A number of lean projects have evolved based on having the major training effort focused on a ranking executive responsible for an individual project, and then subsequently moving down to foremen and trades.

Howell* is wary of the idea that people can be trained in lean; some learning is required but it does not persist without leadership support. This support is usually the limiting resource. They need not be lean experts but they do need to be an active force for studying, learning, and thinking. Simulations such as lean-based games (see lean coaching, p. 140) convey and reinforce very powerfully that collaborative, sharing behaviors hold mutual benefit for stakeholders in contrast to the destructive results of adversarial attitudes.

Matt Horvat,[†] a lean coach at Lean Project Consulting Inc., engages upper management by forming a Lean Leadership Group that meets on a biweekly basis. Meetings usually last one hour, in order not to detract from the busy work schedule. Matt evaluates participants' skill set with the use of a People Development Plan (Toyota Talent and TWI–Job Instruction). A typical agenda is as follows:

1. Defining roles for this meeting (timekeeper, gatekeeper, commitment cop)
2. Reviewing issues of concern from a previous meeting
3. Concerns of management
4. Concerns of project participants
5. LPS roll out schedule
6. Training needed
7. Creating/updating standard work
8. Small group improvement teams report out
9. Other aspects of creating a lean environment
10. Issues to be followed up

Need for Training at Project Inception

The best course of action would be to have a single lean development office coordinating all coaching and training (Howell 2009). Some "all hands" work is needed, where all involved people are brought together. They may work in small groups on specific types of problems, but cross-functional teams are an absolute requirement in order to secure smooth interaction during the project. A fair amount of time—at least a few weeks, should be allowed for people to absorb new concepts. Experience suggests that people adopt lean when they make the connection with what they see, and when it influences how they explain the world. Lean simulations such as the "Airplane Production Simulation http://www.visionaryproducts.biz/" and "Parade of Trades" do this for some people. These simulations are typically used in a classroom setting for lean training, and are especially helpful with a mixed group of design/construction professionals. The Airplane Production Simulation involves having a team of people—often 4, 5, or 6 set up as a series

---

* Comments by Greg Howell, Robert Blakey and Matt Horvat on training the work force are based on e-mails with L. Forbes between December 2008 and January, 2009.
† Self-published report "What Does a Coach Do?" December 9, 2008, pp. 1–8. With permission.

of work stations in an assembly line. Lego blocks are used to assemble a number of model planes. Varied work rules clearly demonstrate the impact of the structure of the work. The Parade of Trades demonstrates the impact of reliable promising on the variability of work flow. (This section on lean coaching is based on e-mails from Howell, Blakey, and Horvat between December 2008 and January 2009 on the subject of initiating a lean construction program.)

Blakey notes the value of coaching/facilitation/training at the project team level. After becoming aware of lean construction, many firms (particularly general contractors or mechanical subcontractors) form their own internal teams to carry forward the coaching/training/facilitation to future projects. However, this approach is limited, as it benefits each firm internally, but not external stakeholders.

Horvat generally starts coaching on lean implementation with a Study Action Team™ (SAT™ is the trademark of Lean Project Consulting), which is very effective in helping participants to understand lean principles and to visualize the possibilities that are available to them. The SATs are most effective with a cross-functional group because team members can see how others work and a relationship can begin to grow. The very act of having people in the same room can have a synergistic effect.

Cross-functional teams benefit especially with explanations of The Last Planner® System. It is helpful for the project engineers and superintendents to work together. This also applies to the project leadership of the general contractor/CM and the leadership of the owner and the architect/engineers. The LPS is built on making and keeping promises, and all team members need to internalize these principles. For small group improvement teams, cross-functional representation is also essential because these teams have a broad impact on a project, and bring a systems approach to finding solutions.

Duration of Coaching

People should be coached until they are able to swim on their own (Howell). Blakey suggests that, at a minimum, the coach should be involved throughout the originally assigned project. Many firms establish an internal position related to lean implementation after becoming aware of it. Typical coaching engagements last between 3 and 5 months, involving about 3 days on-site every other week. With an ambitious team, this is generally enough to set the basic habits in place. Larger organizations such as nationwide builders with many divisions are more complex and generally require a longer time. They often need additional coaching/facilitation to standardize approaches on a companywide basis. As shown in Figure 6.1, with the passage of time, a team's knowledge increases while the coach's input decreases.

Horvat, as a coach, sees his role as building the capability of project staff, instead of doing any administrative work. He familiarizes them with spreadsheets for data capture and tracking and helps them through the daily and weekly meetings to accomplish that critical function. This lean-related role is typically carried out by a junior field superintendent.

Sustaining Lean Initiatives

Some lean consultants usually have a monthly follow-up, at least until procedures are firmly established; at that point the team can begin to innovate on their own. They should be able to carry out PPC measurements, and to formulate improvement strategies based on them. (See Chapter 15 for more information on sustaining lean efforts.)

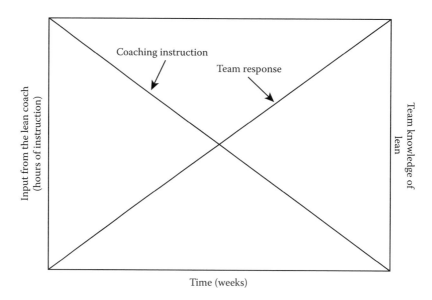

FIGURE 6.1
Impact of lean coaching.

Qualities of a Lean Coach

A lean coach is an unique resource for facilitating the learning of lean principles by project teams. Successful coaches have the following attributes:

- Facilitation skills—the ability to engage a wide variety of people
- Patience
- The ability to not take things personally
- Charisma and presence are great qualities in a coach

Lean Coaching

Lean construction requires that stakeholders adopt new beliefs and behaviors that depart significantly from their experience with traditional construction methods. The adoption of new paradigms occurs most readily with the help of an impartial third party—a facilitator, or more specifically, a lean coach.

Matt Horvat, P.E., a coach with Lean Project Consulting presents the following cases to illustrate the benefits and challenges of lean coaching. Matt is an industrial engineer by training and has gained extensive experience in the construction environment. In Matt's view, people set out on a lean journey for a variety of reasons. In many cases, management sees the positive results of lean projects and wants to derive similar benefits. Some self-education occurs and a manager experiences a heightened interest in the topic; but is unsure of how to get started on a lean journey. Hiring an expert to work with the organization's team for a couple of months is a typical approach.

Lean is best described as a management philosophy based on the concepts of *respect for people* and *continuous improvement*. It requires fundamental change in the attitudes of the organization and this is where an expert coach can help. Three factors must be addressed in order to change the future: (1) our commitments, (2) our self-narrative or story we tell ourselves about ourselves, and (3) our action or practice. Changing any one of these three factors can yield positive results, but the change is most effective when all three have been addressed.

It is also the reason that most implementations fail. For example, changing one's actions is typically the first thing that happens in a lean implementation. Although 5S or the LPS may be initiated, some managers continue to operate in traditional ways. Their story doesn't change. They sustain the belief that the most effective way for getting work done is the traditional approach. They emphasize the use of "carrot and stick," command and control, lowest initial cost, and risk diversion strategies. These strategies contradict the lean philosophy and neutralize attempts at implementing methodologies such as the LPS, which is based on "bottom up," commitment-based planning.

Making a commitment to changing the future is another effective approach for changing the results, but also does not work completely when used alone. When people make strong declarations to change in public, miraculous things happen. People's perspectives change. If a strong leader tells teams that they are going to adopt the LPS and be the best in the industry or region using this approach, the followers will make a point to familiarize themselves with the approach. Even with repeated declarations, this is typically still not enough, although it represents an integral part of the change.

The creation of a lean culture involves having stakeholders embrace the principles of *respect for people* and *continuous improvement,* but this concept is too abstract for most people to begin working on directly. Additional steps are needed to inculcate these values in the hearts and minds of an organization's people, and a certain amount of time has to be invested in the lean journey as it is impossible in one moment to change long-established beliefs and attitudes.

Activating all three of the foregoing approaches is the best method for inducing change. Leaders have to make strong declarations in public to announce their commitment to the lean philosophy. This needs to be followed up with active deployment of a lean process such as the LPS. To accomplish the transformation, stakeholders have to take an objective look at their self-narrative or how they function, and see how it differs from the lean philosophy. It is only when this type of introspection takes place that people can begin to adopt a new perspective and be receptive to the new behaviors that a lean environment requires.

All too often, managers want the benefits of lean without doing the learning themselves. That is a prescription for failure as people in an organization are quick to detect how well their leaders believe in a particular philosophy; subliminal messages convey management's sincerity even in the absence of the spoken word. If leaders do not have their hearts and minds invested in a new initiative, why should workers embrace it?

Lean coaches have been effective in building knowledge with a model termed The three As of Learning: Awareness, Acceptance, and Action. Leaders will do best with a coach that knows a lot about their desired future state; that is, *how* they wish to change and is expert in their industry (Figure 6.2). Beyond imparting a new curriculum, lean coaches

FIGURE 6.2
The three As of learning.

can sometimes help simply by being available for advice while an organization is transitioning to a new operating mode. On typical construction projects, adopting lean represents a quantum leap for most construction professionals, and indicates a new way of life for practically all of the project personnel. For these two reasons a professional coach experienced in lean construction is by far the best resource to have. That individual can help to create an implementation plan and can maintain contact with the organization as lean implementation takes place.

As a practicing coach, Horvat spends time with the leaders of the company or project clarifying their goals. More importantly, he makes an assessment of the world that they live in. He tries to see situations through their eyes and spends time meeting with and becoming familiar with people all across the project.

Details of the Three As

Awareness

With the project team together, a pair of production system simulations are used to give people the experience of changing the rules of how work moves through the system. The Airplane Game (Lean Zone® Production Methodology, (http://www.visionaryproducts.biz/) shows the effect of pull scheduling, work leveling, and other concepts. Traditionally, work gets done in the field by having foremen start their crews as soon as possible. The Airplane Game puts project participants in a simulated manufacturing environment. The participants (often four, five, or six) are asked to represent an assembly line in which one worker assembles a few Lego blocks to represent an airplane section, then passes it to the next person for further assembly. One team member serves as the quality inspector. The last person should complete the airplane and set it aside as finished work.

With each round of the assembly, people are given different work rules to see how they affect the performance and output of the assembly line. In the first round people are told that they are being paid by the piece (build as much as you can—"local optimization"), that they are there to do the work and nothing else (no talking or thinking), and that to reduce set-up time they must maintain integrity of a batch size of five airplanes before moving them to the next work station. Downstream workers wait anxiously to get started while upstream workers struggle to finish a batch. In this round it is typical to see people running to get inventory, and large amounts of inventory accumulate between workstations (unbalanced line). Results are recorded on the following factors: work in process (WIP), rework, time to produce the first "good" plane, and the total number of "good" planes produced. The coach then discusses the impact of the work rules and possible improvements to the system.

The simulation is restarted in the second round, reducing the batch size to one and a queue space of one. That is, a maximum of one assembly can be in the waiting area between workstations. Some thinking and talking are allowed, a bonus is paid for team performance, and the quality inspector remains as an inspector only. Results are recorded, and the coach discusses the experience with the team.

In the third round the line is balanced, communication is completely open, people are given the authority to assist others, the inspector is put to work, and people are responsible for their own quality. Results are recorded. The improvement is dramatic. It is common for a team to complete five good planes in the first round and to complete 30 planes by the third with only changes to the way work moves through the system. In addition, there are interesting safety impacts. Most people say that the first round is the most unsafe.

The Parade of Trades™ (http://www.ce.berkeley.edu/~tommelein/parade.htm) shows the combined affects of dependence and variation in a very visual manner. A team of seven people sits around a table and represent seven building trades. A smaller team can be used, provided that each member represents a building trade.

a. Each team is provided with a die and a cup for tossing. The dies vary—one is a regular die with faces labeled 1–6. A second die has faces labeled 3 and 4, alternately. A third die may have a different numbering sequence.

b. The first player is provided with a number of tokens that represent the number of work assignments to be carried out.

c. The team has a chart subdivided into columns labeled as follows: Production Capacity, Actual Production, Lost Capacity, and Workable Backlog.

One member records values on the chart to indicate respective die rolls; each roll represents the number of people that come on the job for a week (i.e., the production capacity). That member passes tokens to the next trade, in accordance with the value of the toss, and notes that value in the production column. If the die toss indicates a value greater than the number of tokens held by that member, then the shortage is listed as lost capacity. As the game progresses tokens are passed between trades representing the building process: concrete, masonry, façade, carpentry, mechanical, electrical, and paint in that order. Passing a token is equivalent to having one trade complete its work and release that work for continuation by another trade.

If a given trade does not have enough tokens to pass on, then the duration of the project is extended. The game ends when all tokens have been passed through from the first trade to the last. Performance is tracked by noting the number of weeks elapsed, and by totaling each column, especially lost capacity and workable backlog. With more than one team, it is possible to show how the variability of the dice affects the speed at which work moves through. Typically three teams are formed and three different dice are used. One team has a standard six-sided die with the numbers being {1, 2, 3, 4, 5, 6}. The other dice that are used have the same average of 3.5, but the spread is different. Overall the dies have a mean toss of 3.5, but the range of probable values is different, and that introduces less reliability into the project. Typical roll counts are {1, 2, 2, 5, 5, 6} and {2, 3, 3, 4, 4, 5}.

Like the airplane simulation, the results are dramatic. People estimate (correctly) that it will take approximately 10 weeks to move 35 pieces of work through the first workstation. People also generally assume that the painter, workstation 7, will complete in the 17th week. In fact, the team with the higher spread in the numbers {1, 2, 5, 5, 6} takes considerably longer to complete all the work. The concrete workstation can finish much earlier than the other teams, but bottlenecks happen with the higher likelihood of low rolls. This represents the outcome of work that is highly dependent on multiple trades that are coupled together. As is typical of the construction environment, people typically prefer speed to reliable performance. With this simulation, it is easy to see that there is a much greater improvement in throughput when work reliability is given priority over speed.

Performing these simulations with the expanded team of subcontractors, architects, engineers, designers, owners, construction managers, general contractors, project engineers, and so on brings out more conversation. Participants have a great opportunity to learn firsthand how the factors used in the simulation can be implemented and used practically in their projects. Although learning is personal, it is much more effective when it is provided by a peer group—a lean leadership group can have a much greater impact than one teacher with many students.

Acceptance

Through a powerful debriefing, a lean coach can bring out awareness of the current state of affairs in a construction organization as a good starting point. This awareness now needs to be translated into a commitment to change the managing principles in keeping with the model to change the future: commitments, story, and actions.

In initial presentations and through conversations about lean, the leadership learns to say the right things. From there, strong declarations are made about their commitment. A lean coach can continue to work with the lean leadership group throughout the engagement to help them keep in touch with their people and express the needed level of leadership during this transformation.

The lean coach can then be highly effective by introducing the leadership to lean management in a book club style group. With the shared understanding of lean, new possibilities are discussed. Simulations help participants to understand the lean manufacturing concepts of pull, kanban, waste, goal setting, and so on. This is needed to enable leaders to see the difference between current approaches and the lean approach.

Action

Learning is not useful unless it is coupled with action. An effective coach will engage participants in small group improvement activities to develop the habit of continuous improvement. Adoption of The Last Planner® System takes place throughout the beginning stage of transformation. In addition to coaching, someone familiar with lean tools can do training on techniques such as problem solving with sticky notes, group facilitation methods, visual management, five-Why analysis, and A3 thinking.

The objective of the coaching engagement is to enable participants to adopt lean management principles. Project managers typically want to adopt the LPS (they see the results), but are generally not prepared for its requirements and may not fully understand this shortcoming. Engaging a coach is the most effective way to accomplish the needed transformation; it involves gradual discovery as lean students may not be aware of how lacking they are in understanding fundamental concepts, and learning takes time and a lot of energy.

Horvat points out that commitment is essential—with a coach, it can take a couple of months to build the appropriate foundational habits to begin a lean transformation. Then, constant reinforcing and reflection are required to keep learning. The bigger the team is, the longer the transformation will take; the less people are ready to drop their existing habits the longer it will take as well. Shock therapy may work. By the end of a professional coaching engagement, there should be people in place to maintain the new systems and behaviors. But in reality, a successful implementation will reduce waste and result in freed capacity not additional overburdening.

Case of Ready Mechanical

Jack, CEO of subcontractor Ready Mechanical, has heard some of the buzz about lean from trade magazines. He keeps a subscription to ENR and notices an increasing amount of talk related to removing waste and lean scheduling. Jack feels ready to go down a lean path, thinking that at a minimum it won't do any harm.

Ready Mechanical installs HVAC systems in the field and has a prefabrication shop with six administrators and 10 installers doing about $3 million of business per year. Jack buys a popular book about lean, and over one quick weekend thinks a good starting place would be to adopt a 5S program. In the metropolitan area of a city of about 100,000 there are a couple of consultants to choose from, and he picks one from the list to give his team a presentation. Jack discusses this decision with his team and tells them that even though he doesn't know a lot about lean, it seems smart. He explains that although things are working well at present, the team should be prepared for the future and lean seems to be an attractive innovation.

There is some resistance to Jack from Bill, the operations manager. Jack tends to adopt new ideas very impulsively and becomes disheartened when things don't go as planned. Jack assures Bill and the others that whatever happens they'll go slowly and work together. The presentation occurs and small groups begin a 5S activity to sort, straighten, and shine their work area. Jack and Bill stay in touch with their progress. After a couple of weeks Jack personally organizes a discussion between the teams so they can learn from each other. This routine continues and gains momentum for a couple of months. Old and unneeded inventory are recycled, safety at the fabrication shop is visibly improved, and replenishing inventory has been made considerably easier. Previously, the people in the shop didn't see the things lying around or the general disorganization as being a problem. Now they seem to be looking at things with different eyes.

Jack is surprised that the 5S methodology has such a positive impact on the people in the shop. He gets the field crews to take a look at their trucks and begins to work 5S into the vehicles and toolboxes. Installation becomes easier. Comments from the foremen include "it takes time, but it makes for a better place to work," and "we know where each other's stuff is."

Jack is busy with his regular work, but wants to get more serious with Ready Mechanical's lean endeavors. He joins the LCI and reads *The Toyota Way* (Liker 2004). Over the next year, Jack and his team reread the book together and begin to find ways to adopt lean manufacturing techniques. Initially, they begin to try and do more prefabrication in the controlled environment at the shop and to try delivering and making material ready when it is needed at the job site. Long-lead items are still ordered early, but they typically go to the job site and are not his problem until installation.

Ready Mechanical's project managers, Linda and Jeremy, attend the annual congress of the LCI and decide on using the LPS. The LPS is about construction scheduling and is mostly used by GC/CMs, but they believe that the principles are relevant so they begin. To get started they generate a 6-week schedule based on the schedule from the GC. This is updated weekly and submitted to the GC's field superintendent. Work plans are created by the field foremen, reviewed in weekly meetings, and submitted to the GC.

The initial response from the GC is one of confusion, but the field superintendent and the other specialty trades see Ready Mechanical as being unusually prepared and effective at getting their work done. Despite that, there have been several work stoppages and delays. After a couple of months of using the LPS, Linda and Jeremy are able to show the GC what has caused them to stop working the few times that things have gotten in their way. They generate a chart to show the reasons for delays or stoppages so they can fix the problem. There were six work stoppages because the previous trade was not complete when planned, four work stoppages because of unanswered questions, and one work stoppage due to weather. The GC's field superintendent agrees that if everyone would use a weekly work plan, coordination would be much easier.

Debrief of the Ready Mechanical Case

To review this case and learn from Jack's examples, answer these questions with a group of your peers.

1. What did Jack, the CEO, do when he wanted to get started with Lean?
2. Do you think lean needs to come from the bottom or the top of the organization? Why?
3. What could Jack have done if Bill, the operations manager, had been less willing to participate?
4. What elements of the LPS can a subcontractor not carry out by itself?
5. Reread the story of Jack and Ready Mechanical and notice the three As model of learning. What parallels can you draw? What would you do differently?

National Builder

National Builder (NB) is a general contractor located primarily on the East Coast of the United States. They have become familiar with lean principles from working with major manufacturing companies and have completed a couple of pilot projects using the LPS. The champions of the lean movement have been young and energetic junior project managers from the construction management programs at two highly regarded universities. The initial response is that the technique is better than what they had been doing before.

The West Coast division, a relatively small facility based in Portland, Oregon decided to see what they could do with lean scheduling, as they heard it described. A corporate representative came out to Portland and told them of the company's plan to have every project use the LPS within 5 years. Examples of weekly work plans (WWP) and 6 week look-ahead plans, as well as photographs of people in action were on the corporate intranet. The New Sight Optometry Clinic project was the first project that NB selected for use with the LPS out of the West Coast office. The project team engaged a consultant to help them begin to use the LPS.

The consultant came on site and gave a 2 hour presentation on lean principles and left the leadership team tasked with establishing leadership vision.

Two weeks later, the consultant came back to help the team create tactical plans to support leadership direction. The vision was to use lean. Together, the consultant and the leadership team created a plan to adopt the LPS, utilize a problem-solving framework plan, do, check, act (PDCA), and start an idea system (Quick and Easy Kaizen [QnEK]). Over the next month the consultant delivered an introductory session on each of these plans. LPS champions, PDCA team leaders, and QnEK leaders were specified. Workshops were conducted and the results displayed on a bulletin board in the job trailer.

The team modified the format of the existing schedule to fit the needs of a 6-week look-ahead plan (6WLAP). They developed spreadsheets to make WWP and the charts of the PPC, and variance chart. Knowing that the LPS starts with pull planning, the team took the existing schedule from the beginning of the project through Slab on Grade Complete. They wrote out sticky notes—one per activity—and held a meeting with the trades to confirm the schedule. This was very well received. It was one of the first times that foremen from the trades had the opportunity to talk about what they wanted and about the job with everyone before it got started. Everyone agreed that the LPS could work on this project and committed to filling out the WWP.

Although a couple of superintendents were embracing the principles of the LPS, the project leadership did not get in the habit of making improvements to the planning system. Variances were recorded and displayed, but no action was taken to reduce their frequency. The presentations of the small group improvement activities were getting dusty on the wall; no one volunteered for additional work of making improvements. Business was getting back to normal. People were overburdened. After a couple of months the project leadership said that lean scheduling had really helped to improve attitudes, but there was no marked change in productivity.

Debrief of the National Builder Case

To review this case and learn from NB, answer these questions with yourself and a group of your peers.

1. What did the West Coast division do when they wanted to get started?
2. Would you say the focus of the West Coast division was on doing lean things or doing things in a lean way?
3. Do you think that the situation can be salvaged? What would it take?
4. What similarities do you see between this case and your business environment?
5. How was the three As approach violated? How was it aligned?

Examples of Lean Project Delivery Application

The remainder of this chapter has examples of lean construction applications from the perspective of a subcontractor. Chapter 7 will show examples of the application of lean project delivery on entire projects.

Subcontractor Applications of Lean

A tenet of lean construction is its application to an entire project, involving all related parties. Its very essence is the optimization of the project, as opposed to the local optimization of a particular trade or subcontractor. Given the fact that lean construction has yet to be instituted as a standard requirement in construction projects, there are many instances in which a subcontractor that has embarked on the lean journey will be engaged in a project where the prime contract is not based on lean principles. Nevertheless, a significant benefit can still accrue to the overall project and to other stakeholders, including the owner. Not to be overlooked is that lean successes at the subcontractor level can set a powerful example for other project participants. Subcontractors that ascribe to the lean philosophy are well equipped to deliver their work on time, within budget, with high quality, and with few or no safety incidents. They are also inclined to be good business partners as their modus operandi is based on high-schedule reliability, clear communications, clean site habits, and predictable pricing.

The following examples illustrate these principles. They are based on (1) Tweet/Garot Mechanical Inc., a specialty mechanical/plumbing contractor; (2) Belair Excavating, a site services subcontractor; (3) Grunau Company Inc., a specialty mechanical/plumbing/electrical contractor; and (4) Superior Window Corporation Inc., a window manufacturing company.

Case 1: A Specialty Contractor's Lean Journey and Successes: Tweet/Garot's Lean Journey

Tweet/Garot Mechanical Inc. is a company over 100 years old that provides plumbing, HVAC, engineering/design, and energy saving solutions for businesses and government organizations in several regions of Wisconsin, the Upper Peninsula of Michigan, and locations throughout the Midwest. With over 300 employees and more than 70,000 square feet of fabrication space and facilities in Green Bay and Wisconsin Rapids, Wisconsin and Menominee, Michigan. Tweet/Garot's president, Tim Howald, was the 2006 recipient of the Rotary Club of Green Bay's Free Enterprise Award because of his business success and community involvement.

Tweet/Garot Mechanical's mission places the needs of their customers above all else. They have a continuous commitment to quality, safety, and productivity, providing the best value available for the specialized services they offer. They strive to deliver services "on time, on budget, and on quality."

Tweet/Garot has an outstanding safety record—their employees have worked in excess of four million man-hours without a lost time accident; an incredible achievement in a demanding, changing, and hazardous industry. Innovation has been a significant factor in the success of Tweet/Garot. Computers, safety managers, prefabrication, modular units, and lean construction techniques were all adapted and used by the company.

Lean Journey

According to Chris Warren (2009), Director of Risk Management and Lean Continuous Improvement, Tweet/Garot embarked on a lean journey in November 2006 and progressed by degrees. First, they collaborated and partnered with a number of lean advocates. MSOE School of Engineering and Optima were initially involved. WCM Associates became a partner in the process. Later on a consulting firm (quality support services [QSS]) under the guidance of Dennis Sowards assisted in implementing a number of lean initiatives.

About 18 months into its lean journey, Tweet/Garot carefully studied its operations through the process of value stream mapping. It enabled them to observe unnecessary activities and duplication of effort. They eliminated duplicate purchase orders (POs), and implemented better inventory management of their consumables through vendor managed inventory (VMI). In VMI, agreements are made with preferred suppliers to designate specific quantities and categories of materials and supplies to be made available as needed; this obviates the need for Tweet/Garot to maintain inventories of these items and improves their cash flow.

5S implementation

Under Chris Warren's stewardship Tweet/Garot applied the 5S methodology to both their shop and field operations. They cleaned up the shop to make equipment more accessible; one particular machine that was locked into a corner was taken out and placed in an accessible location as part of a "manufacturing cell." They also implemented standard-ization for various job categories. By studying job sequences they have been able to set up storage bins and "gang boxes" for each type of job so the needed tools can be found quickly. As the average worker costs 90¢ per minute a delay of 10 minutes to find a tool costs the company $9.00, hence the improved organization provided by 5S is highly effec-tive in reducing non-value-adding activity.

In the yard, many items that were customarily left on the pavement were retrieved, inventoried, and organized for easy access. Excess items were disposed of as part of a Kaizen event; savings of $105K were realized, in addition to recurring future savings.

Tweet/Garot enhanced their outstanding safety record with a stretching program. The director of risk management and lean continuous improvement established a stretching program that requires workers to do stretches every day in order to reduce injuries. This was done without an incentive from their insurance company, but as a good business prac-tice that was said to reduce lost work hours.

In terms of their work processes, Tweet/Garot adopted behaviors in keeping with lean construction practices. They focused the actions of the work force with a "plan of the day" and they conducted "daily huddles" to discuss the specifics of each job in progress to ensure that they can make reliable promises to their clients and other companies involved with those jobs.

Example of a Lean Project

Tweet/Garot's lean experience was an important factor in winning a multi-million dollar subcontract for a hospital construction project in Wisconsin as the contractors, subcon-tractors, and all management players were required to have lean experience. The facility (ThedaCare Physicians in Shawano) was a 70,000 square foot, two-story medical building. The project was a fast-track project with a 6-month time frame, yet as a medical facility it involved very complex systems such as extensive medical gas piping. It comprised two surgery suites, an MRI, CT Scan, other diagnostic specialty equipment areas, and an oncol-ogy area for cancer treatment.

Lean Construction techniques were critical to the success of this project as the time lines were very demanding. Daily meetings were held to discuss and resolve scheduling issues and conflicts, material delivery schedules, short-term planning, and trade coordination.

Quality control requirements were incorporated in the lean construction processes. The daily meetings included a review of construction progress and quality or scheduling issues that developed subsequent to the previous meeting. These quality or scheduling issues were promptly addressed through action plans.

Tweet/Garot has incorporated information technology (3-D CAD) in designing and fab-ricating pipe and duct work. Through collaboration with the general contractor they were able to use this system to perform clash detection in the routes proposed for the piping. One of their system improvements was to eliminate the practice of having design draw-ings converted to shop drawings that were used for fabrication purposes, a process that required the work of three employees. The improved system converted 3-D CAD drawings directly to fabrication with laser plasma cutting tables, requiring the attention of only one employee, thereby eliminating two positions.

The 3-D CAD methodology also enabled an integration of fabrication work with the construction schedule. The fabrication work was sequenced with the overall master schedule so that fabrication work was done only for the assemblies needed for the immediate installation work. This enabled the use of the JIT approach so that assemblies could be loaded on delivery trucks and taken to the building for direct delivery to the respective floor just prior to installation. Overall significant productivity improvements were derived by reducing the need for measurement and fabrication in the field.

All of the plumbing fixtures and assemblies were prefabricated in Tweet/Garot's plumbing fabrication shop. All components were preassembled in a controlled indoor environment. All fabrications were labeled and tagged with their proposed locations. They were loaded on movable transport racks and shipped to the construction site for final installation. Also, toilet fixtures and associated fittings were staged on a "trolley," rolled down the hall on a floor, and workers could take the fixtures off at each location and bolt them in place using pre-packaged fittings.

In order to provide a high-quality installation, Tweet/Garot installed high-efficiency water heaters and low-flow plumbing fixtures. Utility savings of 5–10% were expected to result, in comparison with standard plumbing equipment.

6S at Tweet/Garot Mechanical, Inc.

Early in their lean journey, Tweet/Garot implemented a cleaning and organizing effort that they refer to as 6S (Safety, Sort, Set in Order, Shine, Standardize, and Sustain). "Take this welding station as an example," says Chris. "The idea came from an employee who said, 'Every time I go to get a welder it takes me a certain amount of time to set it up.' And by the time he gets it set up, the power cord plug may be broken because it was allowed to hit the floor, so then there's down time. And if the plug is cracked, he has to make a decision."

It was also noted that defects in the welding equipment could prove to be harmful to workers, hence there was a need to ensure that all the equipment could be operated safely. As part of the Kaizen event, Tweet/Garot's staff brainstormed about the optimum selection of welding equipment to meet the most frequently occurring jobs, and laid out the equipment in a way that made all needed fittings and equipment such as rods and cables available without hesitation. They also ensured that all this equipment was maintained in safe working order (Figure 6.3).

The following agenda was developed by the Tweet/Garot team:

- Introduction of team members and leaders
- Clarification of goals
- Process representation (SIPOC)*
- Brainstorming on inventory procedures and yard layout
- Changing the work culture
- Eliminating, recycling excess inventory to reduce waste and gain tax benefits
- Establishing VMI

---

* SIPOC represents the flow in a typical operation–Supplier, Input, Process, Output, Customers.

tweel brot

Process or Area Description:	**Kaizen Blitz**	Owner: Chris Warren	Start Date: 8/15/08	Review Date: 10/22/08
6 S of GB Shop Welding Machines		Team: Rod Dax, Eugene Sinitsky, Justin Johnson, Rod DeJardin, Jeff Greening		

Issues:

Welding machines did not have a place to hang all of its cords and miscellaneous items. Power (Damage), ground, whips, bottle caps. All items on the welders were not labeled where to put the Items at the end of the work shift. Time consuming to change welding wire spool, due to unnecessary items on top of machine.

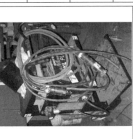

Target Condition:

• Eliminate unused welding machines (9 machines off the floor) (YES)
• Increase additional floor space for fabrication
• Improve ease of access to materials and equipment (YES)
• Increase safety of working in the area (YES)
• Reduce the amount of machine damage; example: power plugs (YES)
• Make equipment mobile for access and positioning (YES)
• Improve use of visual tools (YES)
• Labeling of welding machines (YES)
• 6S markings (YES)
• Parking Lot for extra machines: (Welding machine repair area; 14 of them). (YES)

Current Condition:

First issue was: how many welding machines did we have in the shop? We did not know. Answer: **49.** How many do we really need? About **35** Another challenge was, there are **6** different types/models of machines.

Results to date

• Increased sq. ft of floor space: 135 sq. ft.   X $ 9.50
• A place for everything
• Expectations that nothing is to be on top of machine, at end of the shift.
• A designated parking lot for extra machines

Savings:

• Frustration
• Better Safety
• Less Damage (5 plugs per yr. + labor = $ 330
• Sq. Ft. $ 1283.0
• Time: $ 260 + 500
• Total: $ 2373

Action Plan:

Action	Owner	Due Date
Set budget	Chris Warren	Complete
Obtain funding source for equipment & labor	Chris Warren	Complete
Obtain labor	Rod Dax	Complete
Gather ideas for improvement	Jeff Greening, Eugene Sinitsky	Complete
Schedule to implement	Chris & Rod	Complete
Implement Changes	Eugene S., Justin J., Rod DeJardin	Complete
Team review of gains after changes – Any adjustments?	Chris Warren, Rod Dax	9-26-08
Review after thirty days of use & the new layout	Chris Warren	10/22/08

FIGURE 6.3

A3 report of 6S application to welding equipment storage.

- Communications: Kaizen newsletter
- Kaizen lessons learned
- Future direction

The Kaizen team established the following goals for the first Kaizen exercise:

- All materials are to be identified (so their disposition can be accurately determined)
- Reduce inventory by 25%
- Establish the paved yard as tour ready (i.e., properly organized and free of clutter and excess inventory)
- Explore consignment inventory (through special arrangements with vendors)
- Eliminate personal and job inventories (these represent waste)
- Assign/maintain area ownership

Policies for yard inventory management procedure

1. The yard supervisor shall control all VMI (Vendor Managed Inventory) and all other inventory in the yard and warehouses.
2. No one shall remove any inventory without going through the yard supervisor.
3. The yard supervisor shall put away all inventory upon receipt.
4. The yard supervisor will work together with the financial control to implement and maintain computer managed inventory of all items.
5. No one shall bring any items into the yard without going through the yard supervisor.
6. All items returned to Tweet/Garot shall be tagged with the following information:
   a. Project manager (owner of equipment)
   b. Where item is returned from
   c. Expected removal date
7. All material from job sites shall be dealt with in one of three ways: returned to vendor (preferred), disposed of at site, or returned using item 6 above.

Summary of Improvements Made

The 5S exercise identified a number of materials and supplies that Tweet/Garot considered to be superfluous. These were donated to the Union Local 400 Training Center, for a tax exempt donation of $5,470

1. $13,978.53 worth of consumables returned to the tool room from workers who had a number of them just in case.
2. The purchase order (PO) process was simplified: The number of POs was reduced by six per day.

3. The cost impact was: six POs per day × 200 days × $50 = $60,000 in savings year over year. Plus the reduction of Tweet/Garot inventory and the improvement in cash flow.

4. The management of steel stock was improved. The goal is to have 70% of all steel material on VMI in 12 months.

5. Elimination of trash = six dumpsters; recycle E&O = three dumpsters mild steel; recycle E&O = two dumpsters of SS; estimated value = $4,741.00.

These results indicate a significant reduction in waste! The operation is now much more efficient. As a result of the 5S exercise, excess materials were identified that were redeployed as:

vendor returns ... in process
Total estimated value = $18,700.00.

In addition to the goals, up to 40% of the space in the yard was freed up; only one safety issue was not yet resolved; and there was a huge savings in efficiency when people knew where everything was. The cost benefit was calculated based on a savings of $0.50 per square foot:

9481 square feet × $0.50 = $4,741 year over year.

Overall Summary

Donation to training center	$5470.00
Use of consumables	$13,978.53
Elimination of POs	$60,000.00
Increase yard space	$4741.00
Recycle material	$4865.15
Vendor returns	$18,700.00
	$107,754.68

Ideas were developed for maintaining communication with staff to keep them informed on the progress of 5S and other lean endeavors.

Communication Plan

- Report out at end of Kaizen event (September 29, 2006)
- Send out an e-mail to all employees about Kaizen
- Post photos and a brief overview of the event in the shop
- Publish a brief article with pictures in the lean newsletter
- Present an overview in all associate meetings (fourth quarter)
- Post lean 5S Kaizen event on lean bulletin boards
- Discuss the progress with 5S in departmental meetings

Lessons Learned

- Teams do have power (they have powerful ideas)
- Team accomplished a lot in a week

- *Team diversity ... added to its effectiveness*
- *A Kaizen event is a bigger task than anyone thought*
- We were able to free up more space than we thought
- We save too much stuff in our everyday activities
- We have more space than we thought
- We may not need to add onto the yard
- It was inefficient and costly to have an disorganized yard
- It was hard to get started because of the disorganized yard
- *How undisciplined we are*
- The need for future Kaizen events
- We really started working as a team as the week went on
- *Everyone brought a lot of quality ideas to this event*
- Come in with an open mind
- There is going to be more to do than you thought
- *You are going to find resistance as you perform your tasks ... from other people in the company*
- When in doubt, throw it out
- There are no dumb ideas
- Work together as a team ... not against each other
- *We all have a better understanding what a 5S is*
- *And it really did not hurt as bad as we thought*

Future Direction: What's Next for the Kaizen Team?

- Team meets to finish "Open Action Items"
- Refine metrics and continue to work toward the goal
- Team must be allowed time to implement action items
- Team disbands with celebration when all action items are completed
- Team members then become available for new teams

Case 2: Belair's Lean Experience and Lean Construction Successes

This case portrays the lean activities of Belair Excavating, a multiservice subcontractor. Belair is a family owned site service and excavation company dating back to 1953. They have embraced lean construction principles for a number of years and have significantly improved their competitiveness as a result. Belair's executive leadership team relative to the case comprised CFO Tracy Dabrowski, Belair leader and catalyst for the introduction of lean; CEO Mark Murlowski; and COO Michael Murlowski.

The Lean Journey

In 2005, Belair's CFO Tracy Dabrowski* attended a class entitled Lean Construction at the college of St. Thomas, in St. Paul, Minnesota. She was so impressed with the class and its implications for Belair that she contracted the consulting services of the teacher, Sara Brazilier to implement lean methods in the organization. Paul Pasqua,† former director of development supported Belair's early lean endeavors. Sara led the initial team effort, introducing and institutionalizing lean in three of Belair's branches. This implementation took about 18 months and led to significant improvements in Belair's ability to predict the chances of success with a particular bid. The improved procedures have enabled them to monitor their WIP and not repeat previous mistakes over and over.

Human Relations and Lean

Part of the initial introduction of lean at each branch was a comparison of a triangle and an upside down triangle. The former represented a traditional organizational chart with the CEO at the top. The upside down triangle represented the new Belair philosophy in which management worked to provide the field with processes that enabled the field workers to do their jobs right the first time, and *both the responsibility and the requirement* to stop the job whenever they felt workers were doing something wrong. This of course was a very difficult process to instill in a construction environment that always had very clearly defined "top down" operational hierarchy.

Belair's leadership also introduced the concept of the people that worked in the process having both the right and the responsibility to work on the process. In order to promote pride in the work force, the company held reviews of all critical processes and sought consensus from all stakeholders involved in them. Belair found that employee involvement in various processes helped not only to identify those processes that were critical, but *encouraged and allowed* them to make it better. CFO Drabowski found that agreement on an actual process in construction is more than half of the battle.

Benchmarking Performance

Benchmarking was done internally at Belair's three operating locations. They were naturally competitive and management saw early in the lean introduction process that the problems in one branch were already being addressed in another branch. They determined that by talking about a common issue, the brainstorming of ideas from all the branches produced better results throughout the organization. Internal benchmarking provided the opportunity to improve each branch's deficiency without having to go outside.

Design Approach

Belair has a very forward-thinking design approach that very probably leads the excavation industry. One is the bidding process and the other is preplanning. Belair uses a very sophisticated prequalifying matrix (PQM) for all new bids. Dabrowski points out that site work companies often bid work that does not fit their processes or equipment; this may happen because of their reliance on a subjective tribal knowledge process. Belair,

---

* Belair Excavating documents, Belair's Lean Experience (Paul Pasqua) pp. 1–3, November 25, 2007.
† Belair Excavating documents, Belair Lean Improvements (Tracy Dabrowski) pp. 1, November 21, 2007.

instead, instituted a rigorous objective process that ensures that all the decision makers know enough about the bid to be confident that the company will delight the customer and make a fair profit on the work, if it wins the bid. Belair considers this verification of process capability to be absolutely essential (see Figure 15.3—Scoring of Final Project Grade).

Belair includes in its success criteria an understanding of project risk, to ensure that their efforts to win the bid are justified, and that the outcome will be mutually beneficial to all concerned. Jobs of marginal benefit are avoided, as positive outcomes are unlikely; to avoid that possibility, mandatory pre- and post-bid meetings are held with the customer. Customer satisfaction is placed at such a high priority that bids are presented in person and not by fax or mail. Research shows that pre- and post-bid meetings significantly increase the likelihood of acceptance.

Preplanning

Belair has had great success with its pre-planning process. Their motto is "Measure Twice and Cut Once" to minimize the possibility of errors of any kind. Once Belair wins a job and establishes an internal approach to doing the work, they invite all the stakeholders to an on-site kick off meeting. The participants include the owner, the general contractor, Belair's staff (including equipment operators), and significant subcontractors. They employ the rules of brainstorming to make sure everyone understands each other's viewpoint and expectations. A checklist (attached) is based on past experience and is used to ensure that nothing is overlooked and avoidable rework is reduced.

LEAN PREPLANNING AND MOBILIZATION CHECKLIST FOR JOB INITIATION

by Michael Murlowski (June 22, 2007)

(Reproduced with special permission of Belair Excavating)

PRE-PLANNING ITEMS

- **Signed Contracts:** Negotiate and finalize contract with GC. Establish contracts and or POs for subcontractors and vendors.
- **Funds in Place:** Verify that adequate funds are in place to complete the scope of work. In addition, confirm that Notice to Owner lien process is complete and is recorded.
- **Superintendent Packets:** Work off the Field Superintendent's Packet Table of Contents.
- **Project Manager Folder:** Establish Project Manager Folder to include Project Manager Table of Contents.
- **Established Budgets:** Enter project budgets into viewpoint and obtain project job number. Start project initiation sheet.
- **Safety information:** Put together GC required information regarding Belair Excavating Safety issues including emergency route maps, phone numbers, job specific safety items, GC specific safety items, underground locates, traffic control issues, etc.

- **Identify Production Tracking Issues:** We will need assistance with set-up and tracking to maintain project cost control. BAT files.
- **Develop CPM Schedule:** A critical path schedule and bar chart identifying milestone completion dates will need to be developed or obtained from the GC.
- Identify Equipment and Manpower Needs
- lternative Resource Allocation: Trucking and Dumps
- **Conduct Pre-Planning Meetings:** Schedule weekly in house meetings to identify that the Pre-Planning items as outlined are completed as action items, and schedule project initiation meetings with the GC, Owner and Belair Excavating subcontractors and Vendors to have a complete understanding of project planning and expectations. If needed Belair Excavating should drive these meetings for our own self-preservation relating to this project's success. Include with the Pre-Planning Meetings a partnering meeting with the GC, Belair subs, Owner, Testing Company and Engineers, and all key players in the project.

Mobilization Process

- **Equipment and Manpower Issues:** Resolve any resource allocation issues prior to mobilization.
- **Locates:** Have location plans completed prior to mobilization. Will require continued updating on a minimum 10–14 days relocate.
- **Identify Known Conflicts:** Identify known conflicts to the GC and the Owner and discuss impact prior to mobilization.
- **Job Site Readiness:** Visit the job site to assure excavation area as well as the dump areas are ready for the start of work. (Bus Ride with GC and Staff)
- **Trucking Availability:** Conduct weekly meetings with the trucking vendors to discuss service requirements and get buy-in for project success. Face to face communication so they understand our production requirements. Weigh the advantages of going by the load vs. by the hour for ease of billing and production incentive.
- **Permit and Licensing Requirements:** Identify any special permits, licensing and or Public Ordinances that may affect our productions.
- **Weather Considerations:** Identify what our back-up plans will be if we run into wet weather. How will our loading and dump operations be affected.
- **Safety Compliance:** Inspect operations heavily at the beginning of the project to identify any possible risk items that need to be addressed.
- **Tracking Trucking and Accountancy:** Provide a plan of action to assure trucking and tickets are being accurately accounted for. Establish log sheets on-site to track possible trucking issues on a daily basis.
- **Documentation Requirements:** Discuss with the field staff the documentation requirements that will be required on a daily basis and follow-up continually to make sure these requirements are being met for production analysis. Discuss work order process and Daily Recap requirements with the Supt. and include the GC so they have a full understanding of how we will expect to document pre-approved extra work items.
- **Special Mobilization Requirements:** Take into consideration special oversized equipment and trucking needs that will need to be addressed.

Post-mortem Studies

Belair attributes a part of their success to conducting a mandatory postmortem evaluation at the completion of each job for the purpose of continuous improvement. This evaluation compiles field staff's experience with the job and clarifies whether the outcome was positive for both the customer and Belair. (The process is detailed in Chapter 15.)

Lean 2010 Improvements

Belair's* CFO set a number of critical goals to be accomplished through the lean initiatives. Six initial goals were:

1. Make reliable time and production data easily available to system users, mostly within 24 hours of occurrence
2. Improve equipment utilization at the job level
3. Base job estimates on accurate, accessible historical production cost data
4. Increase field and office productivity
5. Shift time spent by field superintendents and project managers from paperwork and data chase to value-added activities
6. Create a platform on which we can operate at three times our current volume

Field-based improvements:

- Improved field production by 4.5%
- Improved equipment utilization by 6%

Administration improvements the CFO reported:

- Real-time cost and production data were placed in the hands of each field superintendent daily. This provided instant recognition of cost to budget variances.
- Improved timeliness of monthly financial reporting—7 days; posting of committed costs—15 days; posting of equipment costs—30 days.
- Ability to perform job cost projections improved from monthly to daily.
- Eliminated errors between field personnel and accounting.
- Reduced processing of payroll data from eight steps to two steps.
- Reduced worker hours to process redundant data by the equivalent of three full time equivalent (FTE) staff.
- Reduced paper work required to process payroll eliminating over 58,000 sheets of paper on an annual basis.
- Improved accounts receivable days by 10%.
- Paperless storage of accounts payable invoices. This reduced mail and copy costs by 40%.

---

* Based on document Belair Lean Improvements (Tracy Dabrowski), September 21, 2007.

Case 3: Grunau Company Inc.'s Lean Journey

The Grunau Company, Inc. is headquartered in Oak Creek, Wisconsin, with offices in Indianapolis, Indiana; Pittsburgh, Pennsylvania; Orlando, Florida; and Boardman, Ohio. Grunau designs, installs, and services mechanical systems for HVAC, plumbing, fire protection, and specialized electrical systems. The company is also active in specialty metal fabrication under the name Grunau Metals. Four generations have led the company, starting with Paul J. Grunau, the original founder in 1920. He was followed by Paul E. Grunau, Gary Grunau, and Paul W. Grunau who purchased the company in 1999.

Grunau refocused its business strategy in recent years to concentrate on building long-term relationships with customers. Their goal is to truly understand a customer's business, and to get involved earlier in a project's development process. According to Ted Angelo (2009),[*] Executive Vice President, Grunau firmly believes in running their business with an orientation toward building high-trust relationships. They have embraced lean principles to accomplish their strategic goals.

Since April 2003, Grunau Company Inc. has been developing business practices adopted from lean production theories. These theories seek to minimize waste and maximize value by creating better workflow, thereby increasing productivity. When used consistently, these philosophies, which were first implemented in the manufacturing industry, save time, materials, and costs.

Lean Initiative 1. Application of The Last Planner® System (LPS) Technique

Grunau has used The Last Planner® System identify the tasks that need to be completed each week. On simple projects, the foreman serves as the last planner and may keep the schedule by hand. The foreman is a working part of the crew. On big jobs, the foreman is the last planner, but Grunau has a project engineer who works closely with that person. The project engineer attends the weekly meetings and provides input on where the trades are and what they WILL do (i.e., their work commitment) the following week. This approach helps Grunau to make reliable promises on work they are scheduled to perform to other parties that are involved in a given project. Such reliable promises are an essential component of lean construction. Grunau uses software to track their progress.

Grunau uses the project's master schedule to create a look-ahead schedule. In some cases, there may be a project look-ahead schedule, and Grunau uses it to create their own look-ahead schedule. As a sub, Grunau pulls weekly assignments from the look-ahead schedule. Grunau has done this with five trades managed by the company: plumbing, HVAC, pipefitting, sprinkler installation, and sheet metal. They use an Excel spreadsheet to track 1 week and 6 week work assignments pulled from the GC's master schedule. It also shows percentage of plans complete (PPC). Grunau uses a computerized system—a Last Planner® System program for tracking work assignments with regard to time, labor hour, and material requirements for each construction task. The Last Planner® System program was upgraded with five preprogrammed macros that greatly reduce the steps involved in building a work plan for the subsequent week.

---

[*] Angelo, T. 2009. *Lean Case*. The Grunau Company Inc. The description of Grunau's adoption of lean construction, including application of the last planner technique, daily huddle, space reorganization (5S) and supply chain improvements. Self-published notes, March 30, 2009, pp. 1–5. With permission.

Lean Initiative 2

Use of building information modeling (BIM): On big jobs, Grunau uses BIM to keep track of routing for various systems to ensure there are no clashes between lines (pipes, ducts, etc.). Grunau gets a file on a CD in Navis Works from the Architect/Engineer (A/E). It is used to show various lines for services (piping, etc.). Conflicts/clashes can be detected.

Human relations, daily huddle: The daily huddle is a pre-task planning exercise for discussing what is needed for each job for a given day. The huddle helps to reduce accidents, eliminate waste, and improve efficiency. It supports the lean mission: To continually examine processes to provide greater value to customers without waste.

The team members write notes on 3×5 cards to promote discussion, and a volunteer is selected each time to read questions from the card. A typical list of questions for the huddle is as follows:

- What are our productivity goals for today? Answer: 300 feet of pipe, boxes hung, fixtures set, and heads roughed in.
- Do we have what we need? Answer: Yes, all materials, tools, information to get the job done TODAY.
- Are there any obstacles to achieving our goals? Answer: Is another trade in our way? Is the area not completely open to us? Can we move around area?
- Does anyone see a better way? Answer: Is the routing of piping/duct okay, or is there a better way? How about the material handling—any better way? Share a better way to complete a task. If a crew member develops a way to eliminate waste, inform a lean team steering committee member so we can share with others in the company through a future newsletter.
- Were there any near misses yesterday? Answer: This creates safety awareness for team members. Someone else may be subject to the same situation.
- What safety hazards should we be aware of today? Answer: There is going to be a lot of crane activity on the west side of the building. The conditions out there are slippery. Pay special attention to where you are walking.

Remember to sustain, we must keep it fresh since we are reviewing on a daily basis.

Observed Benefits of Daily Huddles

The daily huddles have changed workers' behavior. Even though the work for a given week may have been already planned, workers increasingly go through a thought process before taking action on a project.

Lean Initiative 3

Space reorganization: Grunau's Oak Creek facility was housed in a 22,000 square foot building. Management recognized the need to prepare for expansion, yet they opted to optimize their existing space so they could use the available space more effectively. The CEO directed the Grunau team to reorganize the existing space using a 5S approach and leave unused space for up to 20% growth in staff and operations. The team was also directed not to touch the exterior walls of the facility, but rather to make changes to the use of the interior space only. The underlying purpose of the transformation was to "Reduce storage, increase efficiency, and eliminate waste."

This new requirement was very difficult to accept for many employees, especially for those who had worked with Grunau for several years and held supervisory positions and titles. Paul Grunau himself set the tone by announcing that he would not have a private office. Eight "quiet" rooms were to be established that would allow staff to hold meetings, and so on. These were modeled after the "big room" (*obeya*) concept. The work spaces make it easy for employees to collaborate on the planning of projects, as a CAD operator can talk with a foreman about the configuration of a piping layout and enable an estimator to price a job accurately at the same time. Also, space was allocated that would facilitate personal business such as family matters, medical appointments, and so on.

Electronic filing: Previously, Grunau's service department had organized the storage of large numbers of three ring binders that contained operating & maintenance (O&M) manuals from completed jobs. These binders had been indexed with their respective storage locations in a large bookshelf system. Over 500 such binders were eventually stored in service department cubicles, with the overflow accommodated along a building wall. Electronic filing was seen to represent a significant opportunity for space saving. All the O&M manuals were scanned electronically and placed in computer files, with CDs as back-up copies. The storage space that was occupied by the shelving system has been regained and modified to provide employee work space. Two small CD cabinets have replaced the extensive shelving system that monopolized a fairly large storage area. O&M data can now be retrieved through each employee's computer, using project reference information. There are considerable time savings in eliminating the need to physically walk to the storage cabinets and locate the respective binders.

Vendor supply team: Employees studied the supply chain for materials and took steps to improve it. In the previous state, materials would be purchased from supply houses and stored at Grunau's facilities. In the current state, they have a purchase agreement based on delivering materials when needed, where needed. It includes having a driver do the delivery for each trip, but it provides a major saving, estimated at $70,000 annually.

Case 4: Superior Window Corporation Inc.

This case describes a window manufacturing company that first applied lean manufacturing in their operations, then used the lean supply methodology to synchronize their field operations with factory operations. According to Superior Window Corp's President Jay Berkowitz, the company now defines its TAKT, i.e., its demand, via the needs of its installation department instead of using a batch manufacturing approach to meet a production schedule.

Superior Window Corporation Inc., manufactures windows for the construction industry. Their niche is in the supply of windows for large construction projects such as office and institutional buildings. These projects often require a few hundred windows, some of which are of different shapes and sizes; they are often unique to that specific project.

Superior Window Corporation embarked on a lean journey in their manufacturing facility in 2005. They engaged the services of TBM Consultants, experts in lean manufacturing.

Through the application of lean initiatives, they effected several improvements in production volume and quality in their manufacturing operations. As the quantity of windows for a typical project is moderately sized, many window manufacturers meet these requirements through a batch processing approach. However, Superior Window Corporation was able to improve their productivity and product quality by adopting the concept of one-piece flow. These improvements translated into greater profitability for the manufacturing operation because it eliminated waste as defined by Work In Process (WIP).

While Superior Window Corporation is a highly competitive supplier in the window industry in the southeastern United States, they recognized that they could extend lean initiatives beyond the confines of their manufacturing facilities and into their field operations. Many of Superior Window Corporation's contracts require them to manufacture windows and install them in the shell of a building as a part of the construction process. Their application of lean in supplying windows for a specific project is described as follows.

The Cypress Run facility was built using tilt-up technology. Windows had to be transported from the manufacturing plant to the site and installed by the company's workers. It was most important to dovetail the supply and installation work with the master schedule as it was inadvisable to store uninstalled windows on the site because of possible breakage. The traditional installation method required many steps, and it was typical to fit windows loosely in place and adjust them to the proper fit in the course of several visits to the site.

The president decided to apply lean through a Kaizen event to the window installation work in order to identify and implement reductions to the labor requirements. However, it was recognized that changing from the traditional installation method would mean instituting a cultural change in crews that were accustomed to a great deal of independence. These crews usually spent their days in the field, away from management's oversight, and generally set their own goals and time frames. Management decided that it was first necessary to demonstrate that a new method would work—this led to the implementation of a pilot program.

The Kaizen event was held with the installation crews and their supervisors. The purpose was to:

1. Determine project demand rates and convert this to takt time to pull from the factory.
2. Establish a baseline of cycle times.
3. Identify and measure waste to eliminate it.
4. Create one-piece flow in field installation versus the traditional batch mode.
5. Implement the 5S methodology.

Traditional Method

Windows would be produced in batch mode in the factory to meet the specifications of a construction project. In the days preceding an installation job crews would start at the shop and load windows into a truck for transportation to the construction site. The windows were stored in a trailer and retrieved in subsequent days when installation was carried out. On the day when installation began, crews would start at the shop, selecting and loading tools and equipment in a truck based on each worker's preference. One or more trucks would be driven to the site, depending on the size of the job and the number of workers. As workers selected an opening in the building shell for mounting a window, they applied a bed of sealant all around the opening. A window would be hoisted in place and workers would position and fasten it in place with four Tapcon screws. A follow-up visit would be made on a subsequent day to make final adjustments to each window (i.e., using a spirit level to position it correctly). The workers would drive more screws to have a final installation, and caulk around the entire frame of the window, both at the interior and exterior faces of the building. Major shortcomings of the traditional method are:

1. The batch process used results in several people watching while others work. There is very little formal measurement.

2. Preliminary installation of windows does not encourage "ownership"—while workers are expected to position the windows with the right level and plumbness, their work is not identified with any particular individual.

3. In using the traditional method, many problems are discovered and addressed in the course of the job as best as possible. In some cases, irregularities in the window openings are serious enough to waste several man hours while corrections are made.

There is much waste in non-value-adding activities such as: excessive walking, sorting material, and searching for tools. The lack of a tool check procedure frequently left workers unprepared to handle unexpected situations in the field. They would return to the shop to obtain more tools, literally saying "Bye –bye" as the time for resuming work was very unclear.

Pilot Program

Management made a large investment of time and resources to develop standard procedures for window installation based on "best" methods. TBM Consultants helped to develop these new procedures and train the team.

Windows were produced in the factory using the one-piece flow system. The production schedule was based on takt time established by the needs of the master schedule for a construction project. The pilot program involved using the one-piece flow methodology in field installation as well. Management asked a supervisor to volunteer a team for the pilot program, in order to encourage their "buy in" to the concept. A team was provided consisting of two mechanics, two helpers plus a supervisor. Furthermore, the workers selected for the crew were very inexperienced. Management suspected that the work force was not overly enthused about proving that lean could work in the field.

The pilot program was expected to have windows fully installed, adjusted, and caulked in on a single visit to the site, avoiding repeat work (Figure 6.4). The program was set up as a Kaizen event. To prepare for it, a "Discovery Day" was implemented to review the conditions in the field. A process of waste elimination was established. Kaizen improvements: Over 30 improvements were identified. A few of them are described as follows:

- Significant time waste typically occurred with searching for tools. A standard set of tools was developed that could be used for all types of window installations. Some of these items such as hammer drills were duplicated for each tool kit to minimize the need to search for tools once the job begins or to change drill bits.

- Each crew was provided with a wheeled tool box that was equipped with the standard set of tools. This was estimated to save 2.5 minutes per window, or 125 minutes in a job with 50 windows.

FIGURE 6.4
Window installation: Preparing an opening at left; hoisting a window with a "skyglazer" at right.

- A worker was assigned to the role of "spotter" to visit the job site before the crew and observe and/or correct such deficiencies, before deploying crews to the job. The spotter would record the dimensions of window openings for use by the factory. The time saved from crew assignment was 47 minutes.

- The installation work was standardized to have shims and screws readily available for a wide variety of job conditions.

- Process improvement—workers were provided with miner's hard hats. These hats had a built in battery-powered light that enabled workers to leave both hands free for installing shims around each window. They could examine the fit of the window very closely, and ensure that components such as the drip edge were properly mated with the opening. Improperly fitting drip edges may result in rain water migrating to interior space either at each opening, or to the one below it (in the case of two-story installations).

On the scheduled day, staging for the job was checked at 6 a.m. There was a tool inspection—there should be two copies of each tool or material; should one be mislaid while working, another could be retrieved to continue working without delay. Tapcons (self-tapping concrete screws) were selected to provide 1-½" imbedment in the window openings. Crews were established—two two-man teams to carry out all installation tasks. A truck with a telescopic hoist and platform (called a snorkel) was positioned in the yard and enabled workers to raise each window to its mounting position in the building shell. Once the new procedures were developed with the participation of the crews, everyone was required to follow them "by the book" in order to minimize waste. For example, no site work would begin until a "spotter" had first visited it and resolved any potential problems.

Comparing methods

The traditional method (batch method)	The one-piece method
Crew size: five workers	Two workers per window
Five workers installed four windows in 6200 seconds	Four workers installed four windows in 3900 seconds

Benefits: 36% of the (cycle) time was saved, using one less worker. With the improved method, there is no need for windows to have to be "punched" (i.e., rectified for construction errors). Excess caulk can be wiped with solvents easily when it is wet, "doing it right the first time" is a very cost-effective method. Cleaning up dried caulk several days later can be very time consuming and difficult to do neatly. Very importantly, the crews took pride in implementing "one-piece flow and in successful installations that are 100% right the first time. The policy of one-piece flow has paid dividends—the shortened project duration is less intrusive to site operations, and is also more profitable to all concerned.

It must also be noted that the construction project was not a so-called lean project; it was delivered with a traditional method. However, this supplier's execution of the work clearly demonstrated the viability of the lean methodology; the supplier was also well prepared to operate within the framework of lean project delivery or integrated project delivery in a future project.

Comments

Superior Window Corporation success is underscored by the observations of the President, which are paraphrased below. This success is especially significant because the lean journey encountered resistance at many points in time until people were won over by its momentum and its successes.

"A dream has come true that I never thought possible" stated the President. Over the Kaizen events both in the factory and field, the dream was that the installation department would set the TAKT time for the company. Prior to this economic recession, Superior Window Corporation was building windows and completing orders on one job and shipping to site. The installation crew was working elsewhere on a project that was shipped earlier.

"Here at Superior Window Corporation Corporation we have succeeded in reaching the point where a trade supervisor has identified three project elevations as the sequence for production and, in fact, has set the TAKT time for the company. The customer has not. We have the pull system in place. Instead of installing windows in the field as a batch process we now do it in one-piece flow in the shop and use what is called a "skyglazer" to raise the window assemblies from outside the building to the openings. We no longer bring windows inside a building unless it is necessary. It is truly remarkable that in difficult times we can perform at the highest level."

Questions for Discussion

1. Name at least four prerequisites for lean construction. Why are they important?
2. Explain if a lean coach is needed to begin a lean construction program and why.
3. How does the Parade of Trades simulation illustrate the impact of reliability on project duration?
4. Describe some of the benefits that the Tweet/Garot Company derived through 5S application.
5. What motivated Belair to embark on the lean journey?
6. Describe some of Belair's process improvements that were driven by lean initiatives.

7. Explain how the Grunau Company uses The Last Planner® System.
8. How did the Superior Window Corporation Window Company apply Kaizen, and in what way was it beneficial?

References

Alarcon, L. F., A. Grillo, J. Freire, and S. Diethelm. 2001. Learning from collaborative benchmarking in the construction industry. *Proceedings of the 9th annual conference of the international group for lean construction (IGLC-9)*, August 6–8. Singapore, University of Singapore.

Ballard, G. 1999. The challenge to change. *Selected Articles from the Lean Construction Chronicle* Spring edition.

Bureau of Labor Statistics. 2006. Fatal and non-fatal occupational statistic from 1992–2004. Online at http://www.bls.gov.

Liker, J. K. 2004. *The Toyota way*. New York: McGraw-Hill.

Macomber, H., and G. Howell. 2005. *Using study action teams to propel lean implementations*. Louisville, CO: Lean Project Consulting Inc.

The Parade of Trades (http://www.ce.berkeley.edu/~tommelein/parade.htm).

Tsao, C. C. Y., and I. D. Tommelein. 2001. Integrated product-process development by a light fixture manufacturer. *Proceedings Ninth Annual Conference of the International Group for Lean Construction (IGLC 9)*, August 6–8, Singapore.

Tsao, C. C. Y., I. D. Tommelein, E. Swanlund, and G. A. Howell. 2000. Case study for work structuring: Installation of metal door frames. *Proceedings Eighth Annual Conference of the International Group for Lean Construction (IGLC 8)*, UK, July 17–19.

Polat, G., and G. Ballard. 2004. Waste in Turkish construction: Need for lean construction techniques. *Proceedings of the 12th annual conference of the international group for lean construction (IGLC-12)*, August 3–4. Helsingor, Denmark.

Warren, C. 2009. *A specialty contractor's lean journey*. Green Bay, WI: Tweet/Garot Mechanical.

Bibliography

Ballard, G. and G. A. Howell. 2003. *Lean Project Management Building Research and Information* 31(2): 119–33.

Howell, G. A. 1999. What is lean construction. Proceedings of the 7th annual conference of the international group for lean construction. Berkeley, CA: UCLA.

Womack, J., D. Jones, D. and Roos. 1990. *The machine that changed the world*. New York: Harper Collins.

7

Lean-Based Project Delivery Methods

This chapter addresses more recently developed lean project delivery (LPD) systems based on relational contracting principles. There are several methods for the process of designing and constructing facilities. They include:

1. Design-bid-build
2. Design-build
3. Engineer-procure-construct
4. Design-construction management (CM) contracts
5. Design-agency CM contracts
6. Fast track
7. Partnering/alliances
8. Lean project delivery methods

Methods 1 through 6 have existed for many years and have been used with varying degrees of success, depending on the type of project and the skills required. The pros and cons of these delivery methods were discussed in Chapter 1. Essentially, their effectiveness is limited by the fragmentation that has occurred in the construction industry. Designers minimize construction details in order to shift risk to contractors; so–called constructability reviews attempt to reduce the knowledge gaps between designers and constructors, but in practice tend to occur far too late to make optimal improvements to the design and construction processes. Construction documents often lack adequate information because (a) owners press for reduced budgets, or (b) designers may not fully understand builders' information needs, due to an absence of communication. The design-build method of project delivery integrates design and construction activities as a single responsibility for a scope of work defined by the owners' project requirements (OPR) and basis of design (BOD) through design criteria. However, this method often minimizes the owner's input once the owner has provided these documents and obtained a GMP from the contractor.

A Construction Industry Institute (CII) study by Sanvido and Konchar (1999) identified the primary factors for a construction project success to be:

1. A knowledgeable, trustworthy and decisive facility owner/developer.
2. A team with relevant experience and chemistry assembled as early as possible, preferably before the project design is 25% complete.
3. A contract that encourages and rewards organizations for behaving as a team.

Methods 1 through 6 do not meet these criteria consistently, pointing to the need for improvement. Partnering and alliances are early types of relational contracts and have the potential to meet the foregoing criteria, depending on the specifics of a given project, and the rules that govern the interaction between stakeholders. Lean project delivery methods involve approaches that build on the relational principles included in partnering and alliances by adding lean process methodology to minimize all forms of waste. This chapter presents three cases with different interpretations of the lean concept: integrated project delivery (IPD), integrated project delivery with an integrated agreement, and Lean Production Management. Other interpretations are continually being developed.

Disadvantages of Traditional Contracting Contracts

Traditional construction contracts are adversarial in nature—they typically include penalties for under performance or nonperformance by each party in a project. Owners contract with a general contractor (GC) or construction manager (CM). In turn, these parties contract with subcontractors for different disciplines, an increasing trend in the industry. There are many limitations in this contracting structure and relational contracting seeks to address them. Traditional commercial contracts provide little incentive for subcontractors to collaborate or cooperate with each other, as each is driven by contract language to selfishly focus on completion of their portion of the project to be on time and within budget. In Matthews and Howell (2005), the authors point to four systemic problems with traditional contracting that can be addressed with a relational contracting approach.

Design Ideas Often Lack Field Input: The format and type of construction design documents is a major concern in many projects, as designers are more focused on the form of the finished product such as a pipe section, a room, or a control system, and not so much on the method of how it gets produced. Constructability reviews during the late phases of design seek to address that problem, but with limited success. Even in the case of the construction management form of project delivery (CM) where trade contractors are engaged while design work is in progress, they often reserve their most creative ideas in order to seek an advantage when required to submit a firm price for the contract. If these creative ideas are submitted later in the process when the subcontractors are secure in their contract status, the designers are severely limited in their ability to incorporate fresh ideas.

Cooperation and Innovation Are Inhibited: The traditional contract structure inhibits coordination, discourages cooperation and innovation, often rewarding some of the parties for optimizing their performance at the expense of others. The subcontractors in a project are generally very interdependent, yet they have a very restrained relationship with each other. For example, communications and electrical subcontractors must complete their installation of raceways and outlet boxes before a drywall contractor installs the drywall that covers and surrounds them. If the drywall is installed first, then expensive corrective work is needed. Drywall installers are focused on meeting a schedule that is the foundation for their progress payments, they are loathe to delay a portion of their work if there are any delays in previous steps of the project. In fact, given the communication silos that characterize many projects, the involved parties may simply not be aware of each other's delays. This pursuit of individual gain has resulted in a less efficient industry with lower productivity levels and less innovation. Fault finding and defensiveness create a litigious environment.

Planning Systems are Not Coordinated: Subcontractors in traditional contracts have a legal obligation to the GC or CM, but no clear-cut responsibility to the other subcontractors. Consequently, their planning is guided by the master schedule, with little consideration of each other's schedules. The GC/CM often is not concerned with the detailed planning of its subcontractors, resulting in many random events as field conditions change.

Self-Preservation Is the Subcontractors' Mantra: There are many examples—often anecdotal—of subcontractors being tied to a fixed price contract with the prime contractor, with little protection against the vagaries of unknown site conditions and changing market forces such as rising commodity prices. The saying "knowledge is power" leads to over protectiveness. Subcontractors do not wish to reveal to others more than is legally mandated, lest that knowledge somehow work to their disadvantage. They take an adversarial approach to the contract, defending their legal rights to the hilt. They try to optimize their performance with little concern for the overall project, which they see as the prime contractor's responsibility.

Relational contracting enables the parties to work together for mutual benefit, gain knowledge, and use it creatively within each project. This enables them to reduce risk instead of shifting it to others, and to achieve a successful outcome beyond their self-interest. This outcome benefits all the stakeholders, especially the client, and minimizes the need for troubled participants to improve their compensation by reducing quality and cost. Successful relational models place value on successful project outcomes; they include team-based incentives or reward mechanisms, using collaborative behavior in meetings and reviews as well as project team goals.

The behaviors that facilitate relational contracts are most beneficial in projects of long duration, high uncertainty, and complexity in which transactions become less discrete and more difficult to define in detail. In reality, such projects exceed the capacity of traditional contracts as they involve many unforseens, risks, and contingencies that cannot be covered by contract language alone if optimal project performance is desired. Relational contracts anticipate many scenarios—positive and negative alike—and bind the parties to act in the interest of overall project success.

The Lean Construction Institute has classified projects under two distinct headings: Stodgy (i.e., simple, certain, and slow projects) and Dynamic (i.e., complex, uncertain, and quick or time sensitive projects).

The lean methodology is aimed at the second category of projects. It seeks to make work flow more reliable by reducing the variation in commitments made between stakeholders that is responsible for much of the waste in traditional projects.

Overview of Relational Contracting

An understanding of relational contracting principles is essential for a successful deployment of lean construction. It is a transaction or contracting mechanism that apportions responsibilities and benefits of the contract fairly and transparently, based on trust and partnership between the parties. There is evidence, however, that when stakeholders trust each other explicitly then they will even overcome adversarially worded contracts to work for the good of the project.

Traditional methods of project delivery are based on transactional contracts that govern the exchange of goods and services. These contracts work well in situations where the deliverables are easily defined and a single outcome can be expected, but many projects do not meet those conditions. The vagaries of construction delivery and its susceptibility to chance events render transactional contracts limited in their ability to meet the needs of complex projects with uncertain requirements.

Relational contracting provides a more efficient and effective system for construction delivery in projects that require close collaboration for execution. The relationship between the parties transcends the exchange of goods and services and displays the attributes of a community with shared values and trust-based interaction. Williamson (1979) refers to the relationship as assuming the properties of "a mini-society with a vast array of norms beyond those centered on the exchange and its immediate processes."

Working relationships between the stakeholders are improved, resulting in more efficient work processes and less conflict. A spirit of teamwork leads to better project outcomes, including relatively higher financial returns. Consequently, the most significant factors that underlie relational contracting are cooperation and dependency between the parties for mutual benefit.

Two contract types have some similarity to relational contracting, but have significant differences. Partnering has been used both as an individual project mechanism and as a long-term strategy. Alliances link the owner/client with the contractor and supply chain, usually in a long-term relationship. Partnering emerged to reduce adversarial attitudes and resulting litigation—better relations between participants led to a willingness to maintain dialogue. It has been limited by an erroneous assumption that trust and communication were the missing ingredients for project success. Relational contracting differs by being based on the linking of interdependent parties to improve the value chain. Lean works by designing the underlying planning and logistics system to be more reliable. While partnering speaks of relations and trust. Lean works on establishing and maintaining work processes that improve both the reliability to make and keep commitments, as well as the reliability of the stakeholders to execute the agreed-upon plan. Lean requires that the "HOW-TO" be explicitly defined and agreed to. It is by no means an overly-optimistic dependence on promises that may not be fulfilled.

While the hard dollar contract underlies the partnering approach, relational contracting proactively manages the project to maximize progress and quality while minimizing the disputes that may be caused by adversarial attitudes. Partnering has the shortcoming that it seeks to build relationships but tacitly accepts that a claim may result from a contract. Research in Australia by Rowlinson and Cheung (2004) points to the importance of cultural factors in relational contracting. Professional characteristics of frankness, honesty, and an affinity for positive confrontation are attributes that enhance the probability of successful relational contracting. Australian professionals are said to share these characteristics as a cultural trait. Other cultures that do not share the same attributes may need to make adjustments in order to make relational contracting work effectively.

Characteristics of Relational Contracting

- Team interests have equal or greater weight than the legal agreement.
- There are shared values and common goals.
- Interdependence between the parties is an acknowledged aspect of the business model.
- Trust between the parties is an essential ingredient; team-building activities reassure team members of the others' integrity, character, trustworthiness, and competency.
- Sharing of knowledge and ideas is a requirement.
- Financial benefits and losses are apportioned between the parties.
- Overall, relational contracting not only benefits each specific project, but leads to the adoption of values, behaviors, and actions that extend to the industry as a whole.

Benefits of Relational Contracting

- The ability to work on a face-to-face basis. The collaborative approach eliminates the need to formally document everything. Direct discussion leads to faster decision making. Job satisfaction is higher due to a more pleasant working environment.

- More innovative solutions become possible, and quality is improved as more people come together to discuss problem issues.

- An elevated level of trust in the contractor's competence and trustworthiness relieves inspectors and others of the chore of being on the spot at all times to see every detail of the work. When such a situation arises, quality costs are reduced as the contractor can be trusted to carry out the job correctly; that is, "do it right the first time."

- Better, faster solutions are possible that lead to savings for the contractor as well as the owner/client.

- More harmonious relations free up the parties to focus on work issues and not contractual issues.

- The shared use of ICT, avoids duplication by project team members and significantly reduces informational costs while improving response time.

- There is an atmosphere of goodwill that helps to resolve situations where one party fails.

Obstacles to Relational Contracting

- A reluctance to engage in relational activities at the very outset inhibits relational contracting. Because of entrenched attitudes "This is not how we do business" some people and their companies may be unwilling to carry out the activities prescribed for relational contracting. Team members should be required to sign a detailed memorandum of understanding as part of the relational agreement. This should include requirements for training, as well as monitoring the agreement to ensure that it is being followed.

- The participants may not believe in the efficacy of relational contracting. A remedial approach requires revisiting the initial relational contracting workshop, and refacilitating the team on specific assignments. It is especially helpful to engage the support of champions who can reinforce and sustain the relational process.

- A lack of understanding in the project team, resulting from members who have not attended all (or any) of the relational workshops. In that situation, members may not have fully accepted the relational philosophy and inhibit the effectiveness of the team. Inclusion of relational concepts in craft or tertiary training is suggested for improving the acceptance of relational contracting.

- High staff turnover causes several problems—replacement staff may be unfamiliar with the concept; they have also not "bonded" with the team; the spirit of trust among team members may be diminished. It is recommended to have provisions in the budget for additional relational contracting training meetings.

Limitations of Relational Contracting

- Regardless of the degree of trust between the contracting parties, there will always be areas of difference because the two parties will inevitably have some goals that are different.
- Any element of contestability will always carry some tension between the contracting parties.
- Trust is rarely all-encompassing; each party will trust the other on some things and not others and may feel justified in withholding information.
- Perhaps most of all, there will be challenges for both parties in adopting a trust-based approach. For reasons of accountability, for example, both parties will need to recognize that the other may wish to monitor their trust of the other's actions, including checking on areas of distrust.

Developing a Successful Relational Contracting Culture

- Employing people who are receptive to a new way of doing things. A belief in relational contracting is essential.
- Providing training and development companywide to all staff. Training industry wide would help to increase the pool of people who understand and accept the RC culture.
- A commitment by all involved companies to send their staff to workshops on relational contracting.
- Alignment of team objectives: Common vision/goals and an understanding of each individual's responsibilities and values are identified as two of the most important aspects of relational contracting. The criteria for selecting partners in RC approaches must reflect the client's business objectives and comprise both "hard" and "soft" qualities.
- Team members, along with the partners thus selected, should agree on their mutual objectives, and devise common performance appraisal plans.
- Team members should commit not to do anything that is contrary to the spirit of the relational contract. They should not impede the work of others (for example, by doing work out of sequence).
- Team members should not accept an assignment that cannot be performed correctly or safely. They should not leave the job until they know what they are planning to do the next day.
- Relational contracts should communicate management's commitment to optimizing performance at the level of the project and not push for local optimization of resources.

Relationship Building among Team Members

Mutual trust and open communication are critical factors for relational contracting. The parties in an exchange relationship should have the confidence that others are reliable in fulfilling their obligations. It is essential to "open" the boundaries of the relationship because it can relieve stress, enhance adaptability, smooth information exchange, encourage joint problem solving, maintain transparency, and provide better outcomes.

Relationship management entails stimulating communication and breaking down barriers, and a facilitator is crucial to making these behaviors occur (Rowlinson and Cheung 2004). A facilitator sets the stage at the beginning to promote open communication, willing cooperation, and a brainstorming approach to problem solving. The facilitator is also a critical factor in keeping these "channels" open and maintaining nonadversarial attitudes through ongoing workshops. Relationship management emphasizes communication and problem solving without a focus on the financial aspect of a project; differences in expectations at different staff levels suggest that a facilitator needs to filter the process to serve these needs.

Current Examples of Relational Contracts

Relational contracts were first developed in England for North Sea oil exploration, and have been used extensively for such projects. They were next applied in Australia for major infrastructure and building projects. They were subsequently adopted in the United States by various professional groups and have gained popularity in major construction projects. Examples are:

- Consensus DOCS 300: Issued by a consortium of 22 organizations in the construction industry including the Construction Users' Roundtable (CURT), Associated General Contractors (AGC), and the Associated Specialty Contractors.
- Single Purpose Entity: Issued by the American Institute of Architects (AIA).
- Integrated Form of Agreement for Lean Project Delivery (LPD): Used by Sutter Health System in California for major construction projects.

Organization Structures for Relational Contracting

There are many possible scenarios to describe the interaction between the parties in relational contracting. As the concept of lean design and construction evolves, various configurations of relational contracts will develop (Matthews and Howell 2005). The commercial relationships in relational contracting promote mutual trust and the sharing of knowledge that leads to innovation and reduces the risk of overruns in schedules and budgets while improving the quality of completed construction.

Integrated Project Delivery

The IPD is one type of relational contract. Owen Matthews* and Greg Howell (2005) define IPD as a relational contracting approach that aligns project objectives with the interests of key participants. It creates an organization able to apply the principles and practices of the LPD system. The IPD was developed and trademarked by Westbrook Air Conditioning

---

* Matthews O. and Howell G., *Lean Construction Journal*, 2, 1, 46–61, April 2005. With permission.

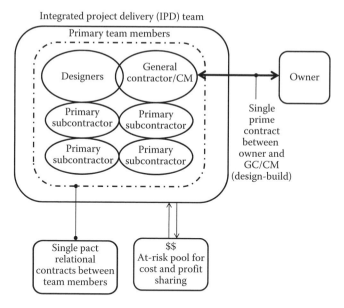

FIGURE 7.1
The integrated project delivery team.

and Plumbing of Orlando, FL. Owen Matthews, CEO, developed the concept to have more effective teamwork in Westbrook's projects and better outcomes as a result.

Integrated project delivery (IPD) is a work in progress as it is continually evolving with each new project (Figure 7.1). The fundamental principle of IPD is the close collaboration of a team that is focused on optimizing the entire project as opposed to seeking the self-interest of their respective organizations. The model adopted in the second case by the Sutter Health System in California involved a legal agreement that mandated the collaborative behavior of the team. It was called the integrated agreement for LPD between owner, architect, and CM/GC.

As described in the first case, Westbrook and the IPD team won a design-build contract worth approximately $6,000,000.00 to design and construct a central chilled water plant. This project is recognized by the Lean Construction Institute as a pioneer example of the IDP process. IPD employs both transactional and relational contracts. The basic requirements of IPD can be met by:

1. Creating a business entity (i.e., the IPD team that incorporates designers, contractors, and subcontractors). A contractual agreement binds the members of the business entity together. In one specific project, Westbrook successfully applied the concept to a design-build team, holding it together by a single "pact."

2. Having the parties serve as business entities in their own right and held together through a relational contract or integrated form of agreement (IFOA). Some of the parties establish a classic transactional contract with the client and suppliers.

As was configured for Westbrook's project, a single business entity contracts with the owner, and the entity's project team comprises the architect/engineer, the GC, and the primary subcontractors. These parties are termed primary team members (PTMs). There may be other subcontractors that are not PTMs. The client has a single prime contract with the IPD team. This contract may be of the traditional type such as design-build, design-bid-build, and so on that defines scope, design and specification requirements, cost, and schedule. PTM members may display their cohesiveness with the IPD team by wearing the same hardhats and the same logo. They operate under the direction of one general superintendent whom they authorize to seek the best overall methods to get the work done.

In the Sutter Health project(s) the architect and the CM or GC had a contract with the owner. Representatives of these three entities formed a Core Group, charged with the responsibility of administering the project.

Westbrook IPD Project: Case A

Matthews and Howell (2005) provide a case study that illustrates the application of the IPD method with Westbrook and an IPD team. Westbrook and their IPD team established two principles for the project: all PTMs were responsible for all provisions of the prime contract with the client and the PTMs shared the risk and profit for total project performance.

One primary team member had a prime contract with the owner/client. The PTM with the prime contract and the other PTMs entered into a single pact; they jointly and severally bound themselves to each other and to carrying out all the conditions, terms, and requirements of the prime contract. The price contracted between the owner and one PTM was the contract price for the IPD team. This scope included all the work required to complete the project.

The PTMs agreed to share the costs incurred in the job and also to share the profits, using a formula that reflected their participation in the project. There was no accounting by these team members to determine who was over or under budget as that would represent a backward step to the traditional approach of "every man for himself." PTMs agreed to open their books for the project to other PTMs and the owner. Each PTM obtained a certificate of insurance based on the prime contract's requirements.

Other key provisions in the agreement consisted of the pact between the PTMs ended when the project was completed and all terms and conditions of the prime contract had been fulfilled, including warranty obligations. At that point, the profits were to be determined, adjusting for any cost overruns that may have occurred. The resulting profits would be distributed to the PTMs based on the previously agreed-upon formula.

Teamwork was practiced by innovation and optimization. Team members were expected to shift work and costs between different companies and trades to improve performance and reduce overall costs. Team members agreed to promptly advise others if there was any impediment that would delay their work. This would enable the team to make necessary logistical adjustments to meet the requirements of the prime contract. Team members agreed to support each other through thick and thin. If a member ran into difficulty, the others would seek to help instead of alienating or even terminating that member. They treated honest mistakes as a "part of the job" and focused on making sure that the overall job was not jeopardized. The fact that this situation deviated so much from traditional construction practice was a testament to the strength of the relational bond between team members. Team members viewed each other as having integrity, character, and competency and who deserved support if they made honest mistakes.

Cultural Factors in IPD Implementation

Matthews and Howell point to barriers that had to be overcome before PTM members could "buy in" to the IPD concept. Prior to joining the project team, the respective subcontractors were accustomed to traditional project delivery methods; they focused on optimizing their own performance and did not feel responsible for overall project outcomes. The IPD principle of shared responsibility is a major departure from the norm; if one PTM makes an error the cost is shared between the entire team. It reduces the profit in the overall project. Similarly, cost savings are shared between team members and with the client; these savings could be considerable, depending on the extent of the collaboration on a project.

One important factor in the cultural change on this project was the PTMs' exposure to training on the principles of lean construction. They became aware of the benefits of lean systems design, and that local optimization would diminish project-level optimization. A team spirit, open communication, and mutual respect were seen as essential ingredients for IPD to succeed. The top executives of the PTMs met twice monthly for breakfast and fellowship, and this interaction fostered a spirit of trust and mutual respect. It also enabled them to discuss and refine the details of the IPD process.

The team created an organization for the duration of the project that was best suited to its needs. This team included a project executive, director of design services, project manager, project superintendent, manager of ICT. The most appropriate team members were selected for these positions. As they performed their functional duties their respective companies were reimbursed for their time and these costs were charged to the project as direct job costs. This also fostered these leaders' identification with the team instead of with their respective companies. Each PTM company was paid monthly to cover its direct costs incurred. At the end of the project, gross profits were determined and paid to the PTMs based on a prearranged formula.

Project Logistics of the IPD Team

During the design process, the IPD team had the benefit of the entire spectrum of skills—architects, engineers, subconsultants, the GC, subcontractors, and so on. That they functioned as members of a common organization contributed greatly to full collaboration between all involved disciplines. Value engineering efforts benefited from the collaboration between all these stakeholders. Constructability was also significantly enhanced as product and process issues were reconciled. Very importantly, a GMP reflected the creativity and innovation provided by this collaboration and was therefore more realistic than would be derived with a standard bid process.

Team members looked for duplication of construction-related activities and avoided it in their cost structure (e.g., rental equipment and other resources were shared by the team and were essentially charged once on the project). This contrasted with traditional construction where several different trades might each rent an air compressor and add its cost to the project.

The IPD team decided to aim for a zero accident rate in the course of the project. Extensive safety compliance procedures were established, with the superintendent having authority over all site safety issues of all trades. The team shared the cost of these safety measures.

The IPD team adopted the concept of "spending more to save more." The mechanical engineer laid out the equipment room (a major part of the project) in detail using object-based 3-D. This layout was immediately usable by the fabrication shop for producing sections of large bore pipe. The engineering cost was increased, but was more than offset by the savings in shop drawing preparation and installation time.

The PTM organization participated in various team-building exercises to promote communication and commonality of interest. Team members typically met twice monthly for breakfast or other social activity and often used the occasion for training on lean concepts to improve their knowledge base. Disputes occurred far less frequently than in traditional contract structures, largely because team members were aligned with a common goal in mind.

Project Specifics

Westbrook and the IPD team won a design-build contract to design and construct a central chilled water plant housed in a steel framed building. The capacity of the plant was 12,000 tons of chilled water, starting initially at 3000 tons. The project comprised a main utility building housing chiller equipment, supported by steel columns with large diameter piping supported by hangers from the roof structure (Figure 7.2). A GMP of $6 million was established with the help of value engineering. The team took several actions that led to job completion much faster than in conventional projects. They recognized the importance of an early finish and cooperated closely to obtain that result.

The steel erector was included in the process and integrated the fabrication of structural steel columns with the construction schedule. Innovative changes were made to the layout of steel columns to accommodate equipment installation. The mechanical contractor used 3-D imaging to identify desired equipment locations, and collaborated with the structural engineers to make the needed changes. Such rapid design changes would not likely have occurred in conventional projects. The team realized that, with an aggressive construction schedule, the location of underground utilities was of paramount importance. Through weekly team design meetings, the column footers were positioned 30" above grade to accommodate electrical ducts with elbows for the main service.

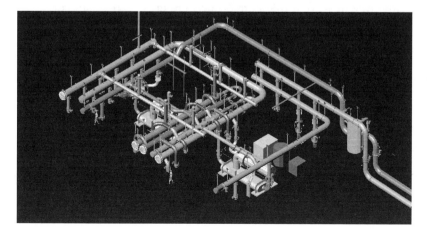

FIGURE 7.2
Three-dimensional view of piping in the equipment room. (From Matthews, O. and Howell, G., *Lean Construction Journal*, 2(1), 46–61, 2005. With permission.)

Another design innovation involved deviating from a power duct system that would have been installed overhead with expensive hangers, and would have taken considerably longer and been more costly. The team devised a method for the underground installation instead, which involved having the GC wash aggregate into a trench around the duct with fire hoses. Prefabrication enabled major savings to be obtained in the project. Through the use of object-based 3-D CAD, the routing of the piping was designed and displayed in 3-D, facilitating integration of the hangers with the supporting roof structure. Figure 7.2 shows the 3-D CAD representation of the piping layout. This technology also provided details for fabrication off-site. The main headers were 30″ and 24″ in diameter; the piping was trucked in from the fabricators on a Just-In-Time basis for installation. The installation itself was simplified to the level of simply bolting components together without the need for field fabrication.

Overall, two 1500 ton chillers were mounted in place and interconnected with piping, all within a period of 10 calendar days. This was a remarkable accomplishment that very probably would not have been accomplished with traditional construction methods at a comparable cost. The schedule was as follows:

Contract award	12/30/03
Design development complete	1/26/04
Demolition complete	1/7/04
Permit Issued	4/14/04
Work start date on site	5/4/04
Plant complete, operational	7/28/04

The effectiveness of the IPD methodology was proven by the fact the team obtained savings of 10% (i.e., $600,000; the final contract price was $5,400,000).

Observations on the IPD Process

The IPD process was seen as a clever solution to tough organization and contracting problems in the contemporary market (Howell 2005). As the contracting incentives and contractual rules reward cooperation and encourage innovation, project transactional costs are reduced and disputes are unlikely to occur—either within a team, or between a team and a client.

Lean Project Delivery (LPD) with an Integrated Agreement

Lean project delivery with an integrated agreement is another version of relational contracting. In a project based on LPD/IA, the parties serve as business entities in their own right and are held together through a relational contract or IFOA. Specifically, an IA for LPD is a modified IFOA designed to govern the relationship between the parties in a construction project and facilitate the steps and processes required for LPD. In effect, these transactional terms represent IPD.

The IFOA is a single contract that the architect and the CM or GC sign with the owner. As opposed to a design-build contract that has a single point of responsibility, the IFOA relies on a Core Group of representatives of the owner, architect, and contractor (GC/CM) to administer the project.

William A. Lichtig (2006), construction attorney with McDonough, Holland & Allen, PC, has done pioneer work with the integrated agreement.* He describes the IFOA as a departure from other project delivery methods and contractual models. It seeks to align the commercial relationships of a construction project's design and construction participants that are assembled as a temporary production system. The Integrated Agreement has been developed in an effort to support the values of LPD that are exemplified in the Toyota Production System—the elimination of system-wide waste and the pursuit of value from the owner's perspective."

Integrated project delivery involves a contractual combination of LPD and an integrated team that is expected to improve project performance in a number of dimensions. If the process is successfully executed, owners benefit from reduced time and cost as well as improved quality and safety. Designers and contractors derive increased profits, improved owner satisfaction, as well as greater employee satisfaction. IPD is designed to support the values and principles of LPD (i.e., to reduce waste and provide value for the owner). Whereas traditional contracts anticipate adverse events and focus on transferring risk, the IFOA seeks to reduce risk by empowering team members to use lean thinking and collaborative approaches. In fact, the collaborative problem-solving skills of the project leadership team (i.e., the owner, designer, and contractor) enable it to transcend unexpected challenges and function as an agile, high performing team that delivers projects successfully.

As Lichtig explains, the IFOA meets the value proposition of the owner (building facilities within expected cost and time) by promoting project efforts to increase the relatedness of the IPD team. It also requires collaborating throughout design and construction, planning and managing the project as a network of commitments. As a result it optimizes the project as a whole, rather than any particular piece. The agreement also calls for combining learning with action, and promoting continuous improvement throughout the life of the project.

Application to the Sutter Project: Case B

Sutter Health is a not-for-profit community-based health care and hospital system headquartered in Sacramento, CA (Lichtig 2005). It is one of the largest health care providers in North California, serving more than one hundred communities. Faced with inadequate facilities, they made a strategic decision to embark on a $6.5 billion design and construction project that would begin in 2004 and be completed by 2012 (see Endnote on page 202). This project included acute care facilities, nonacute outpatient facilities, medical office buildings, and parking structures. Sutter Health's executive management team charged the Facility Planning and Development Department (FPD) to have this multiphase project delivered on time or early, within or below budget, without claims, with maximum safety, and without (FPD) staff burnout.

Starting in 2004, lean project delivery was embraced as an innovative means of reaching those goals. In particular, Sutter Health sought to govern the delivery process as a collective enterprise to meet their needs as owners while designers and contractors could derive

---

* Lichtig W. A. and Sutter Health, Integrated agreement for lean project delivery between owner, architect and CM/CG, 2008. With permission.

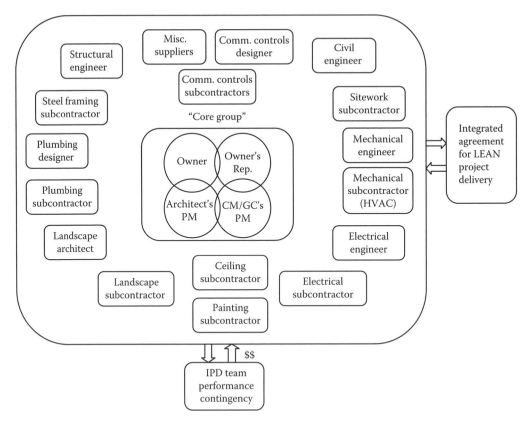

FIGURE 7.3
Diagram of IPD team formation.

increased profit and profit velocity with less risk. They would also experience greater employee satisfaction and better safety performance. The firm of Lean Project Consulting Inc. was engaged to guide the organization in the adoption of LPD. McDonough, Holland & Allen, PC, provided Sutter with legal services, especially with the development and deployment of the IFOA. This effectively led to the adoption of integrated project delivery, IPD, for the project. Ghafari and Associates subsequently provided 3-D (BIM) enabled lean consulting starting in 2007. The following narrative uses this construction program as a model of successful application of the integrated agreement (Figure 7.3).

> The IPD team is comprised of the Architect, Owner, Construction Manager (CM) or General Contractor (GC), subcontractors, and suppliers. As stated previously, representatives of the Owner, Architect, and Construction Manager (CM) form a "Core Group" that manages the project. They select the remaining members of the IPD team including engineers, technical consultants, key subcontractors and suppliers. In order to derive the greatest synergy from team integration, the CM/GC entity is engaged immediately after Architect selection. Major subcontractors are also engaged as early as the schematic design phase. While this approach incurs expenditure for professional services earlier than traditional construction practice, this cost is recouped several times over from the savings and improved constructability that the collaboration provides.
>
> Some relational contracts define a "Core Group" that comprises project representatives of the primary team as well as an Owner's representative. This group manages

the project; it has the power to invite or dis-invite other representatives to its membership. It is recommended that the Owner should have two representatives—one that is knowledgeable about the business and one that is knowledgeable about design and construction.

Team Activities

The purpose of the IPD team is to facilitate collaborative design and construction and commission of the project, in an open and creative learning environment. Team members are required to actively promote harmony, collaboration, and cooperation among all parties performing the project. IPD members sign a joining agreement to document their commitment to meeting the terms and conditions of the integrated agreement, collaborating and cooperating with other team members.

The Core Group performs many functions. It jointly selects additional members of the IPD team, reviewing the recommendations of the owner, A/E, and GC/CM. Team selection is preferably based on requests for proposal (RFPs) with emphasis on past performance, value, and quality, rather than on the low bid. This serves to indicate whether the new team members can work well in a collaborative environment. The Core Group adds major subconsultants and specialty subcontractors to the team during the schematic design phase in order to integrate them at a time when their input and innovation can have the greatest impact on the design and resulting constructability. The IPD team's benefit is greatest when design development is less than 25% complete. The Core Group determines the firm that is assigned the responsibility for a given scope within the project.

The Core Group also develops a joint site/existing condition investigation plan, pointing to the extent of investigative work needed. The group serves in an administrative oversight role, scheduling meetings and problem-solving sessions. The group maintains open information flow, and ensures that all parties work together harmoniously. They promote the network of commitments needed to achieve reliable work flow. This reliability is critical to maintaining a smooth hand-off of work from one crew to another. The Core Group is also required to develop a built-in quality plan, recognizing that 10% of project construction cost is often spent on rework.

Reliable Promising

The integrated form of agreement requires IPD team members to make reliable promises as a basis for planning and executing the project. The conditions for a reliable promise are:

- Satisfaction that is clear to both parties—performer and customer.
- The performer/promissor is competent to perform the task and has access to needed resources (tools, equipment, instructions).
- The performer/promissor has estimated the time to perform the task and has internally allocated adequate resources and blocked the time on its internal schedule.
- The performer/promissor is sincere when making the promise that it can reasonably be fulfilled.
- The performer/promissor is prepared to accept the legal and reasonable consequences of not performing as promised and will promptly advise the team if confidence in meeting the promise is lost.

Trust

The parties recognize that each of their opportunities to succeed in the project is directly tied to the performance of other participants; they agree to work together in the spirit of cooperation, collaboration, and mutual respect, using best efforts to perform the work expeditiously and economically as needed.

The IPD team members are required to provide a Core Group representative to participate in the responsibility to coordinate, manage, and administer the project consistent with LPD principles.

The IFOA defines the creation of an IPD team at-risk pool account termed the "IPD team performance contingency." The purpose of the account is to promote LPD among team members by funding cost overruns. It is set up in the form of an escrow account. Its existence avoids the need to hold any one team member solely accountable for its own scope and price, which would have the negative result of promoting local optimization rather than overall project optimization.

Value Engineering

Value Engineering involves an organized multi-disciplined team effort to analyze the functions of systems, equipment, facilities, processes, products and services to achieve essential functions at the lowest possible cost consistent with the customer's requirements while improving performance and quality requirements. The CM/GC and subcontractors are required to continuously pursue opportunities to create additional value through reducing capital or life-cycle costs while improving constructability and functionality.

Target Value Design (TVD) Process

Will Lichtig of the LCI describes TVD as an essential part of the integrated agreement. As described in Chapter 4, the TVD process is intended to have a project designed within a desired budget in accordance with a detailed estimate. It accomplishes that end by establishing value, cost, schedule, and constructability as basic components of the design criteria. Essentially, TVD focuses on the objective of creating the best design for a facility that can be delivered with the funding available. In projects where TVD is applied, there is a provision in the IFOA that a project's target cost shall never be exceeded without the express approval of the owner. The Core Group is required to develop a TVD plan, and LPD members provide TVD support services throughout design development.

The TVD is facilitated by having cross-functional teams evaluate tradeoffs and opportunities. The teams include contractors who are very experienced in the marketplace and understand the cost implications of various design decisions. These contractors also advise designers on constructability issues—with the expectation that they will listen. Representatives communicate the owner's value proposition in terms of form, function, life-cycle costs, and other criteria. The Core Group develops an evaluation of the existing site conditions to ensure that it is recognized in the TVD, to maintain the integrity of the target value (established as the estimated maximum price proposal). The IPD team establishes a contingency to address design errors and omissions.

Participants in the integrated agreement observe that, during the design phase, the team spends more time interacting than in traditional projects; the added cost is far outweighed by the fact that everyone is in a "common problem-solving mode." Problems and conflicts are resolved early, at minimal cost, whereas traditional projects would have them deferred to a later stage when the cost of correction is often much higher. IPD contracts stipulate conditions for successful operation with regard to the following requirements:

- Meetings are held at least monthly and special meetings are conducted to resolve issues that need immediate attention.
- Decision making is consensus driven.
- Communication protocols are clearly defined.
- The selection of subcontractors, suppliers, and consultants is based on Core Group input.
- Cost and schedule are considered to be design criteria and are required to be monitored closely throughout the project and not just at monthly meetings.

Project Planning and Scheduling

The IFOA requires "pull scheduling" using The Last Planner® System (or an equivalent system); that is, it must be based on requests from the IPD team members to other team members on whom the requestor's work is dependent. Without the LPS, integrated project delivery may be severely compromised, and teams may resort to traditional push scheduling. The upstream performer must make reliable promises about when it will hand off the work to enable the downstream entity to start. The system must be provided with a means for measuring work performance and the reliability of the planning process.

A preliminary milestone schedule should be prepared by the IPD team during the preconstruction phase. The schedule should be updated during the construction phase to reflect key milestone dates for issues such as deadlines for submittals, long lead time procurement items. The phase plan should be created collaboratively, working backward from the milestone schedule. The make-ready look-ahead plan should be created from the phase plan with a minimum duration of 6 weeks (or otherwise as jointly determined) to indicate possible constraints, so they can be addressed. Weekly work planning meetings are required to bring together the disciplines associated with work identified in the look-ahead process. They should review past performance and investigate the reasons for variance in order to improve performance in subsequent weeks. The dissemination of this information is critically necessary for making process improvements during the progress of the project and not later.

Quality of Work and Services

Integrated project delivery emphasizes the goal of producing defect-free work, at the least cost, and in the least time possible. It places the onus for quality at the source, through the people conducting the work, but it ensures that they fully understand the standards for acceptable work. It proposes the use of mock-ups, first-run studies, and early completion of standard work units. Workers are provided with checklists and benchmarks for the purpose of self-evaluation. It is recommended that workers be empowered to "stop the production line" as is done in lean manufacturing if the work handed-off to them falls below the desirable quality standards.

In the construction phase, the IFOA mandates the application by the CM/GC and the subcontractors of the 5S plan with regard to the common work environment. (5-S is described in greater detail in Chapter 4.)

- Sort: Removing all unnecessary items from the work environment when they are not immediately needed, including office items as well as materials and equipment.
- Set in order: Identifying where items need to be maintained consistently, for easy access.
- Shine/sweep: Creating an orderly workspace with continual scheduled cleanup.
- Standardize: Publishing standard practices for implementing 5S.
- Sustain/self discipline: Continuing the first four elements and seeking continual improvement.

In keeping with the 5S concept, the CM/GC is required to use JIT deliveries of materials and equipment and not encumber the site with them. The CM/GC should not damage or endanger a portion of the work or negatively impact the work of others.

Dispute Resolution

The integrated form of agreement requires the Core Group to resolve disputes rather than the architect. If the group is unsuccessful, the senior management executives from the group try to resolve the dispute in question. If their efforts are unsuccessful, an independent expert may be brought in to have closure in an unbiased manner and without animosity.

Learning

The review of "lessons learned" is an important part of the IPD processes, and it occurs at various points in the course of a project. At the weekly meetings the events of the previous week are reviewed to identify reasons for under or over performance. Monthly meetings may capture the reasons for trends in financial expenditure and the use of contingencies.

Logistics of the Sutter Project

Attorney William Lichtig structured the IFOA to support integrated project delivery. Greg Howel and Hal Macomber of Lean Project Consulting sought to align the physics of work (how work gets done) with lean methodology. They provided an important foundation for the project with a philosophy termed the Five Big Ideas. These ideas were acknowledged in a manifesto signed by the stakeholders (i.e., the owner's facilities-related departments and the architects, engineers, contractors, and suppliers.

The Five Big Ideas* for LPD serve as a so-called True North and can be applied to virtually any large, complex project. They are described in detail in Chapter 3. In brief, they are:

1. Collaborate, really collaborate: Collaboration was recommended to overcome the fragmentation that has become a characteristic of the design and building process. There are many gaps between expectations and reality that have developed between designers, builders, and owners. These gaps lead to poor constructability and low quality.

---

* The Five Big Ideas is the intellectual property of Lean Project Consulting.

2. Increase relatedness among all project participants: The learning experience that benefits project team members is enhanced when they develop relationships and interact with trust and openness. Human nature renders people unwilling to try new approaches as they do not wish to make others aware of their mistakes, especially if there is no relationship between them.

3. Projects are networks of commitments: Management's role in the project environment is to activate and articulate unique networks of commitments.

4. Optimize the project, not the pieces: Conventional practice conditions project managers to push aggressively to maximize speed and minimize cost. This approach represents self-interest to obtain high productivity at the task level, but such local optimization reduces the reliability of the work that is moved downstream.

5. Tightly couple action with learning: This big idea proposes that work should be carried out in a manner that facilitates ready observation of the results of specific actions. The lessons so learned should be used for continuous improvement of costs, schedule, and overall project value.

Sutter's Lean Implementation Strategies

To demonstrate their top management's embrace of the lean philosophy, the owner conducted a 3-day lean summit in 2004 to introduce lean concepts to over 200 facilities staff members. The Five Big Ideas were an integral part of the lean structure that was to be adopted. The owner established a program that included formal workshops, weekly conference calls, and monthly initiatives. Staff members were trained in the use of The Last Planner® System and its application was initiated with a variety of construction projects that comprised the building program. A lean executive leadership group was formed with leading industry executives in the organization to have ongoing discussion meetings with regard to the lean program's endeavors. The participants in construction projects—designers and CM/GCs were selected through RFP and detailed interviews. Designers were selected first, followed quickly by CM/GCs, and then by major subcontractors as early as the schematic design phase in order to promote collaboration at a time when the impact of innovation was greatest.

The design process progressed in parallel rather than the conventional serial approach, and value engineering was continually applied. Through ongoing contractor participation constructability was enhanced. While the early engagement of CM/GCs incurred additional cost, better project collaboration was possible, and the improved quality of the design was expected to yield benefits that far outweighed the costs. Target value design was implemented to design the project to a particular estimate based on the owner's funding availability; as a cardinal rule that figure was not to be exceeded without the owner' consent. The CM/GC and subcontractors retained control over and responsibility for construction means, methods, techniques, and safety precautions.

Pull-based design production was stipulated in the contract. A resource-loaded project work plan was required during the preconstruction phase—the design team was discouraged from advancing aspects of the design before the time designated in the plan, although it was possible to address items described as "workable backlog." A milestone schedule was required to be prepared by the CM/GC in conjunction with the IPD team for Core Group approval. This was subjected to monthly updates. Phase planning involved having the IPD team collaboratively work backward from the milestone schedule. A make-ready look-ahead plan was required to have a minimum duration of 6 weeks.

Management of Risk

Rather than focusing on risk transfer, the integrated agreement seeks to establish systems and empower the IPD team to reduce or eliminate risk by employing new conceptual and autonomic approaches to project delivery (Lichtig 2005). In the Sutter Health projects, risks were shared between project participants, rather than being shifted from one to the other, as occurs in traditional projects. Contingency funds were designated to address various categories of risk, including unaccounted for risks; these were collectively managed by the project delivery team. The risk categories included permitting changes, construction deviations, and price escalation. An at-risk pool account was created by the team to promote lean construction and partially fund certain cost overruns. Risk was further reduced by the use of a cost-plus, guaranteed maximum price (GMP) for the CM/GC as well as some subcontractors.

With regard to the designers, the traditional "errors and omissions" insurance requirement was replaced by a deductible as a percentage of construction cost that was funded from a design contingency. It was available without proof of negligence, although errors that exceeded the deductible were recoverable by the owner. This system avoided adversarial relations with the designer, leaving that professional free to maximize design quality.

Dispute Resolution

Disputes were resolved through an escalating series of meetings, with emphasis on settling issues at the lowest possible level. The process started with informal negotiations, which, if not resolved rose to the project manager level. The next resolution level involved a senior executive meeting; beyond that an independent expert would be engaged, and ultimately, mandatory mediation was required. Overall, the process was intended to foster positive relations between team members by resolving disputes quickly, as team solidarity is essential for lean construction.

Benefits of Integrated Project Delivery

A more recent portion of the Sutter Health capital program initiated in 2007 benefited from the technique of 3-D enabled lean processes applied within IPD. The Sutter Medical Center, Castro Valley, is a 233,000 square foot hospital with 130 beds and was proposed to replace an older facility.

The IFOA was instrumental in focusing the design effort—starting with a layout based on best practices for clinical care. Ghafari Associates helped the project team to align information flows across the design and construction supply chain. By using value stream mapping they were able to visualize the sequence of activities that would avoid design rework. Project team members held joint discussions and placed their inputs on a wall with sticky notes to represent real-time occurrences. For example, it was essential to complete clinical space planning before starting the design tasks. Although past practice conditioned everyone to believe that it would take 15 months to produce 100% construction documents, deferring this work until after the clinical space planning made it possible to produce the documents in only eight months.

Lean Production Management: Case C

The LPM methodology is another successful application of lean project delivery. It is illustrated below in a $128 million EPC project with a processing plant for Ultra Low Sulfur Distillate (ULSD). On time, within budget completion was very critical to this project. It had an end date set by the government with extensive regulatory incentives. There were also penalties for late completion.

Peter Gwynn, a lean/facilitator consultant with Lean Implementation Services, and Dave Koester, a project manager with the Regional Contractors Alliance (RCA), documented the following observations from the project (unpublished report). They describe LPM as an integrated set of project execution principles that create significant changes in the behavior interaction among project team members. These major changes include:

- Striving for what is best for the project as a whole, rather than suboptimizing one's own piece of work.
- Developing the discipline and patience to start work only when prerequisites are complete.
- Making detailed, formal, and visible commitments.

The changes occur gradually over a period of time; implementation and execution of the principles require daily support, facilitation, and encouragement. An experienced, expert resource person is an important requirement for successful implementation.

Implementation, training and facilitation require nearly full-time efforts of key staff and management personnel. Established habits and behaviors are so strong that managers and facilitators must consciously and intentionally pull people back into the process on almost a daily basis. The amount of process management, facilitation, and training necessary for success cannot be overemphasized. Gwynn and Koester point out that LPM is not a simple cookbook action list for project execution performance improvement. The degree of implementation and the corresponding improvement in project delivery are greatly enhanced when the project team leaders operate in an inclusive, collaborative manner, and when key project participants such as project engineers are freed up from rote and mundane tasks to concentrate on planning and facilitation. Daily commitment measurement provides all of the input data for more timely and accurate progress measurement, significantly reducing the traditional project controls effort. LPM management involves ongoing meetings and collaboration, weekly interface meetings, and master planning sessions.

Roles and Responsibilities

In order to successfully implement and support Lean Production Management, the roles and responsibilities of project team members must change from traditional duties, and from past practices.

Managers

The project manager and resident manager focus on implementation, training, and adherence to the LPM process. Project plans are built with informed input from knowledgeable general foremen and foremen, not by staff and management deciding and dictating a schedule from the top down. Managing requires ensuring that the right people are doing the right things at the right time.

In the master planning and interactive planning sessions, the managers must function as equal, contributing participants with other team members. Managers must not communicate their ideas for work sequences as their wants and expectations will inhibit further valuable input and ideas from many participants. Managers take their turns as facilitators of planning sessions, and volunteer to address larger, global project constraints. Managers must make decisions, especially during weekly interface meetings, to authorize and enable the amount of resources necessary to support and implement the work. Decisions must support what is best for the project as a whole, and suboptimization must be constantly held at bay.

Many participants find the amount and degree of planning to be far more than they have previously experienced, and require frequent encouragement and support from managers. This support is often successful as brief one-on-one conversations on the job site, reinforcing how the work is progressing because of the planning and the individual's effort. Periodically, managers must refocus the entire team on the LPM process. Many team members have entrenched habits that gradually draw them away from the LPM methods. Managers must watch for project engineers who get discouraged from continually promoting work process change with field crews, and provide encouragement and support for them.

Project Engineers

Project engineers function more as facilitators of Lean Production Management. They facilitate detailed production planning sessions; recommend, gather, and record daily work commitments; and lead the weekly interface meetings. Engineers continue to update and maintain production progress data, pursue issue resolution with design groups and fabricators, and address constraints.

The value and contributions of project engineers on any project are significantly increased by moving the engineers out of the procurement process. Traditionally, many construction staff and supervisors have considered project engineers as the requisitioners, buyers, trackers, and expeditors for materials, consumables, tools, and equipment. This is one of the most harmful habits in the construction industry. This underutilization and misapplication of engineers' skills is easily mitigated by assigning a project materials management person or group from the craft or foreman ranks to perform these procurement functions.

Production Planner

There is no traditional schedule document in the Lean Production Management system; there is only a continually evolving production plan of activities to completion. There is no single- or small-group developer of a plan in LPM. The production plans are developed and captured from the input of team members, particularly from the general foremen and foremen. One individual, who may have functioned as a planner in traditional construction management, is assigned to record and update the production plans in a construction sequence dependency software such as Primavera.

A more accurate title for this person may be production planner. He or she should lead the master production plan reviews, relative to the desired project completion milestones review the plan to completion, facilitate interactive area and detailed production planning sessions, and assist to gather and record daily work commitments.

The production planner also reviews the Gantt charts from the daily commitments to assist in facilitating the weekly master plan review sessions. The emphasis is to compare the remaining master plan activities based on daily commitments with the remaining activities discussed in the weekly master plan sessions.

Lean Facilitator

A lean facilitator is essential for introducing lean concepts to the project delivery team and for serving as a "change agent" to establish new behaviors in place of less productive practices.

Brief Overview of the Whiting Project

This EPC project was executed with the LPM methodology. The constructed facility was designed for the processing of ULSD. Located in Whiting, IN, it consisted of storage tanks, vessels, reactors, piping for fuel distillates, water, and so on, a power distribution center and an electrical power distribution center. It also included buildings for administration offices and laboratories.

Total installed cost was approximately $128,000,000.

Project construction was scheduled for a 14-month period.

Engineering/design was by a well-known firm located in Houston, TX.

The construction was performed by the RCA in the Whiting region. They established operations in 2001 to improve the value of the construction process brought to their clients, especially the Whiting refinery. The RCA consortium consisted of BMW, Superior Construction, Meade Electric Company, Solid Platforms, M & O Insulation, and Manta Industrial. Dave Koester was the RCA project manager, and was actively employed by BMW Contractors, Inc., a partner in the RCA consortium. Rick Tuttle was the assistant RCA project manager.

The compensation for the project was cost-plus—with incentives and penalties. There was a significant penalty for not meeting owner expectations as well as a significant bonus for exceeding expectations. There was no GC or CM. Each contractor had a separate contract with the owner—there was no IFOA. However, the involved contractors had a previous relationship as an alliance. They had worked together before and some had had an exposure to lean. They had intended to implement lean construction in the project, but lacked a systematic methodology. Lacking clear direction, some of the contractors began to revert to their pre-lean behavior, seeking to optimize their individual performance rather than that of the overall project.

The project was seriously behind schedule in several respects—several key items were an average of 74 days delayed:

1. Piping design: 56 days late
2. Electrical design: 45 days late
3. Control system design: 75 days late
4. Towers: 108 days late
5. Drums and vessels: an average of 43 days late
6. Power distribution building: 120 days late

This situation was a serious concern for the alliance. The ULSD project was the single largest project that RCA had ever attempted and they were particularly concerned that new methods would be required to achieve success. They wished to avoid fueling the observation that "the local history has been a string of large capital projects far over budget, with significantly late completions." In fact, RCA wished to show the client—as well as other clients—that they were both progressive and responsive to improving the owner's equity.

All stakeholders were looking for a significantly new project delivery process, and had become aware of the LPD process through Lean Construction Institute seminars and conferences. Through LCI, the RCA was introduced to Peter Gwynn, a lean/facilitator consultant with Lean Implementation Services. At that point the project was already in the construction phase and earthwork was 30% complete.

Gwynn studied the project very carefully before instituting the lean methodology. He determined that the contractors were inhibited by a fear of penalties and uncertainty about how to plan and schedule for lean. Part of their problem was that they lacked specific knowledge in scheduling for reliable flow. In order to institute lean and get the project back on track, several training sessions were conducted with RCA staff. Lean tools were introduced and the team was reinforced in carrying out scheduling exercises. The consultant established a structured and systematic approach to planning for work flow, with special attention to a mechanical engineer with lean experience whose work was on the critical path.

Four distinct levels of lean planning and production control were performed:

1. The master pull schedule: Level 1
2. The look-ahead schedule: Level 2
3. The weekly work plan: Level 3
4. A daily work plan: Level 4

As described in Chapter 4, the first three plan levels are typical of The Last Planner® System (Ballard 2000). The daily workflow plan is appropriate for projects like this one where there is extensive prefabrication and little or no interchangeability of materials/equipment. The diagram is similar to a Level 3 weekly diagram, but the foremen committed daily to completing tasks based on the crew's capability and capacity.

Planning Methodology I: Production Strategy: Master Production Plan

The project manager, superintendents, project engineers, production planner, owner CM, and key general foremen developed the master production plan during a series of six weekly meetings, each lasting about 1.5 hours. During the initial sessions, knowledge of equipment and material deliveries and engineering timing was deliberately not considered, and the group developed the preferred ideal sequence for the major groups of work activities.

Details of Lean Implementation

This plan development included identifying the sequence of working through the different geographic areas of the work site, determining the order of major work activities in each geographic area, estimating the duration of each major activity set, determining what prerequisite work would release each major work activity, and identifying when each work package needed to be complete. After this framework of the master plan was developed over the first few planning sessions, real constraints including equipment and material deliveries and engineering timing were then evaluated (Figure 7.4). Strategies to mitigate constraints were developed, involving initiatives to sequence work by the engineering contractor, prioritize fabrication by vendors, and sequencing deliveries by suppliers. Constraints were specifically assigned to individuals. It was found that by first pursuing the ideal master plan without constraints, the group was able to sequence work to overcome and mitigate more constraints than would be expected, enabling a greater degree of execution of the ideal plan.

This master production plan was reviewed and updated in a 1-hour meeting each week throughout the construction phase of the project. Work sequences and prerequisite work

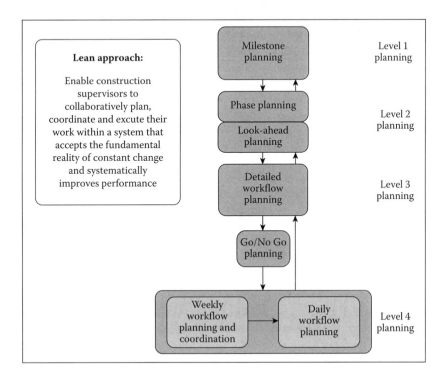

FIGURE 7.4
Lean Production Management planning sequence. (From Gwynn, P., *ULSD Lean Production Management*, Lean Implementation Services, Schereville, IN, self-published, October 2008.)

were reviewed with respect to new constraints and developments, such as engineering and supply deliverables and updated completion forecasts for prerequisite and upcoming work. New constraints were identified and assigned for resolution. The plan was used as a forward-looking tool to collect and present team plans to execute remaining work. Updates concentrated on remaining durations forecast to complete individual work packages, and agreement on what work packages would be released for construction upon completion of prerequisites.

Planning Methodology II: Interactive Area Planning

The project manager, superintendents, project engineers, owner CM, and general foremen developed a more detailed construction plan and activities sequence for each geographic area. One to three meetings lasting about 1.5 hours were required for each geographic area. The owner's CM participated in several of these sessions. The general foremen were key, active participants who particularly identified interfaces among the crews and trades, and identified the degree of completion of prerequisite work that would allow efficient execution of each work package. Each interactive area plan was a more detailed expansion of the corresponding section of the master production plan. Additional sessions to review and update individual area plans were conducted on an as-needed basis, when a significant amount of new constraints or developments were encountered that collectively impacted the planned work sequences.

Planning Methodology III: Detailed Production Planning

The general foremen and foremen for a particular trade developed detailed task and component-level construction and installation sequences for the scopes defined in the interactive area planning sessions to which they were assigned. One or two planning sessions for each area lasting about 1.5 hours were required for each area; the sessions were facilitated by project engineers. The foremen studied the 3-D model and drawings prior to these planning sessions, and developed component-by-component installation sequences for discussion.

   Prior to beginning work on specific piping work packages for geographic areas of the project, it became advantageous for an engineer to conduct a "Go-No Go" meeting. During this meeting, usually held a week before the work was planned to start, the general foreman and other foremen would review a checklist to verify that they had a sufficient amount of engineering, materials, resources, and completed prerequisite work to perform the work efficiently without interruption.

Planning Methodology IV: Daily Production Planning

Each afternoon, individual foremen met with a project engineer or project clerk for about 10 minutes to select and commit to detailed work tasks for their crew to perform the next day. These tasks were consistent with the detailed work flow plan developed in the detailed production plans. The foremen also communicated the status of work committed for the current day, and reasons for work commitments that were not performed.

   Figure 7.5 describes an example of the work activities Establishing a daily commitment enhanced coordination with other trades. Daily huddles with the trades improved the accomplishment of daily tasks. Large EPC projects often require daily work plans

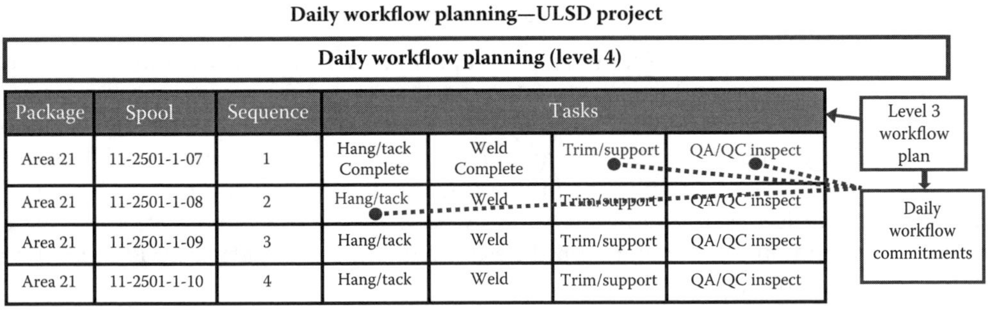

Daily workflow planning—ULSD project

Package	Spool	Sequence	Tasks				Level 3 workflow plan
Area 21	11-2501-1-07	1	Hang/tack Complete	Weld Complete	Trim/support	QA/QC inspect	
Area 21	11-2501-1-08	2	Hang/tack	Weld	Trim/support	QA/QC inspect	Daily workflow commitments
Area 21	11-2501-1-09	3	Hang/tack	Weld	Trim/support	QA/QC inspect	
Area 21	11-2501-1-10	4	Hang/tack	Weld	Trim/support	QA/QC inspect	

Daily workflow planning (level 4)

FIGURE 7.5
Daily workflow planning—ULSD project. (From Gwynn, P., *ULSD Lean Production Management*, Lean Implementation Services, Schereville, IN, self-published, October 2008. With permission.) The involved work package is in Area 21 of the project. This daily activity is pulled from the Level 3 Workflow Plan. The piping spools are listed in the sequence of installation, and the tasks explain the actual work to be carried out. For example, spool 11-250 1-1-07 is to be positioned with a "hang/tack" by the welding crew, then fully welded. The Trim/Support step is followed by QA/QC inspection.

supported by a computerized planning system that is tailored to their needs. This project was provided with extensive software support (SPS/Production Manager). According to Peter Gwynn of Lean Implementation Services, the SPS system facilitated

- reliably pulling material to the work area one day prior to installation.
- coordinating action in real-time across teams with interdependent work.
- measuring commitment reliability.
- consolidating daily workflow plans for material handling teams. and
- productivity reporting.

In preparing daily schedules, foremen-loaded data in the computers. Algorithms used these data to develop daily schedules for the production system. Each day, a print command generated production schedules for the work crews. On a large project involving 30 to 40 foremen, the production schedule is subdivided by area, with coordinates indicating where to find it, and where the systems are to be placed.

(1) Foremen make daily crew assignments following weekly workflow plan and Level 3 detailed workflow plan to the maximum extent possible; when change occurs and makes Level 3 detailed workflow plan either no longer optimal and/or impossible to follow, supervisors rapidly replan the work package. (2) Foreman commit daily to completing tasks they are confident will be completed based on the crew's capability and capacity, status of prerequisite work, weather conditions, having necessary material, equipment and information, satisfactory resolution of conflicts and constraints, and so on. (3) Daily commitment process is supported by SPS/production manager that enables: reliably "pulling" material to work area one day prior to installation; coordinating action in real-time across teams whose work is interdependent; measuring commitment reliability and capturing and acting on root cause reasons for plan failures; capturing and real-time reporting of accurate and objective productivity; consolidating daily workflow plans for material handling teams.

As last planners, foremen must confirm that material handling crews have secured the material/systems and staged them for easy retrieval and installation. Supply chain systems have not yet reached the level of reliability where they can be relied on to operate on a JIT basis. While these items were staged on site for installation, significant improvements were derived—previously, storage on site for long periods of time could result in loss or damage.

Planning Methodology V: Weekly Interface and Commitment Coordination

Each Friday afternoon, the project manager, superintendents, project engineers, production planner, owner CM, and general foremen participated in a 1-hour interface planning meeting, which was conducted and facilitated by a project engineer. Each crew requested specific work tasks to be performed by other crews during the upcoming week that were necessary to support or release planned work. Upon communication of the request, the general foremen discussed and agreed upon what support or prerequisite work would be performed, and on what day it would be completed. Constraints were identified and assigned for resolution.

The general foreman had the final authority to commit to work requests for the next week, or to defer requests to beyond the next week if there were constraints to completion. This resulted in clear agreement and understanding of what support and prerequisite work would be performed the next week, and what work items could not be performed. Each general foreman then communicated and committed to the quantities and work tasks they planned to perform during the upcoming week, upon knowing what commitments were made for support and prerequisite work.

The reasons for noncompletion (RNC) were recorded and tabulated (Figure 7.6). The Pareto Chart reflects a reasons summary. The most frequently occurring RNCs were "Commitment/planning" and "Prerequisite Work." These items were addressed in subsequent weeks, resulting in higher levels of PPC/reliability and reduced variability.

Results

Despite an average late delivery of 74 days of key items as noted below, RCA was able to recover most of the original schedule and the project was delivered 14 days late, but did not impact agreed-to delivery dates of low sulfur products to the client's terminals. In addition, only 7.4% of the budgeted overtime cost was utilized.

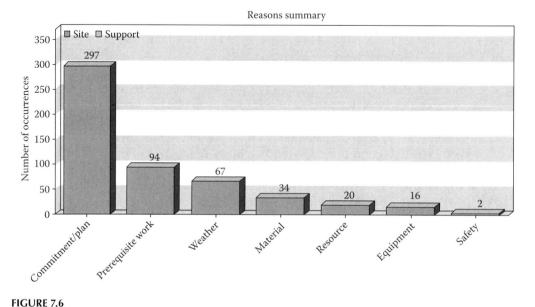

FIGURE 7.6
Reasons for noncompletion. (From Gwynn, P., *ULSD Lean Production Management*, Lean Implementation Services, Schereville, IN, self-published, October 2008. With permission.)

For comparison purposes, the Whiting DHT Unit is very similar to the Joliet Selective Hydrotreater Unit (SHU) erected in 2003 by RCA. Large bore piping on the SHU was installed at an average field labor rate of 2.23 man hours/foot. Large piping on the ULSD project was installed at 1.99 man hours/foot. The actual achieved piping performance data from XOM SHU was used to estimate DHT piping; the performance multiplier for DHT large bore piping was 0.78, which is a field labor improvement of 22%. The man hour parameters include the foreman time with the piping crew.

A definite correlation between the reliability of people to make and keep commitments (planned percent complete), and the field performance multiplier was measured. As the reliability of achieving daily work task commitments improved, the performance multiplier for the piping work decreased.

Figure 7.7 shows the impact of reducing project variability. Between December 2005 and February 2006 there were wide excursions in the PPC values for planned assignments. In some weeks the PPC reached as high as 82%, but fell to zero after several weeks. The mean PPC was in the vicinity of 40–45%. With process improvements, the PPC reached a mean of approximately 80%, with several weeks' values at 100%. The trend line rose from 40% in December 2005 to 85% in March 2006. This project demonstrated that maintaining application of the Lean Production Management principles requires continuous focus, support, and facilitation by management and staff.

With up to 20 field teams simultaneously engaged in the detailed planning and daily commitment processes, the support staff was at times extended, and not all daily work was performed in an optimal manner. When work was begun too early in an area, because not enough prerequisites were completed or constraints removed, the impact was nearly immediately identified by low commitment reliability measurements. Conversely, review of reliability performance quickly pointed to areas where investigation and support or training would improve performance.

With the more efficient performance of work, and the agreed and understood prerequisite work for each work package, field craft overtime was held to 7.5% of total direct straight time man hours. For the first 10 months of construction, Saturday overtime was used only a few times for a small crew to perform a specific prerequisite task that would release a much larger group of construction tasks for multiple crews. The RCA lean production team was

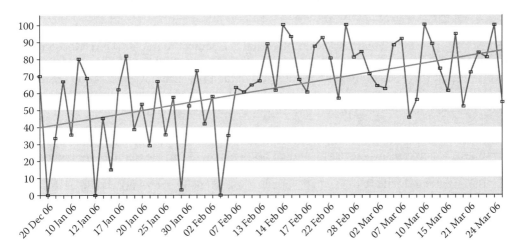

FIGURE 7.7
Diagram of percent project complete showing improvement in variability and reliability.

under considerable pressure to increase overtime as conventional thinking dictates that to reduce a schedule one must increase labor. Instead, the team worked to improve reliability first, and this drove productivity improvement, which resulted in schedule reduction.

During the last 4.5 months of construction, more crews worked on the sixth day to continue the sequence of completing prerequisites to release work packages. This was particularly necessary for electrical work; with the electrical engineering and power distribution center vendor fabrication several months behind schedule, nearly all of the electrical facilities were erected in the last 5.5 months of the project.

Independent Third-Party Audit Results

This project was audited by a well-known, independent third party project analysis company. Their remarks are partially summarized below:

> Based on our analysis, the ULSD project was completed with project outcomes superior to industry: on-budget, on-schedule with an overall capital cost lower than that of industry competitors. Remarkably, the project was able to meet these objectives during a time of unprecedented price volatility and market escalation. The success of the ULSD project was no accident as the project benefited greatly from a number of best practices.

Experiences and Lessons Learned

The most important lessons learned relate to the strategies for overcoming cultural barriers to change and innovation that are so readily found in the construction environment. It was evident that knowledgeable, committed people serve as the best change agents. In this case, the project manager, Dave Koester, was highly focused in bringing about change. The lean consultant, Peter Gwynn, enabled much of Koester's success by being able to provide lean expertise learned in previous projects, and avoid the distractions of day-to-day field activity. Their comments provide an insight into the best practices for success with LPM.

The lean consultant (Peter Gwynn) noted that the project manager (Dave Koester) exercised leadership, which is very critical in the construction industry where people have big egos. Through the force of his personality and leadership, Dave was able to keep contractors in line—he led by example as well. He also viewed it as a mission-critical role to hold the owner's feet to the fire in order to have important decisions made.

> In contracts like this one, anxiety can easily develop over the fear of not meeting deadlines. Even in cases where design and fabrication may use up the contingency, owners and others often overreact, using up high levels of overtime hours. The PM insisted that the contractor should "Go slow to go fast." He pointed out that no work should be done until commitments could be made to have it completed by a specific date. The owner was required to sit in on planning sessions. In traditional projects there is generally highly centralized planning, and extensive reporting to the owner of past events. This is like driving with a rearview mirror. This approach of having the owner involved in various levels, gets them to understand enough to avoid reverting to familiar behaviors when pressure mounts.

The PM was able to play the challenging role of facilitator—this involved asking a lot of questions and bringing out a lot of ideas without telling people what to do. He served as a tie-breaker in situations where different trades would argue over work sequence—who would work next in a particular work area (electricians and pipefitters could be in such a situation). Previously, the choice would go to the trade that was better connected.

It is sometimes necessary to let people fail first in order for them to accept new ideas such as lean construction. A lean facilitator has to serve as a change agent. Project work involves a lot of pressure and there are many technical issues to be addressed. The facilitator must be one who is free to focus on guiding people on a new process without getting caught up in the details of a project. The role calls for someone with passion who will hold people's feet to the fire when pressure builds. One benefit of the facilitator is that he makes it possible for people to hear themselves so people at the foreman level can learn to evaluate projects and learn at weekly review meetings.

The lean consultant's (Peter Gwynn) guidance in implementing lean was directly responsible for the number of lean tools and methods we were able to apply to the Whiting project. While our company model and practice is collaborative planning with the field supervisors actively involved in creating the field work plans and sequences, the consultant guided us in a logical and more disciplined sequence of methods to apply the lean tools. He reinforced the process with suggestions for management techniques to involve and he valued the input of all supervisors. He also provided his company's software, which was effective to capture and link all of the detailed activities in the low-level plans. His recommendations are:

1. Create the master plan (don't call it a schedule) through collaborative sessions with field supervision and the owner. Review the master plan with this group each week, for changes and updates. Once we started the update meeting, we eliminated the traditional weekly "go around the table" construction meeting. The master plan update session more effectively provided all of the information needed by the supervisors regarding plans and interfaces for the next 3–6 weeks.

2. After the master plan is developed, conduct detailed area plans in a similar manner for each geographic area of the project. This takes the plan down to the next level.

3. Do not use the project estimate to back-calculate activity durations for plans— use the guesstimated durations for activities from the supervisors. They are more accurate than we think they are. This was a change for us; we previously back-calculated to quantify durations like the engineers we are!

4. When the foreman makes his detailed activities plan for the next week, don't always jump in and contest the plan if you think it is out of sequence. Let the foreman start on the work; he will see in two to three days that his sequence needs to be revised and he learns this better by this experience rather than by being told to change his plan.

5. Conduct weekly interface planning meetings with the supervisors present—this can be applied to any project whether or not lean is being practiced. What do you need others to do this week to release your work? What do you plan to do this week? Agree on these activities, gain commitment for what will get done when during the week so that all agree with the plan by the end of the meeting. And

nobody can force a commitment on a field supervisor—he must agree that he can achieve the activities by the mutually agreed date. The foreman executing the work gets the final veto if he does not think he can complete the activity when requested. Eliminate unreasonable expectations; eliminate "loudest gets his way."

6. The consultant guided us to further reach upstream into the material receipt and delivery process to include these timings in the plan and work flow. The experience gained in the project suggests that to more successfully implement and apply the lean concepts, project leaders need to be more facilitators and collaborative planners rather than directors. Lean success involves getting field people genuinely involved in planning, showing them that their input is valued by incorporating and executing their ideas, and drawing out the ideas of the more quiet team members in planning sessions and commitment meetings. It is clear that through "command and control" and "nuts and bolts" detail, project leaders would have a difficult time implementing getting much additional value out of lean.

Habits and Previous Practices

Most field supervisors have a strong drive to work productively and to eliminate periods of crew inactivity. Most construction projects have time constraints that are perceived as tight or compressed. There is a strong practice among many levels of construction management and supervision to start erecting work packages as soon as possible to improve the possibility of achieving the completion date. Combined with the usual presence of a varied amount of materials and process equipment on site this can—and does—often lead to premature erection and installation in a sequence that is considerably less than optimal.

Project teams often start installation of work packages too early, erecting incomplete quantities of materials and equipment in a discontinuous sequence because a certain amount of material happens to be on hand. Also, with their desire to meet the completion target, discipline groups tend to optimize the sequence of their work without full coordination with other disciplines. Superintendents spend much of their time mediating competing desires for construction geography and equipment.

Most people in the construction industry (contractor team members, CM, owner representatives, and crafts) believe that working overtime is the solution to many issues. Overtime is the perceived solution to engineering and equipment delays, craft congestion, tight completion targets, foul weather, labor availability, and so on. People join large projects expecting significant overtime. This was not the case on this project, and very little overtime was required to recover from late deliverables.

On conventional/nonlean projects, when barriers and constraints are identified in the field or in planning sessions, they are often verbally communicated as "we need..." or "this should be..." The speaker then assumes that since he stated the constraint to someone else or to a group, someone else will address the constraint. Often, project team members erroneously assume that someone else is addressing the spoken needs or concerns, and constraints linger unaddressed as there is no formal process to capture and remove these constraints, as there is with lean methods.

Interactive and Detailed Planning

Lean Production Management methods expose and capture the best plans and ideas of field supervisors for execution of the work. With all disciplines participating in master production and interactive area planning, work sequences are agreed that benefit the project as a whole, rather than benefiting any individual discipline. The team then understands the path that all disciplines must progress to completion as efficiently as possible for the project to be successful. Team members often contributed ideas and sequences that sacrificed a portion of their preferred work sequence to more significantly enable another discipline to achieve greater efficiency or earlier work package completion. This is counterintuitive and requires extraordinary rethinking and learning by the production teams. External discipline must be brought to bear, usually by an independent third party lean mentor, to assure personnel do not revert to bad habits.

The team understanding and commitment to executing the master production plan eliminates most of the field interferences between discipline work crews, nearly eliminating crew delays, and the need for redirection to other work. Each work group knows who is released to work in each area, and knows the amount of prerequisite work that is agreed to be complete before the next work activities begin. Superintendents had to address competition for site erection access.

Upon development of the current optimum sequence for field construction, the LPM methods channel the supervisors' desire for field efficiency to begin erecting work packages when identified prerequisites are complete, and when sufficient materials, resources, and clear work areas are available to erect the work package efficiently, without interruption through completion. Work on most work packages began 3–4 weeks later than would have been done with traditional project management methods, yet work package completion dates did not extend out by these 3–4 weeks. Work packages started later with much less delay, interference, and interruption and were completed on time because the work could be reliably planned and executed.

Weekly Interface and Commitment Coordination

The weekly interface commitment meeting created a clear understanding among all project participants of what support and installation work would be performed each week. Requested supervisors had the final authority to honestly determine what could be accomplished the next week, and to agree on the day of completion; this reduced the potential for overcommitments and subsequent delays. False expectations about what prerequisite or support work would be complete were nearly eliminated.

In particular, the efficiency and supply and erection of scaffolds was significantly improved, as each general foreman requested his scaffolds for the next week, and the scaffold general foreman made commitments regarding what scaffolds would be erected and the day each scaffold would be performed. These enabled the scaffold general foreman to effectively plan and schedule the scaffold work for the full week, and reduced the amount of "need it now" calls for scaffolds. This also provided a management look ahead at scaffold needs, enabling more proactive measures to provide the required amount of support resources.

Constraint Analysis and Assignment

Identifying, listing, and assigning each constraint for resolution, regardless of how small the constraint may have been, resulted in all constraints being addressed or understood within a known and agreed time frame. Unmet expectations for solutions and continuation of constraining conditions were greatly reduced. The person who volunteered or was assigned to address a constraint had the authority to agree when he would respond with resolution and answers, to allow reasonable time for resolution, and to maintain accurate expectations of the completion of prerequisite work items.

Detailed Planning by Field Supervisors

Having Supervisors perform detailed, component-level planning for the upcoming 5-day period was a very new, challenging experience for nearly every foreman. Each foreman was assigned to determine the specific sequence of pipe spool erection in his geographic area. For each 5-day period, his plan included what spools, welds, supports, and so on were to be installed on each day. The younger, newer foremen adapted more readily to this detailed level of planning, and achieved efficiency results earlier. From the daily commitment reliability curves, it was measured that 3–4 weeks of operation were required for each supervisor to learn, understand, and practice these methods to demonstrate measurable field efficiency improvements. To maintain improved field efficiency, field teams required continued process support and facilitation.

It quickly became evident that before LPM was introduced, planning by field supervisors was actually having a "rough outline" of how groups of components would probably be installed. These planning methods were in most cases a 1-day reaction to the previous day's work; "now that we installed this material today, what do I choose to install tomorrow?"

Summary of LPM Requirements

The implementation of LPM requires an experienced, expert resource person over a series of projects to sustain learning and increase effectiveness. Being a behavioral change methodology, improvements are achieved gradually. While this project achieved measurable construction efficiency and project delivery improvements because of LPM, it successfully implemented only a portion of the available methods and principles.

Proficiency requires more training, building on these successes, and repetition. For one to think that a project team can operate and improve LPM methods after one project without expert resource assistance would not only limit improvements, but would likely result in reductions of performance until LPM is viewed as the previous "management flavor of the month."

At the beginning of the ULSD project, the achievement of the target construction completion date with the demonstrated piping productivity performance improvement and a lower amount of overtime was not believed to be possible. Lean Production Management was the primary contributor to this step-change in achievement. With further application and implementation of the LPM processes, another step-change in performance was achievable.

Overall, LPM requires significant behavioral change among the project team. Implementation requires an experienced, expert resource person. Consistent and sustained focus and energy must be given to adhering to LPM methods and practices. LPM methods achieve a step-change in construction efficiency and productivity.

Questions for Discussion

1. What is Relational Contracting and how is it different from other contracting methods used in the construction industry?

2. Discuss the objectives and the major factors in building a successful Integrated Project Delivery team.

3. Briefly explain the major steps involved in implementing relational contracting and making it successful.

4. What are the Five Big Ideas?

5. How do they impact project delivery?

6. What were the differences between the forms of contract used in the cases on Integrated Project Delivery, Lean Project Delivery with an integrated agreement, and Lean Production Management?

7. How are disputes resolved in Integrated Project Delivery and Lean Production Management?

8. Are there any specific advantages to having a Risk Pool for savings and extra costs in Integrated Project Delivery contracts?

References

Ballard, G. 2000. *The Last Planner® System of production control*. PhD Thesis. Faculty of Engineering. School of Civil Engineering. Birmingham, AL: The University of Birmingham.

Gwynn, P. 2008. *ULSD Lean Production Management*, Lean Implementation Services, Schereville, IN, self published, October, 2008.

Koester, D. and P. Gwynn. Unpublished report. Lean Production Management Case Study—ULSD Project 2005–2006.

Lichtig, W. A. 2005. Sutter Health: Developing a contracting model to support lean project delivery. *Lean Construction Journal* 2(1): 105–12.

Lichtig, W. A. 2006. The integrated agreement for lean project delivery. *Construction Lawyer, American Bar Association* 26(3).

Lichtig, W. A. 2008. Sutter Health. Integrated agreement for lean project delivery between owner, architect and CM/CG, 2008.

Matthews, O., and G. Howell. 2005. Integrated project delivery an example of relational contracting. *Lean Construction Journal* 2(1): 46–61.

Rowlinson, S., and Y. K. F. Cheung. 2004. Relational contracting, culture and globalisation. In International symposium of CIB W107/TG23 joint symposium on globalisation and construction. November 17–19, AIT, Bangkok.

Sanvido, V., and M. Konchar. 1999. Selecting project delivery systems: Comparing design-build, design bid-build and construction management at risk (CII). State College, PA: The Project Delivery Institute.

Williamson, O. 1979. Transaction cost economics: The governance of contractual relations. *Journal of Law and Economics* 22:233–61.

Bibliography

Cheung, Y. K. F., and S. Rowlinson. 2005. Relational contracting: The way forward or just a brand name? In ICCEM 2005 Conference: Globalization and collaboration in construction. October 16–19. Seoul, Korea.

College, B. 2005. Relational contracting—Creating value beyond the project. *Lean Construction Journal* 2(1): 30–45.

Howell, G. A. 1999. What is lean construction. Proceedings of the 7th annual conference of the international group for lean construction. Berkeley, CA: UCLA.

Howell, G. A. 2000. White paper for Berkeley/Stanford CE & M workshop. Proceedings, construction engineering and management workshop. Palo Alto, CA: Stanford University.

Koskela, L. 1992. *Application of the new production philosophy to construction.* Technical Report #72. Stanford University: CIFE.

Macomber, H., and G. Howell. 2005. *Using study action teams to propel lean implementations.* Louisville, CO: Lean Project Consulting Inc.

Mauck, R., W. Lichtig, D. Christian, and J. Darrington. 2009. Integrated project delivery: Different outcomes, different rules. The 48th annual meeting of invited attorneys. Chevy Chase, MD: Victor O. Schinnerer & Company.

End Notes

Sutter Health's Lean Journey

Sutter Health's lean journey was driven by California's regulation (SB1953), requiring seismic upgrades to health care facilities. (CII Research Report 234–11). An aggressive 8-year schedule for the $6.5 billion capital program called for better project delivery. The leaders of the facilities program were: Bob Mitsch, VP, Dave Chambers (Planning), Dave Pixley (Project Management), Morri Graf (Project Controls). Will Lichtig, Sutter Health's legal counsel, introduced Dave Pixley and Morri Graf to lean construction through an LCI workshop in late 2003. Lean Project Consulting provided lean implementation support. Sutter's leadership became committed to lean construction despite its unproven nature, to the extent of funding design and construction providers to "learn lean on the job." David Long, Senior PM, became the program's Lean Coordinator in 2005. Sutter's practice of Integrated Project Delivery has gradually evolved since 2004. Five pilot projects were launched in May, 2004 to implement the Last Planner® System including: Davis medical office building, Modesto 8 story, Roseville emergency department & parking structure. LPS application failed in two projects; the lessons learned continuously improved subsequent projects. Other lean tools were gradually added, including TVD, BIM, A3 reports and set-based design. Some of the later projects are:

- The Fairfield Medical Office Building (Boldt Construction) involved target value design and visual controls—builders collaborated with the designers.

- Camino Medical Office Building (DPR Construction) BIM-enabled Integrated/Lean Project delivery—virtual models helped work teams to visualize work requirements early; 15–30% higher labor productivity; rework < 0.2%.

- Castro Valley 230,000 sq. ft. hospital & MOB (DPR Construction, Devenny Group, Capital Engineering. Digby Christian, Sutter Health Project Manager) LEED-certified "green building," Integrated Form of Agreement, TVD, maximum interoperability of BIM and ICT systems.

8

Information and Communication Technology/ Building Information Modeling

Introduction

New approaches to construction management such as relational contracting and lean design and construction are built on a foundation of team integration and open sharing of project-related information. Information and Communication Technology (ICT) and building information modeling (BIM) are an effective mechanism for providing construction stakeholders with the informational and analytical tools for better management of the construction delivery processes. This chapter includes two cases by Professor Salman Azhar and colleagues that illustrate the benefits of BIM in construction projects.

ICT—a Description

Information and communication technology encompasses computer hardware, software, and communications devices that allow the sharing and access of information conveniently, locally, and worldwide. The rapid growth of the Internet confirms the benefits of global communications and the need for information sharing. This technology and the Internet have affected many aspects of people's lives, including how they work and live. Almost all industries are affected and can benefit from the development of ICT. The A/E/C industry in particular has shown vast interest in adopting the new technologies in the area of 3-D visualization, data analysis, communications and collaboration, information sharing, and BIM. Building information modeling is the process of generating and managing building data during its life cycle. It is a tool as well as a process, and increases productivity and accuracy in the design and construction of buildings.

Impact of ICT on the Construction Industry

ICT first evolved to help in synthesizing or analyzing information. The first two major applications in construction were the finite element analysis program (FEAP) in the 1970s and the drafting software (AutoCAD) in the 1980s (Turk 1997). Since the 1990s, ICT has

TABLE 8.1

Traditional and ICT Supported Technologies in Construction

Needs	Item	Traditional Technology (Becoming obsolete)	ICT Supported Technology
Information processing and management	Project	Drafts, folders	Document management, product and process models
	Company	Archive, microfilm	Data warehouses
	Country	Library, building regulations	National construction information systems
	World	Journals, conferences	Global ICT networks
Interaction facilities	Man with man	Speech, phone, fax, mail	E-mail, video conferences
	Man with application		3-D Visualization, virtual reality, graphical user-interfaces
	Man with machine	Direct contact	Indirect contact using computers
	Application with machine		Robotics, Remote sensors
Time saving	Just-in-time	Book look-up, library look-up, phone call to expert	Database look-up, internet search
	Just-in-case	Reading books, magazines, journals, schools, visiting conferences	Subscriptions to customized content, distance learning

Source: Turk, M., Proceedings of the Workshop on Perceptual User Interfaces, Banff, AB, October 1997.

revolutionized the construction industry. Table 8.1 compares the traditional and ICT supported technologies in construction. At the beginning of 2010, ICT-supported technology has become the norm and traditional technology is rapidly becoming obsolete. ICT tools are being utilized in the construction industry in three distinct areas as will be shown in the following sections.

Information Management and Services

Information management uses all aspects of ICT for capturing, storing, organizing, and retrieving data. Internal (project/company) and external (industry) standards are essential for maximum integration in the process. Shared databases, data normalization techniques, data warehousing, barcode technology, CAD graphics, and BIM are examples of advancement in this area

Communications

All aspects of communicating data and information such as text, graphics, audio, and video are included here. The most common communication tool is electronic mail (e-mail). A variety of software such as ICQ®, IRC®, Net Meeting®, and so on are used to allow seamless discussion between two or more parties. One example is a discussion between a contractor and an engineer trying to resolve an urgent problem encountered on the job site. Hyper text markup language (HTML) can also be used for simple communication such as the use of forms (e.g., to send work progress information from the field), which can then be used for the preparation of progress reports.

Communication tools can help design and construction organizations to coordinate their activities with greater effectiveness and efficiency overcoming the barriers of time and distance. Training tools, such as multimedia, can help workers. However, there is a profound need for uniform and standard data in the construction industry. Without standard data it is difficult to establish a common performance measurement system or a uniform quality assurance program. Uniformity of procedures and standardization of data would greatly enhance the effectiveness of communication among the multiple construction organizations teaming up to build one constructed facility.

Processing and Computing

This includes all systems and models developed for processing data. Technologies supporting the process of developing such systems and models underlie this component of ICT capability. By the advancement of scripting technologies on both the client side and the server side as well as through the availability of plug-in modules, the Internet (including Intranet and Extranet) can be used for engineering and management computing purposes. The most common examples include project scheduling, resource management, and project cost control using shared databases through the Internet.

ICT Tools for Design and Management Processes

Table 8.2 provides an overview of the state-of-the-art ICT tools that are being used in the design and management operations of construction projects.

ICT Tools for Design Processes

Contemporary computer-aided design (CAD) applications that allow 3-D geometric modeling are all based on the same concept (Gero 2000). These systems provide the following types of integration within the architectural and engineering design processes.

- Integration of 2-D drafting and 3-D modeling
- Integration of graphical and nongraphical design information

TABLE 8.2

Tools for Design and Management Processes in Construction Projects

ICT Tools	Design	Management
Information Management and Services	• Integrated CAD systems (Informational databases) • BIM—Building Information Modeling.	• On-line bidding/Permits • On-line building information services • On-line project administration systems Shared project databases
Communications	• Animated 3-D/4-D visualizations • Virtual design studios • Simulation techniques	• On-line Project management and control
Processing and Computing	• Integrated CAD systems (structural analysis and design).	• Model-based cost estimation • Planning and scheduling softwares • E-Commerce Applications • BIM • Virtual Design and Construction (VDC)

- Integration of the data structure and the user-interface
- Integration of two or more applications (e.g., design and analysis, drafting functions, and other applications)

An overview of some of these systems is presented below in terms of concept, technology, application, and implementation requirements.

ICT Tools for Management Processes

Information and communication technology (ICT) offers unique opportunities for managing projects effectively by utilizing automated means to capture, store, and retrieve data; efficient ways to process data into information; and powerful techniques to transmit data/information quickly and in vast quantities. This concept is illustrated in Figure 8.1, which shows an integrated information and control system model for the construction industry. The model provides the mechanism through which information flows to different departments within/outside an organization. To facilitate proper and optimal decision making, the availability of the desired information at its required level of detail is necessary. The flow of this information (through an information system) further facilitates the interaction of all the managers by providing them with a better understanding of what is happening in the project. This knowledge allows them to make decisions proactively.

Construction is a multi-organization process with heavy dependence on the exchange of large complex data and information. The successful completion of the project depends on the accuracy, effectiveness and timing of communication, and exchange of information and data between the project team. In today's ever complex and dynamic environment, most decisions are based on the availability of paper-based information or in some cases computer generated or supported information. However, when a decision maker is not on location or does not have updated information, decisions are based on the information

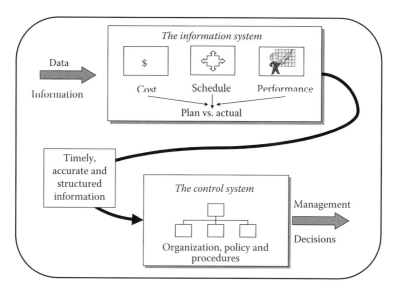

FIGURE 8.1
Model of integrated information and control system environment for construction projects.

available. The available information may not be up-to-date and may also not be sufficient. In such circumstances, if proper information usage is to be made for effective decision making, up-to-date and sufficient information must be accumulated. Technological advancement in ICT with the use of the World Wide Web (WWW) for information exchange management, independent of time and location constraints is fast becoming a medium for distance communications.

Model-Based Cost Estimation and Concept

Model-based cost estimation provides a tool for consulting, contracting, and maintenance corporations to generate quick and accurate cost estimates despite different data formats and standards. The main idea is the transformation of heterogeneous design information into a product model for cost estimation and construction planning. This is made possible through a product model.

Technology/Software and Application/Benefits

Model-based cost estimation could be implemented through a software application COVE® (Cost and Value Engineering). This is a computer-aided engineering (CAE) tool that generates a product model-based on heterogeneous data and information feed-ins. COVE is used through design ++, a knowledge engineering tool (Figure 8.2). All user inputs are passed through a user-interface to Design ++, which then accordingly controls the CAD engine. Heterogeneous data are intelligently transformed to yield an object-oriented product data model (Hannus and Kazi 2000). The application of this technology is primarily for designers and contractors during both cost estimation and tendering stages. Through the developed product model, significant reductions in tender preparation time may be achieved in addition to more accurate cost estimates and the possibility to reuse standard company solutions.

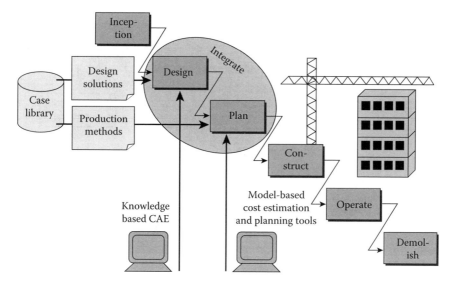

FIGURE 8.2
Conceptual illustration of COVE system. (From http://www.vtt.fi.)

On-Line Bidding Concept

On-line bidding enables bidders to submit and edit their bids thereby resulting in great time and cost savings. The concept behind on-line bidding is the same as behind other on-line applications. The contractors can submit their bids using secured servers. After the closing date, the client can download all submitted bids and make the decision.

Technology/Software and Benefits

A variety of on-line bidding systems are now available in the market such as Invitation to Bid® and Bid Express® and operated through the Web site http://www.Bidx.com. Contractors may subscribe to a provider such as Bidx.com, and must have a digital signature on file at the exchange side. They submit bids as encrypted files to be held in an electronic lockbox until the deadline. At the closing date, the box contents are downloaded by the client, using electronic keys provided by registered contractors to open the bids. Until the moment of delivery, the client does not have access to the bids and the exchange service does not have the service to decode them. This setup ensures the security of the system to stop any act of fraud and hacking (Figure 8.3). The on-line bidding allows the contractors to submit their bids at any time and from any place and have the flexibility to edit them till the closing date.

On-Line Permits and Concept

Before construction actually starts, many permits must be applied for to various government agencies. Engineers then have to evaluate the design for code compliance. Many government agencies worldwide are starting to use the Internet for owners and contractors to submit applications on-line. For example, PermitsNow® in the United States is

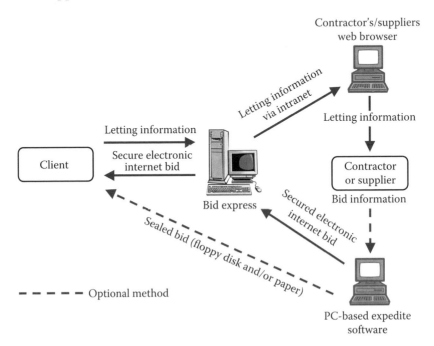

FIGURE 8.3
The on-line bidding system. (Derived from http://www.bidx.com.)

providing permits with on-line services for hundreds of municipal governments. By way of an example, the web may be used to apply for an electrical permit. Once submitted, the permit information and drawings are stored in a database for engineers to evaluate on-line regarding code compliance. Once approved, the permits are issued through e-mail notifications. This kind of on-line permit handling has received a warm welcome from owners and contractors. They can get permits approved faster, the costs of printing numerous copies of drawings are eliminated, and the costs to government agencies are reduced.

Technology/Software and Benefits

Available software, PermitsNow®, can be accessed on-line at www.permitsnow.com. This technology is beneficial for both government agencies as well as general and subcontractors and could result in substantial cost and time savings.

Shared Project Databases and Concept

Shared project databases are gaining wide acceptance in the realm of construction information management. They allow the storage and retrieval of data from a central location independent of time and locational constraints (Figure 8.4). Shared project databases (data warehouses) gather all relevant project information in a central location and allow all project participants to access the same information. This in turn speeds up the decision-making process, reduces data and information redundancy, and reduces costs.

Technology/Software and Application/Benefits

Technology/software requirements for implementation of shared project databases are not complex. A database server is required with which project participants may establish remote connections to retrieve and provide information to the project database. The use

Technology/software

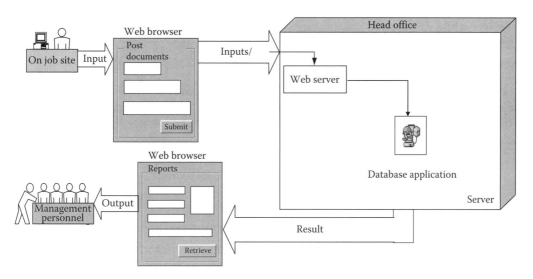

FIGURE 8.4
Schematic view of a web-based shared database system.

of shared project databases extends to virtually all project participants. Usage typically is during the project's life cycle. Upon completion of the project, the information stored within the project database may be used as historical information feed-in for future projects. Participants can contribute to and gain from the shared project databases from any location with a connection to the Internet.

On-Line Project Administration Systems and Concept

On-line project administration systems can provide around the clock information about the project such as status; directory of contractors, consultants, vendors, and suppliers; project drawings and specifications; project control reports and the facility to submit on-line change orders. Those systems facilitate better communication between the head office and remote project sites, which could result in both time and cost savings.

Application/Software and Benefits

Various software titles are available such as Webster for Primavera®, MS Project for Net®. These systems enable dispersed construction groups to work collaboratively together as an effective single team sharing pictures, documents, and real-time videos.

On-Line Project Management, Control, and Concept

On-line project management and control can be done effectively through the use of web cams or handheld and laptop computers that provide continuous data about the project job site regardless of weather and safety conditions for monitoring and control purposes as illustrated in Figure 8.5. Handheld or laptop computers are now standard in the construction field. They are not only to access project information but also to collect important field data, such as manpower and project progress. Many of these handheld computer manufacturers offer attachments such as digital cameras and recording of sound, making these a convenient way of collecting multimedia information for construction projects. This technology provides around-the-clock data and digital views to monitor and control a project while sitting in the head office or other remote location.

Technology/Software and Applications/Benefits

A typical construction site uses three cameras, two stationary and one pan tilt. The cameras typically shoot one picture per hour, but the system can also provide streaming video. The pictures are transferred through the Internet to the company's head office or other remote location. Web cams and/or handheld computers could be used for project monitoring and dispute resolution. They can help resolve critical-path delays caused by other workers or delays because of nature such as floods and earthquakes. They can also be used to stop workers' compensation fraud.

E-Commerce Applications

The growth of the Internet has created new business models that allow buyers and suppliers to conduct business transactions on-line, eliminating the distributors. Many Internet portals, ASPs, and database hosts, facilitate the linkage for a small fee, by allowing buyers to specify goods/products to buy and suppliers with products to sell. The transactions

FIGURE 8.5
On-line project management and control using web cams.

can be directly linked to each company's inventory and pricing, and ordering information can be taken directly with invoices/payments sent electronically (Liu and Erickson 2002). E-commerce applications can be broadly divided into four main categories.

Business-to-Business

An electronic means of carrying out business transactions between two or more businesses. B2B incorporates everything from manufacturing to service providers. An example would be a company that uses the Internet to place an order from the suppliers or retailers, receive electronic invoices, and make payments electronically.

Business-to-Consumer

Similar in concept to the traditional method of retailing, the main difference being the medium used to carry out business—the Internet. Such a method assumes that the consumer has access to the Internet. By selling direct to customers or reducing the number of intermediaries, companies can achieve higher profits while charging low prices.

Business-to-Administration

This category covers all the transactions that are carried out between businesses and government bodies. Currently, this category is in its infancy, but it has the potential to grow with government initiatives.

Consumer-to-Administration

This category has evolved over the last couple of years and provides a platform for consumers to directly contact the government official with matters ranging from buying products to getting permits.

Integrated Computer Aided Design Systems and Concept

These systems provide a complete solution of architectural, drafting, and engineering design problems. The integrated CAD systems combine 2-D drafting with 3-D modeling and provide a complete interface with the engineering design software. This removes the redundancies in the design process and results in both time and cost savings.

Technology/Software and Applications/Benefits

The software titles available for architectural design are updated versions of AutoCAD2002®, 3-D Studio Max®, and Design CAD. For the structural design, SAP 2000® (linear and non-linear versions) and STAAD-III® are popular software. Most of them provide an interface to transfer data. Integrated CAD systems offer the following benefits: reduction in ambiguity, better coordination, lower potential for errors, elimination of the need for physical models, automatic generation of properties, exploded views and sections, and the possibility of extension to other applications.

Animated 3-D/4-D Visualizations Concepts

The use of animation to depict three-dimensional views of a building product/facility provides a useful view of what a finished product may look like as shown in Figure 8.6. The core concept of animated 3-D visualizations is to provide a "near-real" replica of the

FIGURE 8.6
Use of 3-D visualization in construction. (From http://www.vtt.fi.)

finished product/facility from the basic design to the operation of the facility. Not only does this provide a better platform for the architect and more imaginative flexibility, but also a better feel to the customer of their desired product/facility. They also provide a useful visual description to the designer/contractor. The 3-D models have been extended into 4-D (3-D + time), where construction schedules are integrated with 3-D displays to simulate the progress of a construction project. The 4-D provides improved work sequencing. Spatial conflicts between different trades is facilitated and resource utilization is maximized through visual monitoring and 5-D improves the accuracy of virtual estimates.

Technology/Software and Applications/Benefits

Basic technology required for animated 3-D consists of high performance personal computers or workstations connected over a network. Additionally a video studio for digital editing and animation recording graphic cards are required. Software titles available for this purpose include 3-D Studio Max® and Adobe Premiere®. The applications extend from the project development to the facility operation and management stage. All project participants can increase the quality of both the design and construction process. Moreover, instead of having to browse through pages of documents, users can navigate through a model and then through simple mouse clicks, access the relevant information.

Virtual Design Studios and Concept

The virtual design studios allow designers and experts from different locations to interact using audio and video conferencing. Virtual studios allow designers to discuss matters related with design or construction without leaving their office. They can share the same screen and same program thereby entering the same virtual reality space.

Technology/Software and Application/Benefits

The technology includes high-speed computers or workstations with fast Internet access. The main benefit lies in the fact that experts and designers from any location and at any time can discuss a design issue and propose an immediate solution.

Building Information Modeling

Building information modeling (BIM) is the process of generating and managing building data during its life cycle. It is also a tool as well as a process, and increases productivity and accuracy in the design and construction of buildings.

BIM uses three-dimensional, dynamic building modeling software and operates in real-time. It supports the continuous and immediate availability of project design scope, schedule, and cost information that is high quality, reliable, integrated, and fully coordinated.

The process produces the Building Information Model, also known as the BIM, which encompasses building geometry, spatial relationships, geographic information, and quantities and properties of building components. Though it is not itself a technology, it is supported to varying degrees by different technologies. The BIM is a data-rich, object-oriented, intelligent and parametric digital representation of the facility, from which views and data can be extracted and analyzed to generate information that can be used to make decisions and to improve the process of delivering the facility (AGC 2005).

The American Institute of Architects (AIA) has defined BIM as "a model-based technology linked with a database of project information." A BIM carries all information related to a facility, including its physical and functional characteristics and project life cycle information, in a series of "smart objects." For example, an air conditioning unit within a BIM would also contain data about its supplier, operation, and maintenance procedures; flow rates; and clearance requirements.

History of BIM

The term appears to have been adopted by Autodesk to describe "3-D, object-oriented, AEC-specific CAD." This medium was rich in architectural information and distinguished from the traditional 2-D drawing. Professor Charles M. Eastman at Georgia Institute of Technology has done extensive work on the process starting in the 1970s under the heading of a building product model. He has described BIM and documented its application in several publications. (Eastman, Teicholz, Sacks and Liston 2008) Jerry Laiserin is said to have popularized the term BIM as a digital representation of the building process to facilitate exchange and interoperability of information in digital format. In 1987, Graphisoft's ArchiCAD first implemented BIM under the virtual building concept.

BIM Implementation and Technologies

The following technologies are required to implement BIM: CAD, Object CAD, and the most effective technology, parametric building modeling.

CAD Technology

CAD technology has been used by many industry members for several decades. While it lends itself to the automation of certain drafting tasks, it requires much programming effort and high levels of discipline of the users that are responsible for entering data. Layer and naming standards have to be strictly enforced, resulting in higher administrative and management costs. It is therefore considered to lack the sophistication required for BIM.

Autodesk CAD Technology

AutoCAD software by Autodesk is based on CAD technology and has a moderate level of sophistication. Various applications have been installed on the AutoCAD platform, such as cost estimating, structural design, and scheduling that are typically provided in BIM.

Object CAD Technology

Object CAD technology is next on the level of sophistication. It works in a CAD-based environment and accommodates building designs in 3-D geometry—it extracts 2-D documentation from it and can also compile bills of material. However, the quality of its representation for BIM depends very much on the knowledge and versatility of users. Autodesk has a number of products that serve this need: Autodesk Object CAD Technology, ADT, and Autodesk Building Systems. They meet many of the requirements of BIM more easily than AutoCAD. Autodesk Architectural Desktop (ADT) simulates building modeling operation by managing a collection of drawings that each represent a part of the entire BIM model. Because of the loose structure, ADT may be subject to errors if users do not work within its limitations.

Parametric Building Modeling Technology

This technology currently provides the highest level of BIM with the least amount of effort—it is far more sophisticated than CAD and Object CAD technology, and requires a completely different mindset in its users. It is an integrated system that embodies geometry and data representation and the relationships between different elements of the model are user-defined. In a manner analogous to an electronic spreadsheet, as one element is changed, this change is reflected appropriately to all elements of the model. This facility makes the design process (and subsequent construction) faster, more cost effective, less prone to errors, and higher quality.

Autodesk Revit software is based on parametric building modeling technology and is designed specifically for BIM. It has a central project database that contains representations of all building elements. Design revisions are immediately reflected throughout the entire project, and errors are readily detected. A number of software companies have competing products.

ArchiCAD by Graphisoft is based on a virtual building model, and was designed as a BIM system in the 1980s. ArchiCAD behaves as an application peripheral to the model instead of containing all the building data itself.

Bentley Systems is an integrated project model-based on a number of related application modules, and accesses project data by means of DWG and IFC file formats. The modules include Bentley Architecture, Bentley Structures, and Bentley HVAC as well as others. The system is limited by providing interoperability only when proprietary modules are used.

Nemetschek is available primarily in Germany and Europe also providing a BIM platform. Its object interface allows software by other manufacturers to interact with it, unlike the Bentley system.

Practical Applications of BIM

A building information model can be used for the following purposes:

- *Visualization*: 3-D renderings can be readily generated.
- *Fabrication/shop drawings*: It is easy to generate shop drawings for various building systems, for example, the sheet metal ductwork shop drawing can be quickly produced once the model is complete.
- *Automated Fabrication*: In projects that involve technologically advanced suppliers, data from a BIM file can be used as input to program numerically controlled fabrication equipment.
- *Code reviews:* Fire departments and other officials may use these models for their review of building projects.
- *Forensic analysis*: A building information model can easily be adapted to graphically illustrate potential failures, leaks, evacuation plans, and so on.
- *Facilities management*: Facilities management departments can use BIM for renovations, space planning, and maintenance operations.

- *Cost estimating:* BIM software has built-in cost estimating features. Material quantities are automatically extracted and changed when any changes are made in the model.
- *Construction sequencing:* A building information model can be effectively used to create material ordering, fabrication, and delivery schedules for all building components.
- *Conflict, interference and collision detection*: Because BIM models are created to scale in 3-D space, all major systems can be visually checked for interferences. This process can verify that piping does not intersect with steel beams, ducts, or walls.

Building information modeling makes a reliable digital representation of the building available for design decision making, high-quality construction document production, construction planning, performance predictions, and cost estimates. Having the ability to keep information up-to-date and accessible in an integrated digital environment gives architects, engineers, builders, and owners a clear overall vision of their projects, as well as the ability to make faster informed decisions. BIM makes available more design options and alternatives than traditional methods with similar time schedules and budgets. BIM provides two major benefits:

- As important design information is kept in digital format, it can be easily updated and shared with all the involved parties.
- The maintenance of real-time design data in digital form, using parametric building modeling technology, provides major savings of time and money. This leads in turn to an increase in project productivity and quality.

BIM represents a major departure from the traditional computer-aided drafting method of drawing with vector file-based lines that combine to represent objects. It models the actual parts and pieces being used to build a building. While line drawings can be interpreted by people, they cannot be readily interpreted by computers. BIM represents object parametrically in 3-D; these objects reconfigure themselves automatically when software instructions are invoked to show different views in accordance with defined rules. As 3-D objects are machine-readable, errors can be avoided at all phases of design and construction in real-time, as opposed to the time honored method of post facto drawing review. In fact, design or fabrication work can be reviewed remotely through web conferencing tools such as Webex or GoToMeeting.

BIM can be used to demonstrate the entire building life cycle including the processes of construction and facility operation. Quantities and shared properties of materials can easily be extracted. Scopes of work can be isolated and defined. Systems, assemblies, and sequences are shown in a relative scale with the entire facility or group of facilities.

BIM enables a virtual information model to be handed from a design team to the contractor and subcontractors and then to the owner, each adding their own additional discipline-specific knowledge and tracking of changes to the single model. This approach greatly reduces the loss of information that occurs in the transfer of a project from one entity to the other.

These attributes enable BIM to reduce waste and inefficiency in building design and construction. It has proven to be attractive to a number of both public and private users that are concerned about managing completed facilities and wish to reduce or avoid the operational shortcomings that plague many such facilities. In particular, the reduction in design errors and construction errors has a major impact on the users' experience. Whereas the traditional approach involved producing contract drawings that were subsequently converted to more detailed drawings for shop fabrication purposes, a BIM model can readily generate such shop drawings. Furthermore, it provides instructions in machine-readable form that can be used by fabrication equipment for more cost-effective production and greater accuracy than craft labor.

BIM addresses a major shortcoming of the design and construction processes—that of the interoperability of construction documents and information between the various parties to construction projects. This lack of interoperability leads to errors, misinterpretations of building system conflicts, and delays as the design and construction process passes through different phases. Effectively, the owners' project requirements and the designer's intent are often not fully preserved during the transition. BIM improves interoperability by integrating computer models into project coordination, simulation and optimization—the information provided by such models is used to generate feedback.

BIM can serve to prevent the information loss associated with handing a project from the design team to the construction team and to the building owner/operator, by allowing each group to add or reference back to all the information they acquire during their period of contribution to the BIM model. By way of an example, a building owner could use BIM to track down the source of a building problem without leaving a computer terminal. A water valve can be identified in a location of interest, and checked to see whether it is a likely cause of the observed problem. A system user can readily verify the specific valve size, manufacturer, part number, and any other information that resides in the BIM database.

Economic Benefits

Studies by Stanford University's Center for Integrated Facilities Engineering (CIFE) based on several major projects attribute several savings to the use of BIM: (CIFE 2007)

- Up to 40% elimination of unbudgeted change
- Cost estimation accuracy within 3%
- Up to 80% reduction in time taken to generate a cost estimate
- A savings of up to 10% of the contract value through clash detections
- Up to 7% reduction in project time
- An increase in field productivity in the range of 20–30%
- A tenfold or greater reduction in requests for information (RFIs) and change orders

Benefits of BIM at Each Construction Project Phase

Building information modeling supports the continuous and immediate availability of project design scope, schedule, and cost information that is high quality, reliable, integrated, and fully coordinated. Among the many competitive advantages it confers are:

- Faster delivery speed resulting in time savings.
- Better design—design proposals can be analyzed in detail, using simulation and other evaluation tools to identify the best solutions.
- Lower costs, which represents financial savings.

- Better coordination between different disciplines resulting in fewer errors.
- Higher work productivity and quality.
- Provides benefits in the design, construction, and management phases.

BIM Benefits in the Design Phase

During the design phase the scope, schedule, and cost of a project have to be carefully balanced (Figure 8.7). Changes in any of these factors affect the other two. Scope changes directly impact project costs, and are likely to affect the schedule of the project. In the traditional approach, designers readily manipulate information relating to spatial layouts and building geometry. On the other hand, this work is detached from cost and schedule considerations. Cost and schedule information is time-consuming and often difficult to

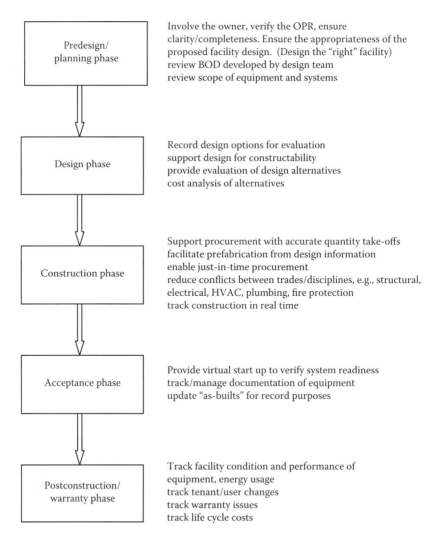

FIGURE 8.7
Process flow for BIM design and construction support.

assemble. It is also based on the design configuration available, and subsequent design changes may easily invalidate cost and schedule estimates.

BIM makes design information readily available to all stakeholders that are involved in the design process. The design team is empowered to make desired changes in real time without the need to do expensive and time-consuming coordination and checking by hand. The design concept is incorporated in the BIM model—as design work progresses its evolution is captured in it. Changes are automatically coordinated instead of requiring manual verification by the design team. The near elimination of errors enables the quality of design work and its documentation to be greatly improved.

The speed of delivering the design and accompanying permit or other documentation is greatly enhanced, with less individual effort. BIM provides the design team with a versatile tool for representing the design concept to the client—renderings can be generated automatically and manipulated electronically for improved client communication.

Three-Dimensional Modeling and Target Cost

This modeling supports the target value method of design. As design work progresses, cost estimating needs to keep pace with the development of the design. The traditional estimating method may prove to be too slow for projects that are fast-paced. Three-dimensional modeling can automatically generate bills of materials, and the associated pricing keeps designers informed of the overall project cost.

Three-Dimensional Enabled Project Delivery Features

- Clash detection
- Construction/installation drawings
- Procurement (bill of materials)
- Structural steel detailing for fabrication

There is no need for data transfer for installation. There is a direct exchange of A/E 3-D model information to steel fabrication. In one project, this eliminated the need for 12,000 drawings. Four-dimensional CAD involves the simulation of construction activities—it provides improved work sequencing, resolution of spatial conflicts between different trades. It also maximizes resource utilization visual monitoring.

BIM Application in the Construction Phase

Constructors can use BIM models to provide far more information than the hard copy construction documents of yesteryear. Take-offs for cost estimating can be readily obtained and work planning can be integrated with the constructor's schedule to see how well it meets the project master schedule. Architectural design details can be readily converted to construction shop drawings and prefabrication instructions for subassemblies such as HVAC ductwork, piping, and so on. Overall, this makes constructors more responsive to the needs of a design in a shorter time frame with less impact on administration costs. Fewer errors are committed in the interpretation of contract documents, and the interaction of various building systems is checked for clash detection long before installation or erection begins. Given the amount of waste that typically occurs in translating the design to the built product, BIM is highly cost-effective for constructors and owners/clients alike.

BIM during Occupancy

Details of building materials, equipment, systems and "as built" information are readily available through BIM to facilitate the management of a facility for optimum performance. As opposed to the hard copy "as built" drawings that are usually maintained for completed facilities, BIM maintains building information in digital form and is more cost-effective for maintaining the platform for keeping current records. In rented space, for example, tenant-related changes can be accurately recorded. The reconfiguration of interior spaces can readily be tracked—not just as repositioned wall panels, but as HVAC, electrical, communications, fire safety, and other layouts. Energy consumption and air quality can be closely monitored and compared with the specified design conditions. Information about tenant-related issues such as square footage, repainting, rental cost per square foot, lighting levels, furniture, and so on are easily tracked and used for financial decision making. Subsequent projects in a facility can be managed for optimum performance by using the accurate records on the facility as well as the lessons learned over a period of time.

Case 1: Using BIM with the Hilton Aquarium

Professor Salman Azhar and his research teams of Auburn University, Alabama describe the following to illustrate the cost and time savings available with BIM on an actual construction project. The teams collaborated with Holder Construction Company, Atlanta, Georgia. (Azhar, Hein, and Sketo 2008a and Azhar et al. 2008b).

The project details are as follows:

Project name: Hilton Aquarium, Atlanta, Georgia

Project scope: $46M, 484,000 SF hotel and parking structure

Delivery method: Construction manager at risk

Contract type: Guaranteed maximum price

Design assist: GC and subcontractors on board at design definition phase

BIM scope: Design coordination, clash detection, and work sequencing

File sharing: Navisworks used as common platform

BIM cost to project: $90,000, 0.2% of project budget ($40,000 paid by owner)

Cost benefit: $600,000 attributed to elimination of clashes

Schedule benefit: 1143 hours saved

The contractor created 3-D models of the architectural, structural, and MEP systems of the proposed building as shown in Figure 8.8. These models were created during the design development phase using detail level information from subcontractors based on drawings from the designers. This method allowed project team members to perform their work in the comfort of their traditional 2-D, drawing-based delivery process and eliminated the potential risk that is often associated with open sharing of digital models across stakeholders. Using the developed models, the project team achieved the following benefits:

(a) Architectural model   (b) Structural model   (c) Plumbing model

  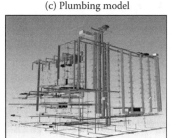

FIGURE 8.8
Building information modeling for Hilton Aquarium, Atlanta, GA. (Courtesy of Holder Construction, Atlanta, GA.)

- Proactively identified 590 conflicts between structural and MEP components and resolved them prior to field installations.
- Design coordination enhanced.
- Additional cost increases avoided.
- Enabled owner to view revisions without issuing change orders.
- Accommodated design changes.

As shown in Table 8.3, through frequent coordination sessions, the project team was able to quickly identify and resolve system conflicts, saving an estimated $800,000 in extras and avoiding months of potential delays. A net saving of $200,000 was calculated using the assumption that 75% of the identified collisions could be pointed out through conventional practices.

During the construction process, non-BIM-savvy stakeholders made use of Holder's visualization models through a free viewer (i.e., Navisworks). The collaborative 3-D viewing sessions also improved communications and trust between stakeholders and enabled rapid decision making early in the process. Finally, Holder's commitment to updating the model to reflect as-built conditions provided the owner, Legacy Pavilion, LLC, a digital 3-D model of the building and its various systems to help aid O&M procedures after occupancy.

Case Study 2: Savannah State Academic Building

This case study illustrates the use of BIM in the project planning phase to perform the options analysis (value analysis). The project details are as follows:

Project: Higher Education Facility, Savannah State University, Georgia

Cost: $12 million

Project delivery: CM-At-Risk/GMP

BIM scope: Planning, preconstruction services, value analysis

BIM cost to project: $5000

Cost benefit: $1,995,000

TABLE 8.3

An Illustration of Cost and Time Savings via BIM in Hilton Aquarium Project

Collision Phase	Collisions	Estimated Cost Avoided	Estimated Crew Hours		Coordination Date
100% Design Development Conflicts Construction (MEP Collisions)	55	$124,500	NIC		June 30, 2006
Basement	41	$21,211	50	hrs	March 28, 2007
Level 1	51	$34,714	79	hrs	April 3, 2007
Level 2	49	$23,250	57	hrs	April 3, 2007
Level 3	72	$40,187	86	hrs	April 12, 2007
Level 4	28	$35,276	68	hrs	May 14, 2007
Level 5	42	$43,351	88	hrs	May 29, 2007
Level 6	70	$57,735	112	hrs	June 19, 2007
Level 7	83	$78,898	162	hrs	April 12, 2007
Level 8	29	$37,397	74	hrs	May 14, 2007
Level 9	30	$37,397	74	hrs	July 3, 2007
Level 10	31	$33,546	67	hrs	July 5, 2007
Level 11	30	$45,144	75	hrs	July 5, 2007
Level 12	28	$36,589	72	hrs	July 5, 2007
Level 13	34	$38,557	77	hrs	July 13, 2007
Level 14	1	$484	1	hrs	July 13, 2007
Level 15	1	$484	1	hrs	July 13, 2007
Subtotal construction labor	590	$564,220	1143	hrs	
20% MEP material value		$112,844			
Subtotal cost avoidance		**$801,565**			
Deduct 75% assumed resolved via conventional methods		($601,173)			
Net adjusted direct cost avoidance		**$200,392**			

Source: Courtesy of Holder Construction, Atlanta, GA.

The contractor coordinated with the architect and the owner in the predesign phase to prepare building information models of three different design options. For each option, the cost estimates were also prepared using three different cost scenarios (normal, mid-range, and high range). The owner was able to walk through all the virtual models to select the option that best fit his requirements. Several collaborative 3-D viewing sessions were arranged for this purpose. These collaborative viewing sessions also improved communications and trust between stakeholders and enabled rapid decision making early in the process. The entire process took two weeks and the owner achieved a $1,995,000 cost savings at the predesign stage by selecting the most economical option.

BIM and Energy/Exergy Analysis in Design

Contemporary design trends place a major emphasis on the energy performance of buildings—facilities that meet the energy and environmental criteria of reference systems such as LEED and Energy Star in the United States. In Europe and Switzerland, building regulations such as The European Buildings Initiative and Minergie, respectively, also specify the requirements for high environmental performance. Buildings configured through these systems have been shown to have a reduced carbon footprint and provide a better working environment. The traditional approach of carrying out simulations is based on a completed design, and provides limited opportunities to make design changes later that improve energy performance. One of the reasons for this practice is that architectural design and energy-related design are performed separately. Architects are usually expert in building design but not in building performance simulation, hence this work is done by other experts.

Building information modeling can facilitate the integration of energy level calculations with the design process—the impact of design changes with these energy levels can be examined in real-time. BIM models store multidisciplinary information within one virtual building representation. This information includes geometric, semantic, and topological information—building shapes are 3-dimensional and are represented as geometric information. Semantic information defines material properties such as thermal resistance. Topological information relates to the dependencies of the components of the building.

Based on the thermoeconomic concept of energy, exergy analysis relates to the quality of energy and provides a holistic representation of a building. It requires an interaction between the factors that represent geometric, semantic, and topological information. The parameters that define this information during the design process can be used for performance calculations. Designers can gauge the impact of design changes by invoking the performance calculations almost in real-time. Adoption of BIM early in the design process facilitates interactivity in the design and performance measurement activities. In the United States, the government's procurement agency—the General Services Administration requires designers to utilize BIM for construction projects at the outset of the design process (Schlueter and Thesseling 2008).

BIM and Lean Construction

BIM supports many initiatives that are critical to lean design and construction. As the fundamental principle of lean is to reduce or eliminate waste, BIM addresses many aspects of waste that occur first in the design phases, and later in the construction phase. Lean design promotes the active participation of construction stakeholders as early as the project definition phase. As the design concept is developed, designers, owners, and constructors can work interactively to make decisions that influence the overall project—in real time, concurrently. Traditional design reviews are treated as sequential events, long after significant design decisions have been made. At that point, changes can quickly become time-consuming and expensive. Furthermore, as many disciplines are involved in the design of a project, changes in some elements of the design may not be fully represented by all the disciplines. In traditional projects that oversight often manifests itself as errors and omissions—classic examples of waste.

The "Big Room" or *obeya* concept is adopted in lean construction from the Toyota Production System, and brings together cross-functional teams under one roof to explore problems. Team members generate synergy by collaborating on not just the design, but on the construction process required to bring it to fruition (Macomber and Howell 2005). BIM provides a critical platform for Big Room meetings. "What-if" games can be played with various design approaches and the results can be evaluated immediately. Target value design is enhanced with BIM. Cost impacts of design are quickly determined in a concurrent manner, instead of relying on the traditional estimating approach.

Clash detection is easily accomplished with BIM—design errors often include having different building systems compete for the same limited space in ceilings and building penetrations from floor to floor. Air conditioning ducts and plumbing/fire protection piping typically compete for that space. When these clashes are detected during field installation, corrective action can have significant consequences. Bends in ducts and pipes that were not part of the original design increase their equivalent length, and may restrict the flow of air or water below the design levels defined by the mechanical engineer(s). That in turn leads to suboptimal building performance.

BIM has features that promote prefabrication—an important component of lean construction. As the BIM data are machine-readable, CNC machines can be readily provided with instructions for making sections of ductwork, piping, or other building elements under controlled conditions in fabrication shops. Such off-site work is more cost-effective and more accurate than on-site work, yet the accuracy provided with BIM literally allows such prefabricated components to be unloaded from a truck and mounted in place with fewer work hours than otherwise possible with traditional methods.

Waste due to lack of interoperability between different design and construction management software systems is improving with BIM. As reported by NIST and the Construction User's Roundtable, this lack of interoperability is a cause of project delays, cost overruns, and construction errors.

Factors to Consider with BIM

While the principle of BIM points to significant improvements in many facets of construction, there are still many opportunities for improvement. This may very likely occur as hardware capabilities improve and BIM software development progresses. Each project should have a written BIM plan to say who draws what and when. As BIM is both a tool and a process the owner needs to be educated on the impact of BIM. BIM systems create very large files, especially for large projects. The technology has not yet caught up with the need. The sharing of a single BIM between many users in real time will be a challenge until hardware capabilities increase. Existing BIM models represent architectural information readily. They do not yet have adequately detailed modeling capabilities for such systems as energy performance, fire safety, structural integrity. The specialized software used by consultants such as mechanical and structural design may not yet be compatible with available BIM software. Detailed analysis and "what if" games may not be feasible with a common BIM. Construction stakeholders have at their disposal a number of specialized or "purpose-driven" software products that enable them to optimize-specific aspects of building systems.

- NavisWorks is used for design coordination.
- Graphisoft's Virtual Construction solutions facilitates construction planning and scheduling.
- Primavera may be used for construction resource planning.
- Integrated Environmental Solutions, Inc. produces software that provides building performance modeling (BPM). It requires users to have an understanding of BPM in order to use it effectively—many users may not have that expertise.

Many contractors develop their own software models that incorporate their historical experience with various project categories.

BIM and Sustainable Design

BIM tools facilitate sustainable design. They provide for:

- Structural design analysis
- Energy analysis
- Day lighting and lighting analysis
- Model viewing tools for design review
- Material database
- Database tools for construction management
- Database tools for facility management
- Monitoring and controls

BIM should therefore be viewed as one aspect of purpose-driven software. Construction stakeholders should understand this current limitation. As BIM is primarily a geometric model with parametric variables, it is capable of operating in tandem with many other purpose-driven software systems. Every effort should be made to ensure that interoperability is established in order to have the best possible results from the BIM technology. The IFC-based model exchange provides an ability to transfer objects, their relationships, and their associated property sets.

Legal Issues Relating to BIM

Rights of Ownership

The first legal risk to determine is ownership of the BIM data and how to protect it through copyright and other laws. For example, if the owner is paying for the design, then the owner may feel entitled to own the data, but if team members are providing proprietary information for use on the project, their propriety information needs to be protected as well. Thus, there is no simple answer to the question of data ownership; it requires a unique response to every project depending on the participants' needs. The goal is to avoid inhibitions or disincentives that discourage participants from fully realizing the model's potential.

Control of Data Entry/Usage

Responsibility for updating BIM data and ensuring its accuracy entails a great deal of risk. As designers are subject to liability claims, such issues need to be clarified before BIM technology is utilized. It also requires more time spent imputing and reviewing BIM data, which is a new cost in the design and project administration process. Although these new costs may be more than offset by efficiency and schedule gains, they are still a cost that someone on the project team will have to bear. The cost of BIM implementation must be paid for as well.

Responsibility for Errors

The integrated concept of BIM blurs the level of responsibility so much that risk and liability will likely be enhanced. Consider the scenario where the owner of the building files suit over a perceived design error. The architect, engineers, and other contributors of the BIM process look to each other in an effort to try to determine who had responsibility for the matter raised. If disagreement ensues, the lead professional will not only be responsible as a matter of law to the claimant but may have difficulty proving fault with others such as the engineers (Rosenburg 2007).

As the dimensions of cost and schedule are layered onto the 3-D model, responsibility for the proper technological interface among various programs becomes an issue. Many sophisticated contracting teams require subcontractors to submit detailed CPM schedules and cost breakdowns itemized by line items of work prior to the start of the project. The general contractor then compiles that data, creating a master schedule and cost breakdown for the entire project. When the subcontractors and prime contractor use the same software, the integration can be fluid. In cases where the data are incomplete or is submitted in a variety of scheduling and costing programs, a team member—usually a general contractor or construction manager—must re-enter and update a master scheduling and costing program. That program may be a BIM module or another program that will be integrated with the 3-D model. At present, most of these project management tools and the 3-D models have been developed in isolation. Responsibility for the accuracy and coordination of cost and scheduling data must be contractually addressed.

Summary

Information and communication (ICT) technology holds tremendous promise and potential to bring greater integration in the construction industry. Significant benefits can be realized by developing appropriate ICT solutions. It is expected that with greater availability of cost-effective ICT tools, the traditional concept of project management will be replaced by revolutionary ideas. This should result in significant savings in time, cost, and resources. Fewer disputes and better coordination are expected as a consequence of increasing use of ICT solutions. Building information modeling (BIM) is the process of generating and managing building data during its life cycle. It is also a tool as well as a process, and increases productivity and accuracy in the design and construction of buildings. BIM uses 3-D, dynamic building modeling software and operates in real time. It supports the continuous and immediate availability of project design scope, schedule, and cost information that is high quality, reliable, integrated, and fully coordinated.

The building information model encompasses building geometry, spatial relationships, geographic information, and quantities and properties of building components. Though it is not itself a technology, it is supported to varying degrees by different technologies. The BIM is a data-rich, object-oriented, intelligent and parametric digital representation of the facility, from which views and data appropriate to various users' needs can be extracted and analyzed. It generates information that can be used to make decisions and to improve the process of delivering the facility. BIM supports many initiatives that are critical to lean design and construction. As the fundamental principle of lean is to reduce or eliminate waste, BIM addresses many aspects of waste that occur first in the design phases and later in the construction phase.

Questions for Discussion

1. How has ICT improved construction since the 1990s?
2. What is Building Information Modeling (BIM) and how does it differ from 3-D modeling?
3. How does BIM differ from 2-D CAD?
4. What are the benefits of BIM?
5. Name three or four practical applications of BIM.
6. What precautions have to be taken when implementing BIM?

References

Associated General Contractors of America. 2005. *The Contractors Guide to BIM*, 1st ed. AGC Research Foundation, Las Vegas, NV.

Azhar, S., A. Nadeem, J. Y. N. Mok, and B. H. Y. Leung. 2008b. Building information modeling (BIM): A new paradigm for visual interactive modeling and simulation for construction projects. *Proceedings of the 1st international conference on construction in developing countries (ICCIDC-I).* August 4–5. Karachi, Pakistan.

Azhar, S., M. Hein, and B. Sketo. 2008a. Building information modeling: Benefits, risks and challenges. *Proceedings of the 44th ASC National Conference.* Auburn: Alabama.

CIFE. November 22, 2007. *CIFE Technical Reports.* Available at http://cife.stanford.edu/Publications/index.html

Eastman, C., P. Teicholz, R. Sacks, K. Liston. 2008. *BIM handbook: A guide to building information modeling for owners, designers, engineers, and contractors.* New York: Wiley.

Hannus, M., and A. S. Kazi. 2000. Construction IT best practices: Lessons learned from the Finnish construction industry. *Proceedings of the 1st conference on implementing IT to obtain a competitive advantage in the 21st century,* eds. H. Li, Q. Shen, D. Scott, and P. E. D. Love, 144–54. Hong Kong: Hong Kong Polytechnic University Press.

Macomber, H. and G. Howell. 2005. Using Study Action Teams to Propel Lean Implementations. Lean Project Consulting inc., Louisville, CO 80027.

Schlueter, A., and F. Thesseling. 2008. Building information model based energy/exergy performance in early design stages. *Automation in Construction* 18 (2): 153–63.

Rosenburg, T. L. 2007. *Building Information Modeling.* Available at http://www.ralaw.com/resources/documents/Building%20Information%20Modeling%20-%20Rosenburg.pdf

Turk, M. 1997. Perceptual user interfaces. *Proceedings of the workshop on perceptual user interfaces.* October, 1997. Banff, AB.

Turk, M., and Y. Takebayashi, eds. 1997. Proceedings of the workshop on perceptual user interfaces. October, 1997. Banff, AB.

Bibliography

Ahmad, I., J. S. Russell, and A. Abou-Zeid. 1995. Information technology (IT) and integration in the construction industry. *Construction Management and Economics* 13:163–71.

Ahmad, I., and S. M. Ahmed. 2001. Integration in the construction industry: Information technology as the driving force. *Proceedings of the 3rd International Conference on Construction*

Project Management, ed. R. L. K. Tiong, 429–34. Singapore: Nanyang Technical University Press.

Akinsola, A., N. Dawood, and B. Hobbs. 2000. Development of an automated communication system using internet technology for managing construction projects. *Proceedings of the 1st conference on implementing IT to obtain a competitive advantage in the 21st century*, eds. H. Li, Q. Shen, D. Scott, and P. E. D. Love, 835–54. Hong Kong: Hong Kong Polytechnic University Press.

Anumba, C. J. 2000. Integrated systems for construction: Challenges for the millennium. *Proceedings of the 1st conference on implementing IT to obtain a competitive advantage in the 21st century*, eds. H. Li, Q. Shen, D. Scott, and P. E. D. Love, 78–92. Hong Kong: Hong Kong Polytechnic University Press.

Anumba, C. J., and K. Ruikar. 2002. Electronic commerce in construction: Trends and prospects. *Automation in Construction* 11: 265–75.

CRC Construction Innovation. 2007. *Adopting BIM for facilities management: Solutions for managing the Sydney opera house*. Brisbane, Australia: Cooperative Research Center for Construction Innovation.

Gero, J. S. 2000. Developments in computer-aided design. *Proceedings of the 1st conference on implementing IT to obtain a competitive advantage in the 21st century*, eds. H. Li, Q. Shen, D. Scott, and P. E. D. Love, 16–24. Hong Kong: Hong Kong Polytechnic University Press.

Latham, M. 1994. *Constructing the team*. London: Her Majesty Stationary Office (HMSO).

Liu, L., and C. Erickson. 2002. Engineering and construction collaboration using information technology. *Proceedings of the 1st international conference on construction in the 21st century—Challenges and opportunities in management and technology*, 521–28. April 25–26. Miami, FL.

Sawyer, T. 2001. States turn onto web for highway bidding. *Engineering News Record* 246(8): 53–54.

Skibniewski, M. J., and M. Abduh. 2000. Web-based project management for construction: Search for utility assessment tools. *Proceedings of the 1st conference on implementing IT to obtain a competitive advantage in the 21st century*, eds. H. Li, Q. Shen, D. Scott, and P. E. D. Love, 56–77. Hong Kong: Hong Kong Polytechnic University Press.

Thompson, D. B., and R. G. Miner. 2007. *Building information modeling—BIM: Contractual risks are changing with technology*. http://www.aepronet.org/ge/no35.html

9

Quality Management in Construction: A Complement to Lean Construction

This chapter addresses several topics that relate to the improvement of quality in construction, learning from projects and using the lessons learned to improve. The assurance of quality throughout the life of a project by means of the commissioning function is also included here. The topics discussed complement the adoption and deployment of lean construction. Built-in quality is a component of lean construction. While lean practices enable project teams to work cost effectively and with efficient hand offs from one discipline to another, quality assurance and commissioning are initiatives that ensure that completed facilities perform as expected and without deficiencies. The adoption of total quality management or other quality management systems prepares an organization for subsequent adoption of lean construction. Many of the tools and techniques that are applied to Total Quality Management and Six Sigma are used by practitioners of lean construction.

Part A: Total Quality Management

Overview

The term *quality* can mean different things to different people, but needs a common definition if the people in an organization can hope to pursue it successfully. This is especially important in the construction industry where there is much variation from one project to another. Dr Joseph Juran's definition has been interpreted as "fitness for use," "fitness for intended use," "conformance to requirements," and "conformance to specifications." Dr. Edwards Deming took the view that: "Quality is defined by the customer; the customer wants products and services that, throughout their lives, meet customers' needs and expectations at a cost that represents value." Deming also added that the quality of a company's output cannot be better than the quality determined at the top (i.e., by an organization's leaders).

Total quality management (TQM) is an approach to doing business that attempts to maximize the competitiveness of an organization through the continual improvement of the quality of its products, services, people, processes, and environments (Goetsch and Davis 2006). The principles of TQM create the foundation for developing an organization's system for planning, controlling, and improving quality.

Dr. Armand Feigenbaum played a major role in the origination of the total quality movement; his landmark text of 1951 was titled *Total Quality Control* and has influenced the quality movement. He defines quality systems as a method of managing organizations to achieve higher customer satisfaction, lower overall costs, higher profits, and greater employee effectiveness and satisfaction. Although organizations have adopted a wide variety of quality improvement programs, these programs are based on the concepts advocated by the total quality pioneers. In addition to Dr. Feigenbaum, the most highly acknowledged pioneers are W. Edwards Deming, Joseph M. Juran, and Philip B. Crosby. A number of Japanese experts including Kaoru Ishikawa and Shigeo Shingo were also major contributors to the quality improvement philosophy. Dr. Deming has emerged as the most influential and durable proponent of quality management in the United States and is best known for the Deming Cycle, his Fourteen Points, and the Seven Deadly Diseases.

TQM is based on involving everyone in an organization in an integrated effort toward improved performance at each level (Goetsch and Davis 2006). It integrates fundamental management techniques and improvement efforts in a disciplined approach toward continual process improvement. TQM emphasizes the understanding of variation, the importance of measurement, the role of the customer, and the commitment and involvement of employees at all levels of an organization. Total quality is driven by an organizational strategy and unity of purpose, an internal and external customer focus, obsession with quality, scientifically based decision making and problem solving, continuous process improvement, long-term commitment, teamwork, employee involvement and empowerment, and education and training.

Quality Management Systems

The American Society For Quality, (ASQ), defines a quality management system (QMS) as a mechanism for managing and continuously improving core processes to "achieve maximum customer satisfaction at the lowest possible cost to the organization." It applies and synthesizes standards, methods, and tools to attain strategically important goals. In the construction project environment, quality management may be defined as the process required to ensure that a project's outcome will satisfy the needs for which it was undertaken. TQM also promotes excellence in customer satisfaction through the total involvement and dedication of each individual who is in any way a part of that product/process.

History of Quality in Construction

Quality has always been with us since recorded history—the code of Hammurabi ensured that a builder would have to provide his clients with good quality. If a house collapsed and hurt its occupants because of inferior quality, the builder was required to be put to death. Fortunately, the twentieth century brought a kinder, gentler approach to quality.

Historically, the Japanese were among the first to apply quality improvement approaches in construction on a large scale, although they did not embrace this concept until the oil crisis of 1973. Previously, they thought that the construction industry was inappropriate for the application of total quality control (TQC), because of the inherent variability in projects, and the difficulty in defining "acceptable quality." Takenaka Komuten Company, the sixth largest in Japan, had their formerly impeccable safety and quality image tarnished by the failure of a sheet piling system in Okinawa, in 1975, and embarked on a quality control (QC) program. They were followed by Shimizu Construction Company, the second largest in Japan, which established a QC program in 1976, and by Kajima Corporation, the third largest, in 1978. Subsequently, several U.S. companies adopted TQC programs and the more familiar TQM programs used by U.S. manufacturers.

In 1992, the Construction Industry Institute (CII) published *Guidelines for Implementing Total Quality Management in the Engineering and Construction Industry*. Their research studies confirmed that TQM resulted in improved customer satisfaction, reduced cycle times, documented cost savings, and more satisfied and productive work forces (Burati and Oswald 1992).

While total quality approaches have been highly beneficial to the manufacturing and service industries, they have had limited application in the construction environment. The construction industry has been heavily steeped in the traditional ways of executing projects, and its constituents—designers and constructors—have been reluctant to make a necessary cultural and behavioral change to adopt total quality approaches. One reason for this unwillingness has been the perception that TQM is for manufacturing only (Chase and Federle 1992). The other major factor inhibiting the implementation of TQM in the construction industry has been the notion that TQM is costly and requires a long-time period for implementation. One aspect of TQM that has frustrated the construction industry the most has been "measurement" (Hayden 1996).

Many construction companies in the United States, Singapore, UK, and other European countries have been using TQM successfully for a number of years. They have been reaping rich rewards in improved client, consultant, and supplier relations. They have also experienced a reduced "cost of quality," on time and within budget project completions, and well-informed and highly motivated employees. On the other hand, many U.S. companies have tried to adopt TQM programs and subsequently abandoned them. Studies indicate the failures can be attributed to a misunderstanding of TQM or to faulty implementation. In many cases top management and senior management are generally preoccupied with short-term, project by project profitability and not with long-term, quality-based strategies.

Attainment of acceptable levels of quality in the construction industry has long been a problem. Great expenditures of time, money, and resources—both human and material—are wasted each year because of inefficient or non-existent quality management procedures. The QC procedures that work effectively in a mass production industry have not been considered suitable for the construction industry. Typical explanations point to the uniqueness of construction projects and the lack of a clear and uniform standard for evaluating overall construction quality. Thus, there is great potential and challenge for quality improvement in the construction industry.

Instead, the construction industry has become increasingly reliant on burdensome specifications, which seldom say exactly what the owner intends them to say. This has led the owners to shift more of the project risks to the contractors (Ahmed and Aoieong 1998). The net outcome is that the construction industry has been bogged down with paperwork and defensive posturing; project participants generally tend to have a hostile attitude toward each other. TQM can help to reverse this trend.

Traditionally, inspection has been one of the key components of a quality assurance/quality control system in the construction industry. Regarding inspection, Deming's view is that "routine 100% inspection is the same thing as planning for defects—acknowledgement that the process cannot make the product correctly, or that the specifications made no sense in the first place. Quality comes not from inspection, but from improvement of the process" (1982). This does not mean that inspection ceases. Instead, it means that more effort is put into preventing errors and deficiencies.

Benefits of TQM

From the viewpoint of the individual company, the strategic implications of TQM include:

- Survival in an increasingly competitive world
- Better service to its customers

- Enhancement of the organization's "shareholder value"
- Improvement of the overall quality and safety of our facilities
- Reduced project durations and costs
- Better utilization of the talents of its people

Quality management is a critical component in the successful management of construction projects; critical success factors include the support of senior management, appropriate leadership styles, cultivation of employee's enthusiasm and participation, and open communication and feedback. The most appropriate strategy for effective QMS in the construction industry is to establish effective management practices.

Quality Costs

In order to quantify the benefits of TQM, quality must be measurable. Although there are numerous tools for measuring quality, the "cost of quality," or "quality costs," as advocated by Crosby (1984) and Juran (1988) represent an important indicator. Oberlender (1993) summarized quality costs as follows:

> Quality costs consist of the cost of prevention, the cost of appraisal, and the cost of failure. Prevention costs are those resulting from quality activities used to avoid deviations or errors, while appraisal costs consist of costs incurred from quality activities used to determine whether a product, process or service conforms to established requirements. Failure costs are those resulting from not meeting the requirements and can be divided into two aspects. Internal failure costs are the costs incurred on the project site due to scrap, rework, failure analysis, re-inspection, supplier error, or price reduction due to nonconformance. External failure costs are costs that are incurred once the project is in the hands of the client. These include costs for adjustments of complaints, repairs, handling and replacement of rejected material, workmanship, correction of errors, and litigation costs.

Principles of TQM

As reflected in Figure 9.1, five core principles are embodied in TQM (Harris 1995). These principles can be achieved in an organization with the aid of seven basic supporting elements. These elements are embodied in the criteria of the Malcolm Baldrige National

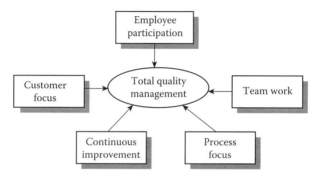

FIGURE 9.1
Principles of total quality management.

Quality Award. Since its inception in 1987, the MBNQA has promoted national competitiveness through demonstrated quality-based achievements. Several outstanding companies have won the award such as: Federal Express, Texas Instruments Inc., Ritz-Carlton Hotel Co., Xerox Corporation, and Merrill Lynch Credit Corporation. It is worthy of note that no construction organizations have yet won the Baldrige Award, although many are undoubtedly adopting quality management techniques as first steps toward that coveted goal.

The Baldrige Criteria are built on seven categories of beliefs and behaviors that are typical of high-performance organizations:

1. Leadership: Senior management must lead this effort by example, by applying the tools and language, by requiring the use of data, and by recognizing those who successfully apply the concepts of TQM.

2. Strategic planning: Senior managers may require support to bring about the change necessary to implement a quality strategy.

3. Customer and market focus

4. Information and analysis: The use of data becomes paramount in installing a quality process. To set the stage for the use of data, external; customer satisfaction must be measured to determine the extent to which customers perceive that their needs are being met.

5. Human resource focus: Communications in a quality environment may need to be addressed differently in order to communicate to all employees a sincere commitment to change. Reward and recognition to teams and individuals who successfully apply the quality process so that the rest of the organization will know what is expected.

6. Process management

7. Organizational results

Organizations are scored out of 1000 points on these categories. Customer and market focus and customer-focused results account for 210 points and are therefore crucial to world-class performance. These categories address how successful organizations listen and learn with customers' key requirements and describe the methods used to satisfy customers, increase repeat business, and generate positive referrals. They evaluate complaint management, its analysis for improvement, determination of customer satisfaction/dissatisfaction, and benchmarking satisfaction with competitors.

Characteristics of the Construction Industry

Construction work is carried out in the form of a project. Projects are becoming progressively larger and more complex in terms of physical size and cost. Currently, the execution of a project requires the management of scarce resources; manpower, material, money, and machines to be managed throughout the life of the project—from conception to completion. The projects have six distinctive objectives to be managed: scope, organization, quality, cost, time, and safety (Figure 9.2). Construction work requires different trades and knowledge but the management, scheduling, and control of those projects utilize the same tools and techniques, and are subject to constraints of time, cost, and quality. There are also unique characteristics of projects, which differ from routine operations.

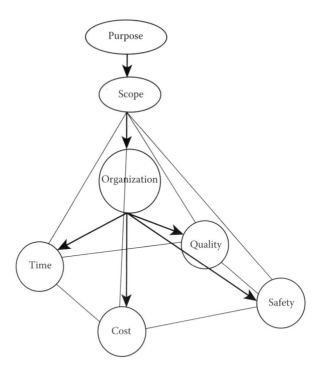

FIGURE 9.2
Six objectives of a construction project.

TQM Principles

There are many approaches available for improving quality: TQMgt, TQS, ISO 9001-2000, QS 9000, Six Sigma, Malcolm Baldrige Criteria, and in lean production the European Quality Award. These approaches are based to varying degrees on the principles established by many of the Quality Gurus mentioned above. Deming in particular, has proven to be the most enduring because of his view of quality as a societal issue. Deming is known for the contribution of 14 Points and Seven Deadly Diseases.

Deming's 14 Points

The 14 Points are summarized in the context of construction as follows:

1. Develop a program of constancy of purpose of improvement of product and service. The organization's strategic plan should include a commitment to improving products and services. Persistence is necessary for quality initiatives to bear fruit in design or construction activities.

2. Adopt this new program and philosophy. Quality needs to be ingrained in the organization's strategic plan—both its mission and vision.

3. Stop depending on inspection to achieve quality—build in quality from the start. Employees need to adopt a quality philosophy to produce to high standards, without relying on inspections.

4. Stop awarding contracts on the basis of low bids. Quality costs may be high with suppliers that have low-initial costs. Low bid awards may lead to failed inspections,

low quality, and subpar performance. Suppliers should demonstrate capable quality systems.

5. Improve continuously and forever the system of production and service. Design and construction organizations should commit to continuous improvement.

6. Institute training on the job. Employees at all levels need to be trained in the best methods for delivering construction-related services. This includes executives, managers, technicians, supervisors/foremen, and workers.

7. Institute leadership. Top management has to lead with quality endeavors and set an example for their employees. Unless top management displays commitment, others will not take quality seriously.

8. Drive out fear so everyone may work efficiently. Construction workers are unwilling to volunteer ideas for improvement in an atmosphere of intimidation. Without their input, inefficient processes are likely to persist.

9. Eliminate barriers between departments so that people can work as a team. Reward teamwork instead of individual accomplishments.

10. Eliminate slogans, targets, and targets for the workforce—they create adversarial relationships.

11. Eliminate quotas and management by objectives. Institute work standards that reflect quality, cost, schedule, and safety.

12. Remove barriers that rob people of pride of workmanship. Provide workers with needed supervision, information, and resources to enable them to do the best job possible.

13. Establish rigorous programs of education and self-improvement. Commit the training and support systems necessary to equip everyone to develop and contribute to a quality environment.

14. Make the transformation everyone's job.

Deming's Seven Deadly Diseases

1. Lack of constancy of purpose
2. Emphasis on short-term profits
3. Evaluation by performance, merit rating, or annual review of performance
4. The mobility of management
5. Running a company on visible figures alone
6. Excessive medical costs
7. Excessive costs of warranties, fueled by lawyers that work on a contingency fee

These "diseases" are considered to be the antithesis of quality.

Crosby's Zero Defects

This process was introduced by Philip H. Crosby in 1979. The Crosby process can help an organization by providing a quality management culture. Lean and Six Sigma can be effective "tools," but a tool becomes more beneficial when it is put to work regularly. To do that a quality "culture" is required to encourage (or insist) that everyone participate in this important process.

Ishikawa

The cause and effect diagram is the brainchild of Kaoru Ishikawa, who pioneered quality management processes in the Kawasaki shipyards, and in the process, became one of the founding fathers of modern management. The cause and effect diagram is used to explore all the potential or real causes (or inputs) that result in a single effect (or output). Causes are arranged according to their level of importance or detail, resulting in a depiction of relationships and hierarchy of events. This can help to search for root causes, identify areas where there may be problems, and compare the relative importance of different causes.

J. Juran

Dr. Joseph Juran is known for several quality contributions: Three Basic Steps to Progress, Ten Steps to Quality Improvement, and The Quality Trilogy. Juran's quality trilogy has three components: namely; (1) quality planning, (2) QC, and (3) quality improvement. This trilogy of quality processes provides a successful framework for achieving quality objectives. The processes must occur in an environment of inspirational leadership and the practices must be strongly supportive of quality. A brief description of the Juran's quality trilogy is given below.

1. Quality planning
    a. Determine who the customers are
    b. Identify customers' needs
    c. Develop products with features that respond to customer needs
    d. Develop systems and processes that allow the organization to produce these features
    e. Deploy the plans to operational levels
2. Quality control
    a. Assess actual quality performance
    b. Compare performance with goals
    c. Act on differences between performance and goals
3. Quality improvement
    a. Develop the infrastructure necessary to make annual quality improvements.
    b. Identify specific areas in need of improvement and implement improvement projects.
    c. Establish a project team with responsibility for completing each improvement project.
    d. Provide teams with what they need to be able to diagnose problems to determine root causes, develop situations, and establish control that will maintain gains made.

Customer Focus

In the TQM philosophy, total customer satisfaction is the goal of an entire system, and a pervasive customer focus is what gets us there. The function of the construction industry is to provide customers with facilities that meet their needs. For a company to remain in business this service must be provided at a competitive cost. TQM is a management philosophy that effectively determines the needs of the customer and provides the framework, environment, and culture for meeting those needs at the lowest possible cost. By ensuring quality at each stage in the construction process and thereby minimizing costly rework—as well as other costs—the quality of the final products should satisfy the final customer.

By definition, customers may be either internal or external. The external customer is the consumer or client; that is, the end user of the products or services being offered. An internal customer is a second process or department within the organization, which depends on the product of the first. For example, the products are plans and specifications for designers, the customers are the owner, and the contractor is responsible for the construction. For the contractor, the product is the completed facility, and the customer is the final user of the facility. There are also customers within the construction organization. These internal customers receive products and information from other groups of individuals within their organization. Thus, satisfying the needs of these internal customers is an essential part of the process of supplying the final external customer with a quality product.

Every party in a process has three roles: supplier, processor, and customer. Juran defined this as the triple role concept. These three roles are carried out at every level of the construction process. The designer is a customer of the owner. The designer produces the design and supplies plans and specifications to the contractor. Thus, the contractor is the designer's customer, who uses the designer's plan and specifications to carry out the construction process and supplies the completed facility to the owner. The owner supplies the requirements to the designer, receives the facility from the contractor, and is responsible for the facilities operation (Burati and Oswald 1993). This clearly illustrates that construction is a process, and that TQM principles that have been applied to other processes are potentially adaptable to the construction industry (Figure 9.3).

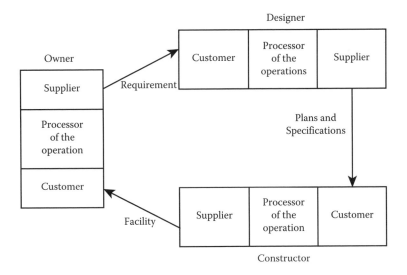

FIGURE 9.3
Juran's triple role concept applied to construction.

Process Improvement

A process is a way of getting things done. A process consists of the tasks, procedures, and policies necessary to carry out an internal or external customer need. According to the TQM philosophy, if the process is correct so will the end result (product). Thus the organization should work to improve the process so as to improve the end product or service.

Three different approaches have emerged for improving the efficiency or effectiveness of a process. Continuous improvement is an approach used on an on-going basis for incremental gains. Benchmarking should be used periodically, and reengineering can be launched occasionally to achieve dramatic breakthrough.

By focusing on a process by measurement and analysis, it can be improved by influencing five Ms of the process; namely, man (people), machine, material, method, and measurement. A sixth factor is that of the operating environment. A strong emphasis on process improvement centers on the measurement of variation, the control of variation, and the knowledge of variation to seek improvement. This analysis is referred to as statistical process control or statistical analysis. It is at the center of process improvement. The objective of measuring the variation in a process is to learn how to control the variation and also how to improve the process by viewing variation as a tool for improvement. The analysis of the positive side (good performance or quality) of the variation of process is referred to as a "breakthrough improvement" or "breakthrough management," which is another key component of TQM.

Continuous Improvement

The goal of continuous improvement is common to many managerial theories; however, what differentiates TQM is that it specifies a specific step-by-step process to achieve it. The steps are as follows:

1. Identify the process
2. Organize a multidisciplinary team to study the process and recommend improvements
3. Define areas where data are needed
4. Collect data on the process
5. Analyze the collected data and brainstorm for improvement
6. Determine recommendations and methods of implementation
7. Implement the recommendations outlined in step 6
8. Collect new data on the process after the proposed changes have been implemented to verify their effectiveness
9. Circle back to step five and again analyze the data and brainstorm for further improvement

The nine-step cycle emphasizes: focusing the progress, measuring the process, brainstorming for improvement, verification, and remeasurement. These four elements are further illustrated in Deming's Plan-Do-Check-Action (PDCA) diagram shown in Figure 9.4. The PDCA diagram stresses removing the root cause of problems and continually establishing and revising new standards or goals (Deming 1986).

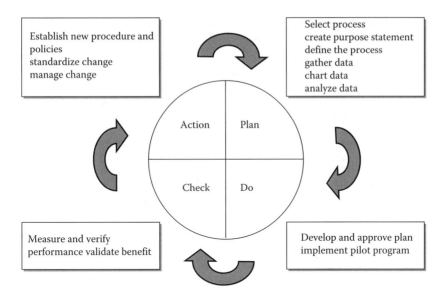

FIGURE 9.4
The PDCA diagram.

Under TQM, management in the construction industry has two functions: (1) to maintain and improve current methods and procedures through process control and (2) to direct efforts to achieve, through innovation, major technological advances in construction processes. The incremental improvement of the process is achieved through process improvement and control. In every construction organization there are major processes by which all the work is accomplished. However, there are innumerable parts in the construction process. Through the use of flow diagrams, every process can be broken down into stages. Within each stage, input changes to output, and the methods and procedures directing the change of state (i.e., the construction procedures) can be constantly improved to better satisfy the customer at the next stage. During each stage the employees should communicate closely with their supplier and customer to optimize the work process for that stage. This requires each employee to recognize their place in the process and their respective supplier and customer.

Quality Improvement Techniques

Total quality management mainly demands a process of continued improvement aimed at reducing variability. An organization wishing to support and develop such a process needs to use quality management tools and techniques. It is prudent to start with the more simple tools and techniques: The check sheet, the checklist, the histogram, the Pareto diagram, the cause and effect diagram (Fishbone diagram), the scatter diagram, and the flowchart. They are described briefly below. Readers may refer to many available texts on the subject of TQM to obtain more detailed information on the use of these tools.

Check Sheet

Used to record events, or nonevents (nonconformances). They can also include information such as the position where the event occurred and any known causes. They are usually prepared in advance and are completed by those who are carrying out the operations or monitoring their progress. The value of a check sheet can be retrospective analysis, so they help with problem identification and problem solving.

Checklists

A list of things that need to be done and when completed are checked off. It can be used in the auditing of quality assurance and to follow the steps in a particular process.

Histogram

Provides a graphical representation of the individual measured values in a data set according to the frequency of occurrence. It helps users to visualize the distribution of data and can display the amount of variation within a process.

Pareto Analysis

A technique employed to prioritize the problems so that attention is initially focused on those having the greatest effect. It was discovered by an Italian economist, named Vilfredo Pareto, who observed how the vast majority of wealth (80%) was owned by relatively few of the population (20%). As a generalized rule for considering solutions to problems, Pareto analysis aims to identify the critical 20% of causes and to solve them as a priority. (See the appendix at the end of this chapter for an example.)

Cause-and-Effect Diagram (Fishbone Diagram)

The diagram, which was developed by Karoa Ishikawa, is useful in breaking down the major causes of a particular problem. The shape of the diagram looks like the skeleton of a fish. This is because a process often has a multitude of tasks running into it, any one of which may be a cause. If a problem occurs, it will have an effect on the process, so it will be necessary to consider the whole multitude of tasks when searching for a solution. **(See the appendix at the end of this chapter for an example.)**

Scatter Diagram

The relationship of two variables can be plotted in the scatter diagrams. They are easy to complete, and an obvious linear pattern reveals a strong correlation between the variables.

Flowcharts

A chart providing a diagrammatic picture using a set of symbols. They are used to show all the steps or stages in a process project or sequence of events. A flowchart assists in documenting and describing a process so that it can be examined and improved. Analyzing the data collected on a flowchart can help to uncover irregularities and potential problem points.

Customer Focus and Quality Gaps

Brown (1992, 1995) defines gap analysis as quantifying customer perceptions of a product or service, and comparing them with what management believes to be the customer's view of the product or services. Gap analysis facilitates an evaluation of the internal barriers in meeting consumer expectations, determining whether the current standards inhibit employees' ability to give the expected service; whether the organization is communicating honestly to its customers; and if the company's vision is at odds with its capability to deliver that vision. Several gaps exist in the relations between the parties to construction projects.

1. Service quality: The gap between customer expectations of service quality and customer perceptions of the organization's performance.
2. Understanding: The gap between customer views of service quality and the organization's view of its service quality.
3. Design: The gap between the organization's perception of customer views and of service quality and the design of the organization's service delivery system.
4. Delivery: The gap between how the organization's service delivery systems should operate and how they actually operate.
5. Communications: The gap between the service delivered and the level of service being promised to the customer.

Gap analysis involves posing three questions: What do you need from me? What do you do with what I give you? What are the gaps between what I give you and what you need? The answers to these questions help to pinpoint service problems, and to implement a strategy to close the gaps, resulting in improved service quality. The gap analysis procedure quantifies the differences between importance and performance, through a survey of both internal and external customers. The largest gaps show the elements of service (or product) quality that merit the strongest corrective action.

Construction, on the other hand combines elements of both the manufacturing and service environments—the manufacturing process is protracted and involves many ongoing interactions between the parties. While a completed facility is analogous to a manufactured item, the interactions involving status information, problem resolution, dispute resolution, and so on embody transactions that emulate the service environment.

Quality gaps between the parties to construction are represented by dotted lines that intersect with the lines of communication in Figure 9.5. Quality gaps exist between owners and designers, between designers and contractors, between owners and contractors. Similarly, gaps exist between designers and code compliance agencies and between code compliance agencies and contractors.

Forbes (1999) quantified the gaps between owners, designers, and contractors for health care facility projects. Surveys were conducted to determine the rankings of the respective parties for issues such as owner satisfaction and performance evaluation. The survey was stratified for both public and private sector facility owners. Comparisons were made using non-parametric analysis of variance calculations. In the case of owner satisfaction, for example, questions were asked for a ranking of the following factors:

- Timeliness
- Adherence to cost estimates
- Clear understanding of job scope
- On-going communication
- Prompt, adequate response to customer complaints
- Attractive aesthetics

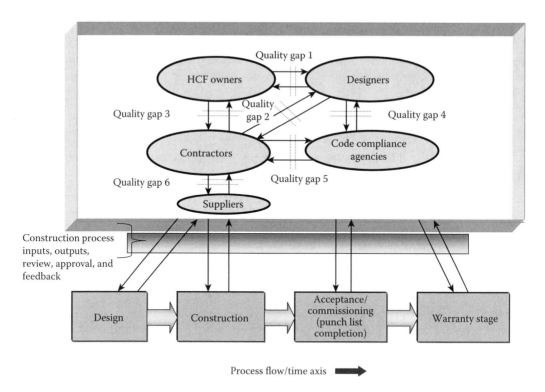

FIGURE 9.5
Health care facility construction process map with quality gaps. (Adapted from Forbes, L., *An engineering management-based investigation of owner satisfaction, quality and performance variables in health care facilities construction.* Ph.D. Dissertation. Coral Gables, FL: University of Miami, 1999.)

- High quality—fit and finish
- Minimum disruption to on-going facility operations

The observations were as follows:

- Public owners and designers disagreed on the ranking of seven of nine criteria.
- Owners and contractors disagreed on six of nine criteria.
- Designers and contractors disagreed on three of nine criteria.

The extent of the quality gaps suggests that many differences persist between the parties to construction projects. Owners generally expect designers—such as architects—to synthesize their needs into owners' project requirements (OPR) and to represent their interests. That they differed in the study by seven of nine criteria supports the general observation that customer satisfaction has much room for improvement in construction projects. Furthermore, designers and contractors seemed to have fewer quality gaps between them than owners and designers as well as owners and contractors.

Barriers to the Implementation of TQM

Employees generally show resistance to the introduction of TQM for a host of reasons, which include fear of the unknown, perceived loss of control, the "it may cost more" syndrome, and an unwillingness to take "ownership" and be committed to change. Other barriers include:

- Perceived threats to foreman and project manager roles
- Disinterest at the site level
- Lack of understanding of what TQM was, particularly on-site
- Geographically dispersed sites
- Fear of job losses
- Inadequate training
- Plan not clearly defined
- Employee skepticism
- Resistance to data collection (e.g., rework costs, nonconformances, material waste, etc.).

TQM efforts should be implemented to increase productivity of the organization (quantity of performance); increase quality (decrease error and defect rate); increase effectiveness of all efforts; increase efficiency (decrease time requirements while increasing productivity).

TQM Implementation

TQM Deployment Structure

A formal structure is necessary in order for TQM to be implemented effectively in an organization. Such a structure is even more important in the construction environment where stakeholders are preoccupied with meeting schedules and responding to daily events. TQM calls for persistent analysis, evaluation, and learning; but this may be nigh impossible to accomplish by people in their traditional roles.

The proposed structure is aligned with the principles embodied in the Malcolm Baldrige award criteria and facilitates the deployment of activities necessary to secure the high level of quality in relation to the design and construction processes. It must be pointed out that the intent of this structure is not to compete with other organizations for the award; rather the purpose is to take advantage of a well-integrated structure that connects the behaviors of stakeholders with the criteria that have proven to be successful for many years.

The proposed TQM structure (Figure 9.6) shows the relationships between the functions that are essential for its deployment. Top management is the driving force behind quality endeavors. They supervise the steering committee, and this committee supervises the project teams. A consultant provides the expertise needed to initiate an effective TQM program. This function may be temporary if its scope is limited to getting the program established. TQM facilitators performs many functions that are critical to the deployment of quality-related activities, and serves as a multiskilled resource for the project teams, the steering committee, and top management. The Malcolm Baldrige Criteria gives direction to all quality-related activities.

Quality Implementation

There are three distinct phases to quality deployment: preparation, planning, and execution.

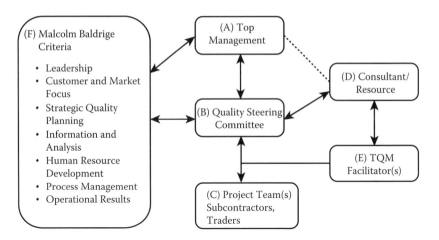

FIGURE 9.6
Proposed TQM deployment structure.

The process starts with a firm's top management embracing the idea of incorporating quality as a value of the organization. They need to be committed enough to generate enthusiasm in the company's employees; everyone should wish to produce quality workmanship, by "doing it right the first time." top management appoints a steering committee to oversee quality efforts. The members should be advocates of quality and placed at an adequate level in the organization where they can make the decisions needed for a program to be effective.

Limitations of TQM Deployment

In traditional manufacturing and service industries, providers have had excellent results with TQM. They have worked closely with suppliers to deliver products and services to high levels of customer satisfaction. They are well equipped to optimize the generation of those products and services, most of which are discrete and/or transaction-driven. A manufacturer takes components and subassemblies from a number of suppliers, and can optimize the consolidation of these items to generate a finished product. A service operation such as a garage is responsible in total for the service delivered and can optimize service delivery as well.

Construction, on the other hand, is a protracted process that involves interaction between many providers. While a single provider can optimize its portion of the work by improving its internal operations, project-wide optimization requires the close collaboration of other providers. The difficulty in establishing that collaboration is the Achilles heel of most traditional project delivery methods. The effect of top management leadership, therefore, may be limited to a specific firm that wishes to improve quality. It may have limited impact on the desires of another provider of construction (or design) services that is a member of the project delivery team. For example, a general contractor may embrace the TQM philosophy, but cannot reasonably expect a mechanical subcontractor to embrace these concepts and demonstrate genuine commitment to its employees with the degree of credibility needed to engage them.

Quality Improvement Concepts

A mission statement should exemplify an organization's focus on exceeding customers' expectations with products of high value and quality, and motivating/empowering its employees.

The organizations that have won the Malcolm Baldrige award have all shared the common attribute of having leaders that were committed to quality and demonstrated it by example through their daily activities. Such leaders were active participants or members of quality councils that were set up to promote quality endeavors. They guided and led quality improvement efforts, established and reviewed performance measures, kept quality as a major topic of all meetings and/or on-going reviews.

Measurement is critical to ensuring that an organization is meeting its goals and expectations. These goals/expectations should be delineated in a strategic plan developed with the participation of all stakeholders to ensure that they are an integral part of the quality endeavor. For quality performance to be achieved there must be specific, meaningful measures of performance based on key drivers. The strategic plan should also delineate the responsibilities of specific individuals and assign time frames for accomplishment.

Key drivers include customer satisfaction, operational performance, financial performance, employee satisfaction, and community service. By way of an example, the indicators of customer satisfaction are: customer satisfaction index, referral rates, percentage of on-time delivery, defects detected by the builder at delivery, defects determined by homeowner inspection, warranty call rates, warranty costs, and response time on warranty calls.

Quality Strategies

1. Establish top management priorities
   - Visionary
   - Set aggressive goals
   - Walk the talk

   First, top management of the organization must establish that total quality is the highest priority of the organization. Executives must provide a clear and reasonable vision, set aggressive goals for the organization and each unit, and most importantly demonstrate their commitment to TQM through their actions.

2. Initiate Cultural change
   - Paradigm change
   - Credibility development
   - Time

   Second, the culture of the organization must be changed so that everyone and every process embrace the concept of TQM. The organization must change its paradigm to adapt to a customer focus emphasis where everything done in the organization is aligned with exceeding customer expectations. This becomes an ongoing way of life for the organization, continually improving and adapting. As part of the culture change, credibility among the employees must be built through rewarding positive steps toward the vision of TQM. The organization must also allow time for the change to occur and be indoctrinated into the everyday aspects of the organization.

3. Establish small teams—provide overall goals
   - Define quality
   - Identify customers wants
   - Measure and change

   Third, small teams need to be developed throughout the organization to define quality, identify customer wants, and measure progress and quality. These teams will be responsible for creating their own goals, given the organization's overall goals.

4. Execute change and continuous improvement

Fourth, change and continuous improvement must be implemented, monitored, and adjusted based on an analysis of the measurements.

A Study of TQM in the United States (Florida) Construction Industry

Syed Ahmed and Salman Azhar conducted a 2002 study of the Florida construction industry; the study observations are presented as guidelines for successful TQM implementation in the industry. The main objectives of this study were to (1) investigate the adoption and implementation of TQM in the construction industry; (2) determine the processes ("what to measure") that are most suitable and appropriate for measurement during the construction project life-cycle; and (3) develop a model ("how to measure") for the measurement and evaluation of the quality performances of the construction processes identified in step 2 above as a tool for continuous improvement. When it comes to measuring work processes, the construction industry does not enjoy a good reputation. The problem, however, can be attributed to the nature of the industry, which lacks solid data gathering and experiences wide fluctuations in productivity.

Phase I

The first stage of this study identified the prevailing implementation and adoption of TQM principles in the construction industry through an in-depth questionnaire. The questionnaire was divided into six parts; namely, their knowledge of TQM, their perception of quality, the data acquisition methods used by them, quality in their organization, the degree of training provided to their employees toward TQM, and the obstacles faced by them in implementing TQM in their businesses. The conclusions from these six sections are briefly discussed below. (Observations are reported in real time, for the sake of convenience)

Knowledge of TQM

The results show that the majority of the contractors agreed that if a contractor satisfies clients, the profits would increase in the long run. They feel that TQM will work very well in their organizations, and that it will be beneficial to all stakeholders. They are, however, not aware of any implementation programs. Most of them feel that TQM is a philosophy used to improve cost estimating and warranty claims. This shows their lack of knowledge about the TQM and the potential benefits in implementing it in their organizations.

Perception of Quality

The analysis indicates that the majority of the contractors perceive quality as a competitive advantage next to elimination of defects. They feel that product/service quality is very important for them in gaining customer satisfaction because it ultimately translates to higher profits. They feel that customer satisfaction is their main goal. Interestingly however, when asked to rank quality, safety, time, cost, and scope in the order of importance, they ranked scope and cost as the more important considerations followed by timeliness, safety, and quality.

Data Acquisition Method

The results suggest that most of the companies in the study collect data to measure performance. They address problems by assigning an individual to solve them. On the other hand, 52% of the companies have a system for capturing customer suggestions, but only 28% measure customer satisfaction through questionnaires. Customer suggestions are gathered by the number of complaints or other methods in 20% of the companies. In most cases, the suppliers and the subcontractor are rated (52%) and when defects in services are identified, they are required to pay for or correct them.

Quality in Their Organization

Although only about 50% of the contractors surveyed have a clear definition of quality in their organizations, 86% are aware of the importance of quality. A majority of they respondents say that they do not have a formal quality improvement program (QIP) in place. Those that do, however, have the full support of their top management. They use a mix of QC, TQM, and ISO 9000 principles in their QIP. Demanding customers, CEO commitment, and competitive pressures are identified as the key reasons for implementing the quality improvement programs. The main objectives of the QIP are employee involvement followed by increasing productivity and cost reduction. There were 40% of the contractors that felt the quality of their products and services improved after implementing such a program.

Training

In the majority of the companies, employees are not given formal training in TQM or other quality improvement programs. Of these companies, 44% reported that managerial/supervisory staff had undergone quality improvement training while 29% of the companies provided training on quality management philosophies to nonmanagerial and nontechnical staff. Training programs mostly emphasize customer satisfaction as a primary goal followed by teamwork and communication.

Barriers to Implementing Total Quality Management

The following list gives a breakdown from the most important obstacle to the least important one.

1. Difficulty in changing behavior and attitudes
2. Lack of expertise/resources in TQM
3. Lack of employee commitment/understanding
4. Lack of education and training to drive the improvement process
5. Schedule and cost treated as the main priorities
6. Emphasis on short-term objectives
7. Tendency to cure symptoms rather than getting to the root cause of a problem
8. Too many documents are required (poor documentation)

It is easy to infer from the above that although TQM has been a magic word in the construction industry for the past few years, methods and techniques to implement the quality management program still need to be developed. The basic problem attributed to a lack of expertise or resources for implementing quality improvement programs is the difficulty in assessing what to measure and how to measure it—particularly the intangible aspects of quality. Without measurement, the notion of continuous improvement is hard to follow.

To address the issue of measurement, an attempt is made to determine the client satisfaction index. This provides a direct reference point from where quality improvement steps can be initiated in the construction industry. Various possibilities of client satisfaction were listed and rated. This gives us a measure or an index of client satisfaction or dissatisfaction that is discussed in Phase II of the study.

Phase II

In the second phase of this study, a second questionnaire focused on the customer was created to identify the processes for improvement. From the analysis of this questionnaire, the client satisfaction index was developed that lists the major causes of dissatisfaction for the clients. They are:

- Lack of attention to client priorities
- Poor planning
- Poor scheduling
- Inadequacy in processing change orders
- Poor delivery schedules and methods

Next, a third questionnaire was developed using the cause and effect diagram to identify the subcauses for the main reasons for client dissatisfaction. This was presented to contractors and their feedback was sought through structured interviews.

The major subcauses for the lack of attention to client priorities were:

- Lack of personnel training
- Lack of quality and cost control
- Inadequacy in contractor–subcontractor coordination
- Lack of conformance to specifications

The main subcauses for poor planning in the construction industry were:

- Low quality of material and workmanship
- Poor management of change orders
- Poor cash flow analysis
- Construction underestimation
- Poor equipment management

The subcauses for poor scheduling were:

- Incomplete design
- Lack of site condition supervision

- Inadequacy in project management coordination
- Improper network model selection

The main factors for the inadequacy in processing change orders were:

Changes in design by clients
Errors in construction design
Defective materials/equipment
Weather delays

The subcauses for the poor delivery schedules and methods in the construction industry were:

- Ambiguity in methods
- Change orders from procurement department
- The availability of materials and equipment

Phase III

This analysis led to the final phase of the research study where the most important identified subcauses by the contractors were presented to the same owners who had participated in questionnaire #2. In questionnaire #4 they were asked to indicate a measure of their satisfaction if the contractors made improvements to the identified subcauses. We call this the improvement index. Following are areas that would increase customer satisfaction the most if the contractors improved them.

- Construction underestimation
- Conformance to specifications
- Project management coordination
- Design changes by clients
- Change orders from procurement department

Customer satisfaction can be greatly enhanced by improving construction estimating, conformance to specifications, project management coordination, and by reducing design changes by clients, and change orders from the procurement department. The above areas were identified after analyzing the results of the last questionnaire in which we had asked the clients to identify the most important processes that need improvement.

This process can be repeated until different areas of improvement are identified through another cycle of the development of the client satisfaction and improvement index. The key to understanding is that clients are now a moving target—their expectations and requirements are constantly changing. To keep up with their ever-changing goals, contractors need to have a system of identifying, measuring, and continuously improving their tangible and intangible products and services in place.

Given the limited size of the sample, it was not possible to confirm that the causes and subcauses of client satisfaction or dissatisfaction identified in the study were statistically significant. The objective of this study was to develop and demonstrate how a system of continuous improvement can be put in place by measuring different processes.

Recommendations for Improvement

Success in TQM deployment is very much driven by the adage: "What gets measured gets done." It is of paramount importance to develop metrics related to key performance indicators (KPIs) and to report on-going measurement to all parties—top management and workers alike. Exception reports may be used to flag nonconformances. Top management should be attuned to having corrective measures implemented to address anomalies.

Factors to Be Measured

The following factors are examples of measurements to be applied in the application of TQM.
Designer performance:

 Number of design errors

 Processing time for shop drawings

 Number of RFIs submitted

 Response time for RFIs

 Customer satisfaction ratings

Contractor performance:

 Accidents, days without an accident

 Number of warranty calls

 Number of punch list items

 Number of failed inspections

 Percentages of delay in schedule

 Customer satisfaction ratings

 Estimated cost versus actual cost—by trade

Best Practices

The following TQM-based best practices are recommended for improving performance by building customer loyalty.

- Develop a clear strategy for enhancing customer loyalty, emphasizing activities that represent high value. Resource limitations often make it impossible to be all things to all customers. Responsiveness and proficiency of customer service operations may need to be weighed against the quality and timeliness of delivery, for example.
- Capture the voice of the customer (VOC). Have customers define the criteria for quality, price, and value.
- Develop measurement systems to collect and analyze customer feedback—information on won/lost customers or projects is critical. Track problem resolution and the percentage of highly satisfied and highly dissatisfied customers.

- Base measurable performance goals on the VOC so those performance thresholds can be compared with customers' needs.

- Establish a clear understanding of market segments in order to deliver value appropriately to each segment. Public sector customers have needs that differ from private sector companies—housing developers and office building investors have different needs as well. Value does not mean the same thing to each of these customers.

- Establish competitive benchmarks—know how well your competitors are delivering value to their customers. You may increase the satisfaction of your customers, but lose them to a competitor who can satisfy them even more.

- Maintain your company's image by such "soft" issues as commitment, partnership, and integrity. If a crisis occurs that jeopardizes the value delivered, responsiveness and communication with customers can keep their loyalty intact.

- Do root cause analysis to see where problems originate; institute corrective action.

- Discard the mistaken notion that construction is a project specific proposition; remember that each satisfied customer provides a building block for the value and corporate image that forge long-term business success.

Above all, embrace change in order to adopt the new paradigm; like the dinosaur, those who do not change are doomed to extinction.

Summary for Part A

Total quality management (TQM) is an approach to doing business that attempts to maximize the competitiveness of an organization through the continual improvement of the quality of its products, services, people, processes, and environments. There is no simple, comprehensive definition of quality. Deming's view is that the customer is the most important part of the production line. Quality should be aimed at the needs of the customer, both present and future. By ensuring quality at each stage in the construction process, and thereby minimizing costly rework, as well as other costs, the quality of the final products should satisfy the final customer.

Part B: Six Sigma in Construction

Definition of Six Sigma

Six Sigma management may be described as the "relentless and rigorous pursuit of the reduction of variation in all critical processes" (Gitlow 2006). It is an organizational imperative to create an order of magnitude change in processes. The Six Sigma approach has two versions. One version is a managerial initiative for measurably improving processes to better meet customer requirements; the other version includes another approach based on statistical tools and techniques that reduce product defects with a target of 3.4 defects per million opportunities (DPMO). Both versions are based on tools and techniques that were previously developed through Shewhart, Deming, Juran, Ishikawa, and others. Six

Sigma management originated in the high-volume manufacturing environment with an emphasis on measuring and reducing defects. Some construction applications may involve repetitive activities that may lend themselves to this numerical target. As much of construction is nonrepetitive, it can benefit from the more general process improvement approach.

Six Sigma is not focused on quality although quality generally improves with its application as a result of reducing the variation in target processes. As opposed to TQM, it is not concerned with changing the culture of an enterprise, but this culture tends to change as Six Sigma thinking permeates it. It facilitates "breakthrough" business improvement as opposed to gradual or incremental improvement. Six Sigma projects typically yield process performance improvements in the vicinity of 30–60% in 6 months or less, with attendant improvements in financial and business performance.

According to the Six Sigma Academy, Black Belts save companies approximately $230,000 per project and can complete four to six projects per year. General Electric, one of the most successful companies implementing Six Sigma, has estimated benefits on the order of $10 billion during the first 5 years of implementation. In 1995, GE first began Six Sigma after Motorola and Allied Signal blazed the Six Sigma trail. Since then, many organizations worldwide have discovered the far reaching benefits of Six Sigma.

Whereas Quality Management Systems such as TQM and ISO 9001:2000 generally require sustained effort for several years, this period of time is longer than a typical construction project cycle and limits application to the construction environment. Six Sigma projects have a much shorter cycle, making the approach more adaptable to the construction environment.

History of Six Sigma

Prior to 1978, Motorola was primarily a manufacturer of car radios and television sets. Bill Smith, a senior engineer and scientist in Motorola's communications division, introduced/developed the Six Sigma concept. He brought these ideas to CEO Bob Galvin who was deeply concerned with the quality problems that had been reported by Motorola's customers. Galvin set a target to reduce manufacturing defects by a factor of 10 in the first 5 years. By 1987, Motorola had reduced its rate of defects by that factor, then revised these improvement criteria to a factor of 10 in 2 years. Subsequently, they established a target of a hundredfold in 4 years and 3.4 defects per million opportunities (DPMO) in 5 years.

Smith and Galvin went on to establish Six Sigma as a primary component of Motorola's culture in order to provide customers with outstanding products. By the late 1980s, Six Sigma had been expanded to reducing concept to market cycle time. It also became the primary means to reduce field failures and increase customer satisfaction. Motorola's Six Sigma methodology evolved—continuous improvement became a main goal of the organization. In 1988, Motorola was awarded the MBNQ Award in its first year. The award is administered through the U.S. Department of Commerce and recognizes U.S. organizations for excellence in quality and overall performance.

Between 1989 and 1993, Motorola was joined by ABB and Kodak to support the Six Sigma Institute. During the mid-1990s, General Electric and Allied Signal implemented Six Sigma with outstanding, well-publicized results. A large number of other organizations have adopted the Six Sigma philosophy in recent years. They include American Express, Ford, Dupont, Dow Chemical and 3M. The Six Sigma philosophy has expanded beyond the original concept of 3.4 DPMO. It has become a business strategy and methodology for improving process performance and customer satisfaction.

Six Sigma Benefits

Motorola views the Six Sigma methodology as a proven way of accomplishing transformational change within an organization. It serves as a business improvement process that aligns an organization closely with customer requirements, applying tools that seek near perfection. Organizations that have employed the Six Sigma methodology have achieved significant benefits. Typical examples are:

- Better process flows
- Fewer defects
- Higher productivity
- Greater customer satisfaction (both internal and external)
- Shorter cycle time
- Faster delivery time
- Reduced work in process (WIP)
- Lower inventory levels
- Greater capacity
- Less waste
- Higher quality and reliability
- Lower unit costs
- Greater profitability

Stakeholders typically experience the following benefits:

- Greater customer satisfaction with products and services
- Better employee morale
- Greater stability for suppliers
- Stockholder satisfaction with lower costs and increased profits

Process Basics (Voice of the Process)

A process is a collection of interacting components that transform inputs into outputs toward a common aim, called a mission statement (Gitlow 2006). It is management's job to optimize the entire process toward its aim, directing resources as necessary to accomplish its aim, even if a particular department or function has to suboptimize. The important thing is for the entire organization to secure maximum profit. This poses a particular challenge for the construction environment as the organization in that situation comprises separate actors (such as subcontractors) who each seek to maximize their individual profits.

Statistically, the term Six Sigma refers to a process in which the range between the mean of a process quality measurement and the nearest specification limit is at least six times the standard deviation of the process. Six Sigma seeks to center the process on the target and reduce process variation. Table 9.1 cross-references the number of standard deviations and the corresponding defect occurrence. (Note: the values indicated assume a

TABLE 9.1

Sigma Limits Vs. Defect Occurrence

Specific Limit	Percentage of Accuracy	Defects (Per Million Opportunities)
1 sigma	30.23	697700
2 sigma	69.13	308700
3 sigma	93.32	66810
4 sigma	99.3790	6210
5 sigma	99.97670	233
6 sigma	99.999660	3.4

process shift of 1.5 standard deviations.) The Six Sigma concept is built on the assumption that:

- Continuous efforts to achieve stable and predictable process results (i.e., reduce process variation) are of vital importance to business success.
- Manufacturing and business processes have characteristics that can be measured, analyzed, improved, and controlled.
- Achieving sustained quality improvement requires commitment from the entire organization, particularly from top-level management.

Features that set Six Sigma apart from previous quality improvement initiatives include:

- A clear focus on achieving measurable and quantifiable financial returns from any Six Sigma project.
- An increased emphasis on strong and passionate management leadership and support.
- A special infrastructure of "Champions," "Master Black Belts," "Black Belts," and so on to lead and implement the Six Sigma approach.
- A clear commitment to making decisions on the basis of verifiable data, rather than assumptions and guesswork.

Methods

Six Sigma has two key methods: define, measure, analyze, improve, control (DMAIC) and define, measure, analyze, design, verify (DMADV), both inspired by Deming's PDCA cycle. The DMAIC process is an improvement system for existing processes falling below specification and looking for incremental improvement. The DMADV process is an improvement system used to develop new processes or products at Six Sigma quality levels. It can also be employed if a current process requires more than just incremental improvement. In the construction context, Six Sigma quality levels are not generally attainable due to the lack of repetition in work tasks. The DMAIC submethodology may be more appropriate (Figure 9.7).

DMAIC Steps

Define: It is the first step in our six sigma approach. DMAIC first asks leaders to define core processes. The critical to quality (CTQ) characteristics must be identified to determine

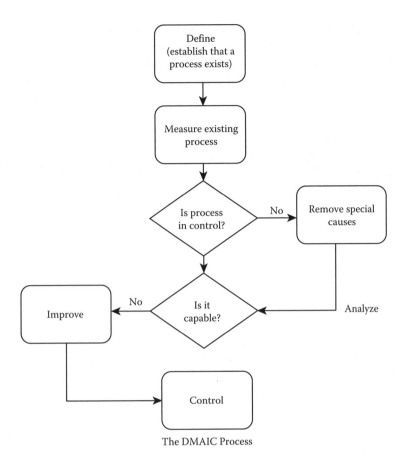

The DMAIC Process

FIGURE 9.7
Flowchart of the DMAIC process.

what is important to customers. It is important to define the selected project scope, expectations, resources, and timelines. The rationale for the project must be established, relative to other potential projects. The first efforts at process documentation are often at a general level. Additional work is often required to adequately understand and correctly document the processes.

Measure: This phase involves developing operational definitions for each CTQ variable, ensuring the measurement system is valid, and establishing baseline capabilities. The Six Sigma approach asks the Black Belt project manager to quantify and benchmark the process using actual data. At a minimum, consider the mean or average performance and some estimate of the dispersion or variation (maybe even calculate the standard deviation). Trends and cycles can also be very revealing. Process capabilities can be calculated once there is performance data.

Analyze: The upstream variables (Xs) for each CTQ are identified with a flowchart. Once the project is understood, the baseline performance documented, and opportunities for improvement identified, an analysis of the process is appropriate. In this step, the Six Sigma approach applies statistical tools to validate root causes of problems. Any number of tools and tests can be used. The objective is to understand the process at a level sufficient to formulate options for improvement. The available options should be compared to determine the most promising alternatives. As with many activities, balance must be achieved. The Six Sigma Black Belt is invaluable in striking the right balance by determining the extent to which each upstream variable influences each CTQ variable. This relationship may be expressed as:

$$CTQ = f(X_1, X_2, X_3, ... X_k).$$

Where CTQ is the characteristic important to customers as clarified in the measure phase. $X =$ ith variable that is hypothesized to influence the CTQ.

Improve: During the improve step of the Six Sigma approach, ideas and solutions are put to work. The Six Sigma Black Belt has discovered and validated all known root causes for the existing opportunity. The Six Sigma approach requires Black Belts to identify solutions. Few ideas or opportunities are so good that all are an instant success. As part of the Six Sigma approach, there must be checks to assure that the desired results are being achieved. Some designed experiments and trials may be required in order to find the best solution (i.e., the levels of the Xs that optimize the CTQ). When making trials and experiments, it is important that all project associates understand that these are trials and are part of the Six Sigma approach.

Control: Successful process revisions are formalized in conjunction with risk management and mistake proofing. Processes are documented and process owners are fully trained to minimize the possibility of variation returning to earlier levels. Process owners are taught the Kaizen approach that seeks to make everything incrementally better on a continuous basis. The sum of all these incremental improvements can be quite large. As part of the Six Sigma approach, performance tracking mechanisms and measurements are in place to assure that, at a minimum, the gains made in the project are not lost over a period of time. As part of the control step we encourage sharing with others in the organization. With this, the Six Sigma approach really starts to create phenomenal returns, ideas and projects in one part of the organization are translated in a very rapid fashion to implementation in another part of the organization.

DMADV Steps

- Define the goals of the design activity. What is being designed? Why? Use QFD or analytic hierarchical process to assure that the goals are consistent with customer demands and enterprise strategy.
- Measure: Determine critical to stakeholder metrics. Translate customer requirements into project goals.
- Analyze the options available for meeting the goals. Determine the performance of similar best-in-class designs.
- Design the new product, service, or process. Use predictive models, simulation, prototypes, pilot runs to validate the design concept's effectiveness in meeting goals.
- Verify the design's effectiveness in the real world.

Six Sigma Tools

As Six Sigma teams define and study business-related problems by the DMAIC or DMADV methodology, they use a variety of analytical tools and techniques. They include applications of:

- Mean, mode, median
- Measurements of central tendency
- Range, variance, and standard deviation
- Graphical analysis
- Histogram, run charts, Pareto charts
- Correlation studies
- Process mapping
- The XY matrix
- Measurement system analysis
- Process capability tools
- Multivariative study
- Hypothesis testing
- Failure mode effect analysis
- Design of experiments

Available software titles for statistical or process analysis are:

- Minitab
- JMP
- SigmaXL
- RapAnalyst or Statgraphics
- iGrafx
- Microsoft Visio
- Telelogic System Architect
- IBM WebSphere Business Modeler
- Proforma Corporation ProVision

Program management tools to manage and track a corporation's entire Six Sigma program:

- Instantis
- PowerSteering
- iNexus and SixNet
- Proxima Technology Centauri
- HP Mercury
- BMC Remedy

Roles in Six Sigma Leadership

Six Sigma has several specific roles and responsibilities (Figure 9.8): The senior executive provides the impetus, direction, and alignment necessary for Six Sigma's ultimate success; must lead the executive committee in linking strategies to Six Sigma projects using

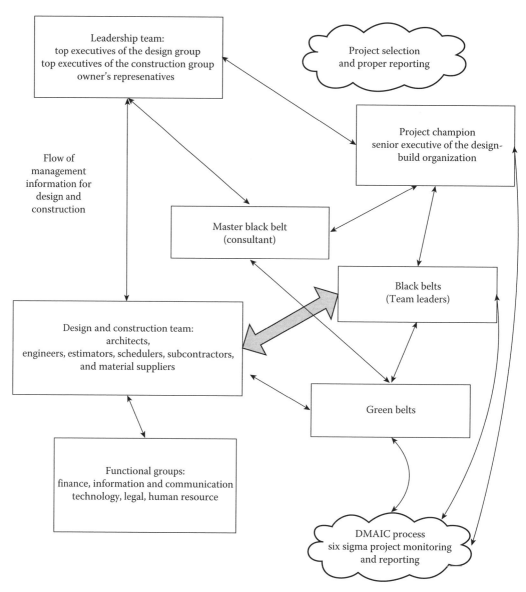

FIGURE 9.8
Six sigma structure for a construction project.

a "dashboard". He/she has the all-important responsibility of leading the overall initiative; must provide the resources necessary for success, and must approve the projects that are to be pursued by Black Belts and their teams. Very importantly, the senior executive must maintain an overview of the system to avoid suboptimization and conduct project reviews.

The Executive Committee members must deploy Six Sigma throughout the organization, using dashboards; Prioritize and manage the Six Sigma portfolio; assign champions, black belts, and green belts to projects; and remove barriers and provide resources.

Champions are responsible for Six Sigma implementation across the organization in an integrated manner. The executive leadership draws them from upper management.

Champions act as mentors to Black Belts, identify projects' impact on the organization, develop and negotiate project objectives and charters with the executive committee; remove political barriers or resource constraints, and keep the team focused on the project.

Master Black Belts, identified by champions, act as in-house coaches on Six Sigma. They devote 100% of their time to Six Sigma. They assist Champions and guide Black Belts and Green Belts. Apart from statistical tasks, their time is spent on ensuring consistent application of Six Sigma across various functions and departments.

Black Belts operate under Master Black Belts to apply Six Sigma methodology to specific projects. They devote 100% of their time to Six Sigma. They primarily focus on Six Sigma project execution, whereas Champions and Master Black Belts focus on identifying projects/functions for Six Sigma.

Green Belts are the employees who take up Six Sigma implementation along with their other job responsibilities. They operate under the guidance of Black Belts. Their assignments last for the duration of a Six Sigma project—typically 3–6 months, then return to their regular job responsibilities.

Process owners are the manager of a process. He or she has the responsibility for the process and has the authority to change it on his or her signature. The process owner must be accountable for the monitoring, managing, and output of his/her process; should empower the employees who work in the process to follow and improve best practices; and should accept, manage, and sustain the improved processes after the completion of the Six Sigma project.

Construction Applications

The factors listed in Table 9.2 are examples of variables in the construction environment that can be evaluated and improved in a Six Sigma program. For example, under the heading of lost time for accidents and delays, the Six Sigma team would collect data on the level of occurrence and the number of lost hours. Brainstorming could identify possible causes within the respective factors. Black belts can use statistical techniques to verify causality between upstream variables and these CTQ variables. Corrective measures would be implemented in operational procedures. Improved communication between upstream crews and successor crews could increase downstream awareness of potentially unsafe conditions. Procedural improvements would have upstream crews place safety barriers at critical areas and send emails to successor crews.

Deploying Six Sigma

There are a variety of strategies for implementing Six Sigma. The adoption of the Six Sigma methodology has to start with the leadership of a design or construction organization. As stated previously, the senior executive provides the impetus, direction, and alignment necessary for Six Sigma's ultimate success. The adoption of any major initiative such as Six Sigma, TQM, or lean construction must be integrated with the organization's strategic goals. These goals are typically documented in an organization's strategic plan; it serves as a road map to executives, managers, and employees for the actions they need to take to secure superior business performance.

Six Sigma adoption requires a long-term focus (although it produces results faster than TQM or ISO 9000) as it requires commitment to the cost of training and consultant support. As with other components of the strategic planning process, there should be goals,

TABLE 9.2

Summary of Typical Performance Factors and Measures

	Performance Factors	Performance Measures
A	Construction quality	Customer satisfaction index Customer complaints Defects, failures, failed inspections Number of punch list items Warranty costs
B	Production issues	Lost time (accidents, delays) Rework costs (time, material) Material waste costs Percentage of projects complete (PPC) Labor productivity (PAR values)
C	Project delivery	Actual versus projected completion Actual cost versus projected cost
D	Project administration	Direct costs Rework labor hours Labor costs Change orders (number by source)
E	Human resources	Employee satisfaction index Training costs Number of employee suggestions Labor evaluation scores

objectives, and expectations for a Six Sigma program. Responsibilities and time frames should be established and clearly communicated to all stakeholders. The following steps need to be taken in relation to the strategic plan:

1. Align executives and staff to Six Sigma objectives, deliverables, and time frames.
2. Mobilize by selecting project teams. Select projects based on importance and/or cost-benefit criteria.
3. Prepare and educate team members by conducting training (through consultant support).
4. Set up reporting systems to track team performance and ensure expectations are met.
5. Manage the Six Sigma process with an ongoing review at the senior executive level.

Some organizations have a corporate group that provides analytical skills—people are trained as black belts, green belts, and so on and are deployed to projects as needed. In some cases, people may be drawn from various business units, receive training, serve for a number of Six Sigma projects and then return to the respective business units. The Six Sigma culture can be gradually spread through an organization in this manner.

According to Gitlow (2006) there are five stages in the life of Six Sigma teams:

1. Forming: Team members get to know each other and establish ground rules.
2. Storming: Reflects efforts by the team to align goals and priorities in order to move ahead. Conflicts may arise and the direction of the team leader may be resisted. Team building exercises overcome these obstacles.

3. Norming occurs when team members display interdependence and develop cohesiveness. When common expectations are agreed for the team's activities.

4. Performing: The team works smoothly and directs combined efforts toward solving problems.

5. Adjourning: Teams disband voluntarily as their assignments have been completed.

The most successful teams direct their energies to the performing phase and lose very little time in the other four phases.

Examples of Six Sigma Application to Construction

A nationally ranked construction organization provides premier engineering, design, construction, and maintenance services to government and private-sector clients in a wide array of industries, including the energy, environmental, infrastructure, and emergency response markets. They are a vertically integrated provider of comprehensive engineering, consulting, procurement, pipe fabrication, construction, and maintenance services to the power and process industries. The company has over 20,000 employees in strategic locations around the world, and includes engineer, procure, and construct (EPC) projects in its areas of specialization. The organization has 30–40 professionals trained in the Six Sigma methodology. Given the demanding requirements of EPC projects, the company ensures that completed projects have systems that meet OPR; hence, a premium is placed on having minimal variation in performance.

Total variation is a combination of product variation and measurement variation. Measurement variation is called GR&R (i.e., gauge repeatability and reproducibility). Repeatability is due to equipment variation and reproducibility is due to inspector variation. There is emphasis on GR&R to evaluate measurement error in inspection activities, essentially to make sure everybody gets the same answers. In EPC projects, the certainty of outcome is critical and deviation from a planned outcome may result in the less gross margin. The financial viability of a project is closely linked to its predictability; if a high contingency has to be added to project costs in order to make a profit, then the company's competitiveness may be reduced.

The six sigma methodology is used to track process variation so that stakeholders can take action to counteract it. EPC outcomes are analyzed by examining the components of a project to see where variability has occurred. For example, in the design of a water treatment installation, original estimates are compared with actual expenditures of linear feet of material versus man hours for rigging and handling.

Overall View on Six Sigma Applications

The organization observed that many of the benefits of the Six Sigma methodology can be derived without having an army of black belts. Much can be achieved by training people to the level of yellow belts, especially if they are process owners. They should be trained to understand specification and control limits, and they should have the knowledge needed to manage those processes that are most important. With regard to lean, Six Sigma is an important complement. Lean reduces waste, but Six Sigma reduces the variability in the involved processes. The training of staff also pays dividends in effecting cultural change in the organization. There is a strategy for performance improvement in engineering—primarily on-time delivery and meeting or exceeding specification requirements. Six Sigma promotes behaviors that facilitate this strategy.

Project Selection

The organization is not highly focused on a particular return on investment. Generally, a business case has to be made for a particular process to be targeted for Six Sigma improvement. There is cost consciousness from the standpoint of not engaging highly paid black belts until there is a real need for them in a project. If a process can be improved through the efforts of a yellow belt, they would facilitate that action. It is recognized that using a high return on investment as a criterion for performance improvement can be highly misdirected. A focus on large dollar values for improvement efforts can sidetrack one from bottlenecks and can be devastating to the overall process. At the same time, it is recognized that as engineering may be only 5% of project costs, it would be less productive to study it first, when procurement may account for as much as 50% of EPC project costs.

There is a strong emphasis on identifying bottlenecks in core processes and using them as candidates for Six Sigma application. Vendor submittal reviews are a typical example. In a project, vendor submittals have to be processed promptly in order to confirm to vendors that they can proceed with carrying out their respective contract obligations. Failure to scrutinize submittals carefully may result in the unintended approval of nonconforming material or activity. If submittal reviews are delayed, then critical path activities may be delayed; hence, reviews are a bottleneck process that impact up to 50% of project costs.

Team Formation

A Six Sigma team is established with previously trained in-house professionals—usually yellow belts in conjunction with process owners; that is, staff members who are involved in the processes that are targeted for improvement. Other contractors are not added to a team as formal Six Sigma participants. Six Sigma efforts are internal, but in situations where other stakeholders impact the company's processes, they are included in process improvement initiatives. For example, the turnaround of vendor submittals is a critical part of EPC projects. To improve this turnaround, Six Sigma teams will get input from vendors and suppliers.

Relationship with Clients/Owners

Six Sigma is established as an internal best practice for optimizing performance. Owners can drive construction behavior with regard to safety, for example, but usually they do not seem to care about a contractor's processes. Owners are typically focused on the price and schedule of a project. They do not generally call for Six Sigma as part of the delivery method.

Implementing Improvements

A business plan is developed around improved processes. A process improvement is developed using Six Sigma and the improvement is implemented on a pilot basis. Validation is carried out to ensure the pilot plan works as intended and the improvement is then institutionalized with core processes.

Process owners are included as much as possible with Six Sigma teams to develop process improvement and ensure that the improved processes operate in a stable manner. While this approach is evolving, the desired "future state" is to have process owners become familiar with the upper and lower limits of target processes, and be empowered to take corrective action when it is needed.

Six Sigma Application to a Major Civil Engineering Project

The Six Sigma process was applied successfully in the construction of the New Doha International Airport (NDIA) project in Qatar. The first phase of the project included the preparation of 22 square kilometers of land, of which approximately half was reclaimed from the sea. The scope included having 60 million cubic meters of sand and rockfill placed within 24 months for a contract value of E337 million. The hydraulically reclaimed fill was stabilized through a combination of compaction techniques, including high energy impact compaction (HEIC). The process of this novel compaction technique was optimized using the Six Sigma quality improvement scheme. It was first documented in Terra et Aqua Number 103, June 2006, by four engineers affiliated with the project (Avsar et al. 2006). Four companies partnered in the project. The major challenges of this project were an extremely short-schedule, requiring over 65 million cubic meters of fill to be placed in less than 24 months, with many important early milestone handover dates.

Why Six Sigma

The reclamation fill had many variables: quality, gradation, water content, and layer thickness; each condition requiring a varying number of passes by compaction rollers. In order to attain a uniform compaction in a non-uniform environment on an extremely congested site, a high level of QC was needed. FAA standards were incorporated in values to be met by a cone penetration test, in-situ dry density, and settlement at a specified design bearing pressure.

Overview of the Process

Three techniques were used in the soil compaction process:

1. Hydraulic compaction, by the action of bulldozers driving over deposited material.
2. High energy impact compaction (HEIC) involves a non-circular module that is towed by a tractor and rises and falls repeatedly, striking the ground in the process. This method has a low frequency but high amplitude, generally leading to a greater penetration depth.
3. Conventional vibratory rollers: The impact rollers were driven in fixed patterns, changing from a clockwise direction to counterclockwise every 10 passes in order to achieve greater uniformity over the compaction area. The optimum speed of the impact rollers was determined to be 10–12 kph.

Implementing Six Sigma

The **DMAIC** process was used as a guideline to ensure that relevant data were collected, analyzed, and converted into information. The tools used in the framework of the Six Sigma implementation for the compaction process included: the compaction process map (or flowchart; see Figure 9.9) to identify potential causes, cause and effect diagram to generate a list of root causes, prioritization matrix of the most important root causes. The process map comprised four stages: initial placement with hydraulic compaction; watering, to provide optimal moisture content; rolling and compaction;and assessment, using geotechnical testing. Each stage involved field-based QC staff in direct contact with the operations foremen.

Root Cause Analysis

As a part of the Six Sigma deployment, a cause and effect diagram was prepared based on brainstorming by project staff from several operational functions. This was intended to identify possible root causes of variation in the compaction process in order to focus mitigating strategies on reducing it. Six categories were used to classify the ideas from the brainstorming

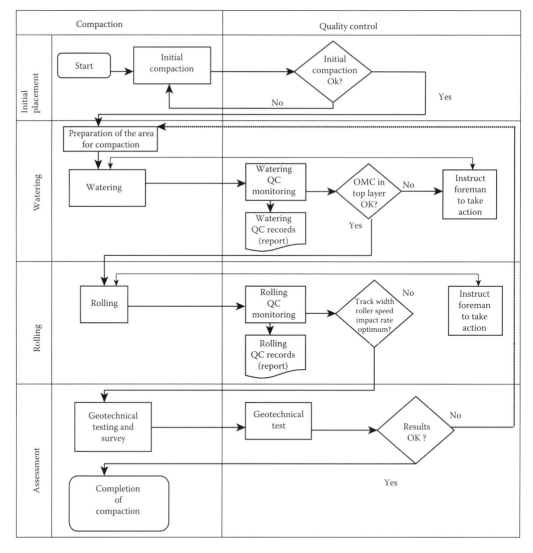

FIGURE 9.9
Map of the NDIA compaction process.

exercise: manpower, methods, machines, measurements, materials, and geo-environment. The most important causes were identified as: manpower, optimal driving speed (for operator-driven equipment); machines, impact rate; and methods, roller track width, timing of added water. The in-situ dry density was the bottom line "effect" of the cause and effect diagram (Figure 9.10). Since, it was also a contractual requirement, it was the best variable to illustrate the influence of Six Sigma on the compaction process. It is noted that here, the in-situ dry density is expressed as a percentage of the maximum dry density (% MDD).

Prioritization Matrix

The prioritized root causes were designated "upstream process indicators" (denoted by the symbol Xs). Optimum values and specification ranges were determined, as shown in Table 9.3. A procedure was established whereby quality controllers took ongoing readings and reported anomalies to the respective foremen. This facilitated prompt process adjustments, thus avoiding rework to achieve the required compaction levels. Two test areas were designated—an uncontrolled HEIC test area, and a Six-Sigma controlled area. In the controlled area, the variability of the sample results was significantly reduced and the process average increased, representing desirably higher values of dry density. As can

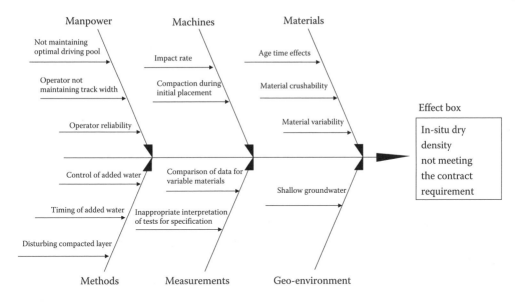

FIGURE 9.10
NDIA compaction cause and effect diagram.

TABLE 9.3

Optimum Values of Process Indicators Xs

Upstream Process Indicators Xs	Optimum Value	Specification Range
Roller speed	11 kph	10–12 kph
Impact rate	2 impact/sec	1.8–2.2 impact/sec
Roller track width	2.6 m/track	2.4–2.8 m/track
Timing of water	20 min/watering session	15–25 min/watering session

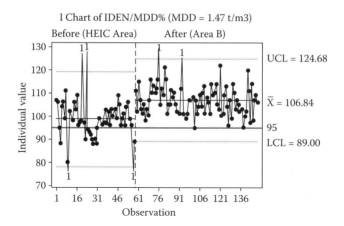

FIGURE 9.11
Values of dry density.

be seen from Figure 9.11 the observations in the uncontrolled test area show more variability than the observations for Area B where the Six Sigma approaches were applied. The two areas are demarcated by a vertical dotted line. The mean compaction was higher for Area B—at 106.84% as compared with a value of 99% for the uncontrolled area. A direct result of the Six Sigma campaign on compaction was that the average in-situ dry density increased by 8% (from 99 to 107% MDD), and the variance decreased by 56% (from 83 to 36).

Conclusion

Both TQM and Six Sigma represent change in the way a design or construction organization carries out its mission. It is change for the better, but will come only to those who want to change and are willing to make the necessary commitment. Purists may take the position that TQM or its variants should not be attempted unless all the necessary ingredients are present, especially top management commitment. But such a textbook scenario is more the exception than the norm. It is better for an organization to make an incremental step toward improving quality than to maintain the status quo. Examples abound of quality initiatives that began at the middle management level, with small successes that were eventually sold upward and embraced by champions at the top management level (Forbes 1994). Organizations that have adopted Quality Management systems with some success will have acquired knowledge and tools that prepare them better than others to venture into lean construction.

Questions for Discussion

1. Many construction projects are awarded based on the lowest bid. How is that practice viewed by the Deming philosophy?

2. Does having more inspections improve construction quality? Discuss.

3. Provide examples of three quality costs in construction projects.

4. What are the main barriers to TQM implementation?

5. How can these barriers be reduced, if not eliminated?

6. Discuss the similarities and differences between the objectives of TQM and the objectives of Lean Construction.

7. Does the prior existence of a TQM program in a construction organization inhibit the adoption of Lean Construction? Discuss.

8. Describe an organizational structure that is recommended for TQM deployment in construction projects.

9. Is Six Sigma easier to deploy in construction than TQM?

10. How are Six Sigma projects selected? What are the main criteria?

11. What are the life stages of Six Sigma teams?

12. Briefly explain how you would apply the Six Sigma concept to a construction project of approximately 12 months' duration. Describe the roles that are usually assigned in Six Sigma projects and explain who would fill them. Explain specifically how the approach could improve cost and schedule performance.

13. How would the application of Total Quality Management (TQM) differ?

14. In what way does Six Sigma vary from the lean construction philosophy?

15. How can Six Sigma support lean construction applications?

Appendix: Quality Tools

Fishbone Diagram

The Fishbone Diagram (Figure 9.12) or cause and effect diagram was developed by Dr. Kaoru Ishikawa. The diagram also bears his name. It serves as a tool to find the sources of quality problems, or in fact, almost any type of problem; it focuses attention on causes, not just symptoms. The diagram looks like the skeleton of a fish, with the problem represented by the head of the fish. The ribs represent possible causes, and the smaller bones are subcauses. The process works best with a group of people who are familiar with the problem, or who may be involved with it. A facilitator works with the group, helping them to work backward from the observed effect caused by the problem, asking "why?" in a manner similar to five-Why Analysis. The diagram serves as an important communications tool as users examine target processes in great detail in order to find a solution to the problem that is associated with it.

A generic form of the chart may be constructed with four bones radiating from the central spine. They are typically labeled machines, materials, methods, and people, as many problems can be traced to the influence of one or more of those factors. Guidelines for problem solving with the diagram are:

1. State the problem clearly in the head of the fish.

2. Draw the backbone and the ribs. If the team can think of possible causes, other than machines, materials, and so on they can label the ribs accordingly.

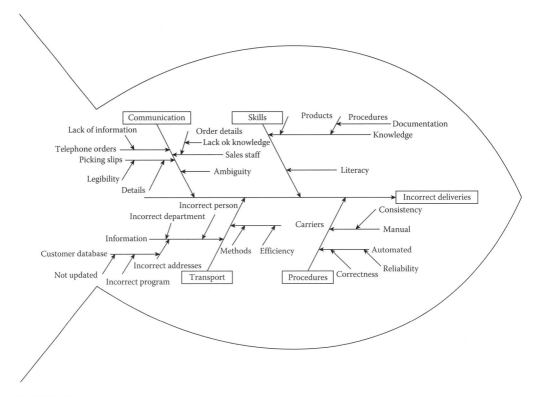

FIGURE 9.12
The fishbone diagram.

3. Working with the facilitator, work backward from the head of the fish, and along each rib to identify subcauses that branch from each rib. Asking "why" five times often helps to accomplish that.

4. Have the team agree on the most significant possible causes.

5. Identify strategies and set goals to address the core causes.

In the example above, a construction project is affected by many incorrect deliveries. The project team has a fair idea of the sources of the problem, and has selected the ribs or branches communications, skills, transport, and procedures. Brainstorming by the team has identified few subcauses on the skills rib such as knowledge, procedures, and documentation. On the other hand, many subcauses have been identified on the communication and transportation ribs. Communications seem to be affected by such issues as lack of information, ambiguity, and lack of knowledge in placing material orders. Remedies could include more training for staff and verification of any orders placed.

The Pareto Chart

The Pareto chart is named for an Italian economist, Vilfredo Pareto (1848–1923), who studied the distribution of wealth in Europe. He proposed an economic theory based on the observation that a few people were very wealthy, while large numbers of people had relatively little means. Essentially, 20% of the people in Italy held 80% of the wealth, hence the "80-20" rule. Dr. Joseph Juran applied this concept to the field of quality, distinguishing between the "vital few" and the "useful many." The Pareto chart helps users to identify those 20% attributes that represent 80% of the benefit. Faced with a variety of problems, a user can find the 20% that, when addressed, provide 80% of the benefit.

The example shown below represents a window installation operation. The data provided, list the number of problems relating to window installation and their unit cost. For example, when a window is installed in a manner that violates the code, it costs $1,100. Four occurrences are priced at $4,400. In performing Pareto analysis, one has to decide whether to use the number of occurrences or the costs of these occurrences. In this case, the analysis was based on total costs. The procedure for drawing the chart is as follows:

1. Set up a matrix similar to the one below, with the respective headings as shown.

2. List the values of the attributes in descending order of magnitude (i.e., the largest cost first and the smallest last). The category "Not finished on time" costs $19,625. By contrast, the category "crew very rude" costs $800 and is listed last.

3. The cost column is totaled to $44,824.10.

4. For each category, divide its cost by this total to obtain a percentage.

5. Use the right-hand column to create a cumulative total.

6. Draw the chart with two "y" axes. The left-hand vertical scale is used for the cost of the transactions. The right-hand vertical axis is used for the cumulative value.

7. Draw the cumulative curve, using the right-hand scale and terminating at a value of 100%.

The completed chart (Figure 9.13) shows that the first four categories represent 80% of the cost of complaints.

Analysis of Customer Complaints

Item	Category	Quantity	Cost per Occurrence	Total Cost	%	Cumulative %
c	Not finished on time	25	785.00	19,625.00	44%	44%
g	Wrong window installed	2	3500.00	7000.00	16%	59%
f	Job not done properly	4	1250.00	5000.00	11%	71%
a	Code infraction	4	1100.00	4400.00	10%	80%
h	Damage during window installation	1	3500.00	3500.00	8%	88%
e	Not starting on time	15	150.00	2250.00	5%	93%
b	Leave job site dirty	18	124.95	2249.10	5%	98%
d	Crew very rude	8	100.00	800.00	2%	100%
		77		44,824.10	100%	

Exercises to Test Your Understanding:

1. A contractor organization is concerned about the quality of its housing construction and does an analysis of the sources of failure in order to take corrective action. The failures are categorized by the affected components. The cost of each failure is listed below.

   a. Develop a Pareto chart by hand based on this information.

   b. Show what building elements that should be targeted for corrective action, based on the 80/20 rule. Show your calculations, including cumulative percentages. You may use the reverse side of each page. (The neatness of the chart is not critical.)

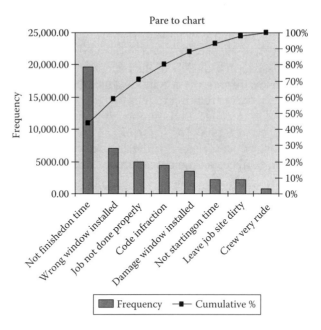

FIGURE 9.13
Pareto chart of customer complaints.

Element	Number of Failures	Cost Per Failure ($)
(A) Electrical	4	90
(B) Woodwork	2	47
(C) Plumbing	8	78
(D) Roof	1	83
(E) HVAC	2	55
(F) Doors	3	11
(G) Damaged walls	12	8

2. Using an on-site materials storage trailer as an example, draw a cause and effect diagram to represent a problem that is reducing productivity. (Trades staff have to queue up to get materials before starting a task, and lose a lot of time in the process.) Sketch the structure of the diagram. Start with an observed effect "Long waiting time." Show possible causes on four bones on the chart labeled Materials, Workers, Methods, and Environment. To save time, show only two causes on each bone.

Pareto Chart: Draw a chart using the following information. Construct a cumulative curve to show the categories that account for 80% of the occurrences.

Item	Number of Occurrences	$ Cost Per Occurrence
A. Roof	8	30
B. Doors	7	9
C. Windows	10	90
D. Ceiling	5	16
E. Carpet	20	5
F. Paint	30	7

References

Ahmed, S. M., and R. T. Aoieong. 1998. Analysis of quality management systems in the Hong Kong construction industry. *Proceedings of the 1st South African international conference on total quality management in construction*, 37–49. November 22–25. Cape Town, South Africa.

Avsar, S., M. Bakker, G. Bartholomeeusen, and J. Vanmechelen. 2006. *Six sigma quality improvement of compaction at the new Doha International Airport Project/Terra et Aqua*, 14–22. Number 103. June.

Brown, S. 1992. *Total quality service*. Scarborough, Ontario: Prentice Hall Canada Inc.

Brown, S. 1995. *What customers value most*. New York: John Wiley and Sons.

Burati, Jr., J. L., and T. H. Oswald. 1992. Implementing TQM in engineering and construction. *Journal of Management in Engineering, ASCE* 9(4): 456–70.

Burati, Jr., J. L., and T. H. Oswald. 1993. Implementing TQM in engineering and construction. *Journal of Management in Engineering, ASCE* 9(4): 456–70.

Chase, G. W., and M. O. Federle. 1992. Implementation of TQM in building design and construction. *Journal of Management in Engineering, ASCE* 9(4): 329–39.

Crosby, P. B. 1984. *Quality without Tears*, New York: New American Library.

Deming, W. E. 1986. *Out of the crisis*. Cambridge, MA: Massachusetts Institute of Technology.

Deming, W. E. 1993. *The new economics*. Cambridge, MA: Massachusetts Institute of Technology.

Forbes, L. H. 1994. *What do you do when your organization isn't ready for TQM?* National Productivity Review, Autumn, 467–78.

Forbes, L. 1999. *An engineering management-based investigation of owner satisfaction, quality and performance variables in health care facilities construction*. Ph.D. Dissertation. Coral Gables, FL: University of Miami.

Gitlow, H. 2006. *Design for six sigma for green belts and champions.*, Upper Saddle River, NJ: Prentice Hall.

Goetsch, D. L., and S. B. Davis. 2006. *Quality management: Introduction to total quality management for production, processing, and services*. New Jersey, Pearson Prentice Hall.

Harris, C. R. 1995. The evolution of quality management: An overview of the TQM literature. *Canadian Journal of Administrative Sciences* 12(2): 95–105.

Hayden, W. M. 1996. Connecting random acts of quality: Global system standard. *Journal of Management in Engineering, ASCE* 12(3): 34–44.

Juran, J. M. 1988. *Juran on planning for quality*, New York: Free Press.

Oberlender, G. D. 1993. *Project management for engineering and construction*. McGraw-Hill.

Bibliography

Ahmed, S. M. 1993. *An integrated total quality management (TQM) model for the construction process*. PhD Dissertation. School of Civil & Environmental Engineering. Atlanta, GA: Georgia Institute of Technology.

Ahmed, S. M., L. P. Sang, and Z. M. Torbica. 2003. Use of quality function deployment in civil engineering capital project planning. *Journal of Construction Engineering & Management, ASCE* 129(4): 358–68.

Ahmed, S. M., P. Tang, S. Azhar, and I. Ahmad. 2002. An evaluation of safety measures in the Hong Kong construction industry based on total quality management principals. CIB-W65/W55 10th international symposium construction innovation and global competitiveness, 1214–27. September 6–10. Cincinnati, OH.

Aoieong, R. T., S. L. Tang, and S. M. Ahmed. 2002. A process approach in measuring quality costs of construction projects: Model development. *Construction Management and Economics* 20(2): 179–92.

Arditi, D., and H. M. Gunaydin. 1998. Factors that affect process quality in the life cycle of building projects. *Journal of Construction Engineering and Management*, *ASCE* 124(3): 194–203.

Associated General Contractors of America. 1993. *Implementing TQM in a construction company.* Publication No. 1211. Alexandria , VA: The Associated General Contractors of America.

Besterfield, D. H. 1994. *Quality control.* Upper Saddle River, NJ: Prentice Hall.

Dale, H. B., B. Carol, H. B. Glen, and B. Mary. 2003. *Total quality management.* Upper Saddle River, NJ: Prentice Hall.

Deming, W. E. 1982. *Quality, productivity, and competitive position.* Cambridge, MA: Massachusetts Institute of Technology.

Omachonu, V. K. 1991. *Total quality and productivity management in health care organizations.* Norcross, GA: Institute of Industrial Engineers; and Milwaukee, WI: American Society for Quality Control.

Snee, R., and R. Hoerl. 2003. Leading six sigma. *Financial times.* Upper Saddle River, NJ: Prentice Hall.

Tang, S. L., and C. W. Kam. 1999. A survey of ISO 9001 implementation in engineering consultancies in Hong Kong. *International Journal of Quality and Reliability Management* 16(6): 562–74.

Tenner, A. R., and I. J. DeToro. 1992. *Total quality management.* Reading, MA: Addison-Wesley.

TQM Concept Map. 2003. http://soeweb.syr.edu/faculty/takoszal/TQM.html

Web Sites

http://buildnet.csir.co.za/cdcproc/docs/2nd/love_ped.pdf
http://en.widipedia.org/wiki/Six_Sigma
http://www.asq.org/sixsigma/
www.isixsigma.com

10

Sustainable Construction: Sustainability and Commissioning

Part A: Sustainability

The First International Conference on Sustainable Construction, in Tampa, Florida, in 1994, was defined as "the creation and responsible management of a healthy built environmental base on resources efficient and ecological principles" (Kibert 1994). The Brundland Commission defines a sustainable condition for this planet as one in which there is stability for both social and physical systems, achieved through meeting the needs of the present without compromising the ability of future generations to meet their own needs.

A green building (or "sustainable building") is a structure that is designed, built, renovated, operated, or reused in an ecological and resource-efficient manner. Green buildings are designed to meet such objectives as protecting the health of occupants; improving worker productivity; using energy, water, and other resources more efficiently; and reducing the overall impact on the environment.

The design, construction, and operation of facilities have a major impact on the proliferation of greenhouse gases. Inefficient energy buildings consume far more energy than is necessary for a comfortable working environment. In turn, this waste in energy escalates the use of fossil fuels and contributes to the proliferation of greenhouse gases. The materials used in buildings such as wall and floor surfaces, paint, laminates, manufactured boards, and refrigerants are contributors to pollution as well, to the extent that they are designed to be environmentally friendly.

The worldwide consumption of fossil fuels is escalating rapidly, as is the presence of greenhouse gases. In 2008, the world consumption of energy was 15 terawatts (1 terawatt = 1000 gigawatts). The world demand for oil was 31 billion barrels of oil per year. The growth in demand is forecasted to be 2% per year to 2025. Although recessionary trends may dampen growth somewhat, the demand for fuel and energy in Asia has been escalating as its economies experience rapid growth. For example, in 2008 China added 14,000 new vehicles to their roads each day; India added 10,000 new vehicles.

Buildings account for one-sixth of the world's fresh water usage, one-quarter of its wood harvest, and two-fifths of its material and energy flows (Roodman and Lenssen 1995). In the European Union, buildings are responsible for more than 40% of total energy consumption and the construction sector is estimated to generate approximately 40% of all man-made waste (CIB 1999).

Initiatives such as Leadership in Energy and Environmental Design (LEED) established by the U.S. Green Building Council (USGBC), Energy Star and others promote green architecture and sustainable construction. These initiatives serve as a guide and reference for building owners, designers, and builders to ensure that future buildings increasingly incorporate the best practices that will in time, lead to a sustainable environment for all.

Importance of Sustainable Construction

In the United States, buildings and their landscapes are a major source of pollution accounting for a third of all energy consumed, more than a third of the carbon dioxide emissions, and almost half the sulfur dioxide emissions produced in the country. According to studies by the U.S. Green Building Council (USGBC 2007), in the United States buildings account for:

- 65.2% of total U.S. electricity consumption
- >36% of total U.S. primary energy use
- 30% of total U.S. greenhouse gas emissions
- 136 million tons of construction and demolition waste (approximately 2.8 lbs/person/day)
- 12% of potable water
- 40% (3 billion tons annually) of raw materials used globally

Lean Construction and Green Buildings

Green buildings promote and support the concept of lean construction. While lean construction involves the minimization of waste, ensuring that the resources used in the building process enhance the value chain for a completed facility, green building design is based on design principles that promote environmentally beneficial long-term operation. Overall, a combination of these approaches significantly enhances sustainability. Not only is a lean building assembled using fewer resources (human, financial, energy, material, etc.), but a green building has a lower environmental impact in its construction continues in its on-going operation as well. <u>Green practices may even reduce initial construction costs</u>. The recycling of construction material reduces the amount of physical waste generated in a project, and the amount of pollution associated with the process is reduced as well. Sorting construction waste on a site is a recommended practice that qualifies for LEED credits. It can also reduce haulage costs for waste disposal as some of that material can be reused.

LEED requires clear documentation of design strategies and construction practices—as credits involve both stages of a project. This leads to better collaboration, including designers and constructors as well as owners. Sustainable design, therefore can dovetail very effectively with lean construction—as designers and contractors collaborate on constructability issues.

Sustainability Practices

- Using resources efficiently and minimizing raw material resource consumption, including energy, water, land, and materials, both during the construction as well as throughout the life of the facility.
- Maximizing resource reuse
- Utilizing renewable energy sources
- Creating a healthy working environment

- Building facilities of long-term value
- Protecting and/or restoring the natural environment

There are several systems that provide a framework for green building initiatives: They include:

- Energy Star
- Green Globes
- Florida Green Building Coalition
- LEED

In terms of the degree of adoption, Energy Star is managed by the U.S. Department of Energy and is applicable nationwide. Green Globes is growing nationwide, as is LEED, but the Florida Green Building Coalition applies in the state of Florida only. These systems vary in complexity and in the degree of verification required in order to have a building certified.

The complexity of Energy Star is moderate. A professional is required to verify the design intent, and follow this up 1 year after occupancy. Green Globes is also moderate in its requirements and includes the Energy Star criteria. The LEED system is complex—it uses a system of credits based on highly multifaceted and structured criteria. A formal application identifies the intent to have a project considered for certification and is preliminarily reviewed, but it is reviewed in detail at completion before the appropriate level of certification is granted (Table 10.1). The Florida Green Building Coalition has a similar two-stage review, but has a single certification level—"certified" or "not certified." Green Globes and LEED both have four levels of certification. Green Globes awards projects 1, 2, 3, or 4 globes based on a point scoring system. LEED has four levels: certified, silver, gold, and platinum. Energy Star has a pass/fail system with a threshold of 75 out of a possible 100 points. All of these systems require a formal assessment by the rating agency for certification. The Energy Star system requires a third-party assessment by a professional engineer. Green Globe certification also requires a professional engineer's review.

The use of any of the foregoing standards will provide building designers with strategies for improving a building's environmental performance on several fronts, including energy efficiency, indoor air quality, potable water usage, and life-cycle cost. However, the LEED standards developed by the USGBC evidently accomplish this objective with a broader scope of influence and greater rigor. Furthermore, LEED certification involves the services of a LEED-accredited professional (LEED® AP) in various aspects of the design and/or construction. It also requires the formal process of commissioning to ensure that certain critical systems in a facility (such as HVAC) perform as required and meet the facility owner's project requirements (OPR). Consequently, LEED is proposed as the most appropriate standard for enhancing the lean performance of buildings on a life-cycle basis.

Sustainability and the U.S. Green Building Council

The USGBC is a non-profit membership organization whose vision is "a sustainable built environment within a generation." Its membership includes corporations, builders, universities, government agencies, and other non-profit organizations. Its mission is: "The USGBC is the nation's foremost coalition of leaders from across the building

TABLE 10.1
Summary of Selected Green Building Rating Systems

Green Building Rating System Comparison	Building Initiatives Green Globe Rating	Florida Green Building Coalition (FLGBC)	Leadership in Energy and Environmental Design (LEED)	Energy Star
Online assessment	Yes	No	Yes	Yes
Third-party assessment (separate from rating agency)	Required only for Green Globes certification	No	No	Yes (by a professional engineer)
Fundamental commissioning	Best practice but not required for globe award	Required for certification MEP designer can do	Yes, required	No
Enhanced commissioning	No	No	Yes	No
Rating agency required?	Third-party verification for globe award	Yes, review by FLGBC Self-certification not possible	Yes	Yes (Energy Star)
Complexity	On-line system may improve efficiency and reduce certification cost. Includes Energy Star criteria	Similarity to LEED Two-stage review, project proposal with points planned, then final plan on completion	Complex weighted average calculation needed	Moderate. Professional verify design intent, follow-up 1 year after occupancy
Degree of adoption	Growing nationwide	Florida only	Nationwide	Nationwide—Dept. of Energy
Certification levels	Four Globe levels (1000 points): 1 globe = 35–54% 2 = 55–69% 3 = 70–84% 4 = 85% and over	Certified or not	Four levels: (NC 2.2) Certified = 29–36 Silver = 37–43 Gold = 44–57 Platinum = 58–79	Pass/fail 75/100 points or more

industry working to promote buildings that are environmentally responsible, profitable and healthy places to live and work." Since USGBC's founding in 1993, the council has grown to more than 15,000 member companies and organizations by the year 2008, a comprehensive family of LEED® green building rating systems, and an expansive educational offering. There is also the industry's popular Greenbuild International Conference and Expo (www.greenbuildexpo.org), and a network of 74 local chapters, affiliates, and organizing groups.

The USGBC has led a national consensus to produce a new generation of buildings that deliver high performance through measurable systems. Council members collaborate to develop LEED products and resources. For more information, visit www.usgbc.org.

LEED (Leadership in Energy and Environmental Design)

Members of the USGBC from all segments of the building industry developed the LEED system starting in 1995 to provide a national standard and continue to contribute to its on-going evolution and growth. Its development included representatives from all sectors of the building industry with funding from the U.S. Department of Energy, and unique in having had its development open to both industry and public scrutiny.

According to the USGBC, the LEED green building rating system is a voluntary, consensus-based national standard for developing high-performance, sustainable buildings. The LEED® certification program is a feature-oriented rating system that awards buildings points for satisfying specified green building criteria. The six major environmental categories of review include: Sustainable Sites, Water Efficiency, Energy and Atmosphere, Materials and Resources, Indoor Environmental Quality, and Innovation and Design.

LEED provides a framework for assessing building performance and meeting goals for sustainability. It is based on proven scientific standards and promotes the use of state-of-the-art techniques for sustainable site development, energy efficiency, water conservation, materials selection, and indoor environmental quality. LEED promotes the proliferation of green building knowledge and expertise by recognizing achievements, and through a program of project certification, professional accreditation. The program also provides training and technical resources. LEED:

- Establishes a common standard of measurement to define "Green" building
- Promotes integrated, whole-building design practices
- Raises consumer awareness of green building benefits
- Encourages competition among industry members
- Recognizes environmental leadership in the construction industry
- Transforms the construction industry/market

LEED can be applied to all building types including new construction, commercial interiors, core and shell developments, existing buildings, homes, neighborhood developments, schools, and retail facilities. There are several subsets of LEED standards:

- LEED-NC for new construction
- LEED-EB for existing buildings

- LEED-CI for commercial interiors
- LEED-CS for core and shell
- LEED-H for homes
- LEED-S for schools
- LEED-ND for neighborhood development.

Facilities are certified to a variety of levels based on a credit system that reflects the degree of compliance with environmentally based standards. LEED standards are continually being upgraded. LEED-NC was revised to NC2.2 in 2005 and subsequently to LEED-NC 3.0. The levels prevailing in 2009 were: Certified (the basic level), Silver, Gold, and Platinum levels of LEED green building certification and are awarded based on the total number of credits points earned within each LEED category. The distribution of credits was changed with each update.

LEED performance areas are highly detailed and can only be presented in outline form in this publication. Readers are encouraged to contact the USGBC for the most up-to-date documentation of the LEED standards (Table 10.2, Figure 10.1). A brief overview of the six categories in the LEED rating system is as follows:

1. **Sustainable sites (SS): By selecting an appropriate project location, designers can reduce urban sprawl, promote the use of mass transportation, and reduce the dependence on personal transportation. Urban infill areas are examples of this practice, as they may already have the needed infrastructure. Site sustainability can also be enhanced by retaining existing trees and landscaping. Plants that have minimal water and pesticide requirements are highly desirable. Recycling of paving material materials and mulches promotes the concept of sustainability.**

2. **Water efficiency: Water usage can be reduced by 30% or more in commercial buildings. In a typical 100,000 square foot office building with 650 occupants, low-consumption fixtures equipped with sensors can save over one million gallons of water per year, assuming that each occupant uses 20 gallons per day. The strategies used include water efficient landscaping, irrigation efficiency, use of recovered rainwater and recycled wastewater, and non-potable water treated by a public agency.**

3. **Energy and atmosphere (EA): Improvements in the energy efficiency of buildings reduce operating costs and enhance comfort. Greater energy efficiency**

TABLE 10.2

Certification Points for LEED NC 2.2

	Levels of LEED Certification		
	Level	LEED (New Construction)	LEED (Existing Buildings)
1	LEED-certified	26–32 points	28–35 points
2	Silver level	33–38 points	36–42 points
3	Gold level	39–51 points	43–56 points
4	Platinum level	52 + points	57 + points

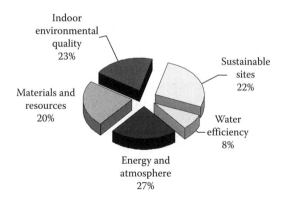

FIGURE 10.1

Distribution of categories for LEED NC 2.2 standards. (From U.S. Green Building Council, *New construction and major renovation reference guide, version 2.2,* 3rd ed. Washington, DC: USGBC, 2007. Reproduced with permission from the USGBC.)

also reduces pollution from power generation equipment. Passive design strategies such as building shape and orientation, passive solar design and the use of natural lighting have measurable benefits. Natural lighting, in particular, has a positive impact on the well-being and productivity of building users. High-efficiency lighting systems provide the maximum amount of lighting per kilowatt; task lighting is more cost effective than area lighting. Advanced lighting controls such as motion sensors and dimmable systems are also highly cost effective. The heating or cooling system for the facility should be properly sized and as energy efficient as possible to maintain adequate temperature and humidity conditions. The careful use of building orientation with reduced glass exposure on the east and west building faces reduces the solar gain of facilities. High R-value insulation and reflective roof and wall colors reduce the cooling load of facilities. Alternative energy sources such as solar cells and fuel cells should be considered.

4. **Materials and resources (MR):** Recycled materials reduce the impact of waste in landfills, and also reduce the impact of natural resource consumption to produce new construction materials. Reuse and recycling of materials from demolition are incentivized by the LEED credit structure. The approved strategies include: storage and collection of recyclables such as metal, glass, plastic, paper, and so on; use of salvaged materials from other locations; sourcing construction materials within 500 miles to minimize transportation-related pollution; using wood certified to come from well-managed forests; in one project 3966 cubic yards of waste material resulted from the initial demolition (over 99% of this waste was reused or recycled); and 9.23 tons of obsolete window glass and 3.25 tons of concrete waste dust were converted into over 8,000 concrete blocks.

5. **Indoor air quality (EQ):** According to the Environmental Protection Agency (EPA) the levels of pollutants indoors are often many times greater than in outdoor air. Buildings that provide for increased levels of filtered outdoor air have lower levels of pollutants and appear to provide a better environment for users. Fewer cases of illness are documented. Hospitals that are designed with lower

levels of pollutants and that have been built based on green principles have observed lower levels of site-borne illnesses. The average length of hospital stays (ALOS) has been shown to be shorter than with non-green facilities.

6. **Innovation and design process (ID):** This category of the LEED standards assigns additional credits for innovative applications and so-called exemplary performance for exceeding specific thresholds for compliance. For example, diverting more than 90% of construction waste from a landfill would be recognized with an innovation credit.

Benefits of Sustainable Construction

According to the USGBC, there are several benefits of sustainable construction including improved health, safety, comfort, and/or lower building operating costs. Other factors of green building include being more environmentally friendly than other building practices, creating structures that are more durable and less polluting, as well as making substantially better use of natural resources. Furthermore, the cost of sustainable construction is often no greater than the cost of a conventional building, and savings can frequently be realized in heating, cooling, and fuel costs over the long-term.

Several factors influence the growth and development of sustainable construction internationally. The United Kingdom based public–private partnership, Partners in Innovation project developed a guide, *Sustainable Construction: Practical Guidance for Planners and Developers.* The guide articulates several drivers for sustainable design and construction, including meeting governmental legislative requirements for a green building rating system (such as the USGBC LEED program in the United States), for reducing carbon dioxide emissions. The drivers include:

- Increased funding and partnership opportunities for sustainable construction;
- Improved workplace or academic performance in green buildings; public recognition of sustainable construction methods; and
- Long-term financial savings through environmentally friendly building practices.

Examples of Operational Benefits

Energy- and water-efficient buildings have been able to reduce their operating costs significantly. Use can be cut to less than half than that of a traditional building by employing aggressive and well-integrated green design concepts.

- Employees in buildings with healthy interiors have less absenteeism and tend to stay in their jobs. The Internationale Nederlanden (ING) Bank headquarters in Amsterdam uses only 10% of the energy of its predecessor and has cut worker absenteeism by 15%. The combined savings equal $3.4 million per year.
- More than 17 million Americans suffer from asthma, and 4.8 million of them are children. 10 million school days are missed by children each year because of asthma, which is exacerbated by poor Indoor Air Quality (IAQ).
- A healthy indoor environment can reduce the likelihood of lawsuits and insurance claims. In *Bloomquist v. Wapello* (500 N.W.2d 1, Iowa, 1993), plaintiffs successfully

sued employers and builders for creating an unsafe work environment due to inadequate ventilation and pesticide applications.

- Insurance companies are using climate change protection activities as a means to manage risk and maintain profitability.

The USGBC reports that energy efficiency measures reduced the operating expenses of the Denver Dry Goods building by approximately $75,000 per year. At a grocery store in Spokane, Washington, 48 tons of waste was recycled during construction. It would otherwise have incurred high-disposal costs. Students in day-lit schools in North Carolina have consistently scored higher than students in conventionally lit classrooms.

A number of studies point to the energy savings derived by sustainable construction based on LEED principles and the ENERGY Star system. Two studies, one by the New Buildings Institute (NBI) and one by CoStar Group, have verified that third-party certified buildings outperform their conventional counterparts across a wide variety of metrics, including energy savings, occupancy rates, sale price, and rental rates.

In the NBI study,* the results indicate that new buildings certified under the USGBC's LEED certification system are, on average, performing 25–30% better than non-LEED certified buildings in terms of energy use. The study also demonstrates that there is a correlation between increasing levels of LEED certification and increased energy savings. Gold and Platinum LEED certified buildings have average energy savings approaching 50%. The NBI Study confirms that newly constructed LEED certified buildings use significantly less energy than their conventional counterparts and that they perform better overall. The report also underscores that monitoring a building's on-going operations and maintenance, as required in LEED for existing buildings: operations and maintenance and Energy Star, are equally important. LEED and Energy Star provide building owners and operators with a valuable structure to maintain high performance and deliver savings over time.

Energy savings under the EPA's Energy Star program are equally impressive: buildings that have earned the Energy Star label use an average of almost 40% less energy than average buildings, and emit 35% less carbon. But beyond the obvious implications of reduced energy use and reduced carbon emissions, the results from both studies strengthen the business case for green buildings as financially sound investments.

According to the CoStar study,† LEED buildings command rent premiums of $11.24 per square foot (2008 dollars) over their non-LEED peers and have 3.8% higher occupancy. Rental rates in Energy Star buildings represent a $2.38 per square foot premium over comparable non-Energy Star buildings and have 3.6% higher occupancy. And, in a trend that could signal greater attention from institutional investors, Energy Star buildings are selling for an average of $61 per square foot more than their peers, while LEED buildings command a remarkable $171 more per square foot. The group analyzed more than 1300 LEED Certified and Energy Star buildings representing about 351 million square feet in CoStar's commercial property database of roughly 44 billion square feet, and assessed those buildings against non-green properties with similar size, location, class, tenancy and year-built characteristics to generate the results. The USGBC cites studies of green

---

* The NBI study was funded by USGBC with support from the U.S. Environmental Protection Agency and can be accessed at: http://www.usgbc.org/DisplayPage.aspx?CMSPageID=77#usgbc_publications.
† For more information on the CoStar study: http://www.costar.com/News/Article.aspx?id=D968F1E0DCF737 12B03A099E0E99C679.

TABLE 10.3

Financial Benefits of Green Buildings: Cost per Square Foot Based on 20-Year Net Present Value (NPV) Calculation

	Category	20-yr NPV (U.S. Dollars)
1	Energy value	5.79
2	Emissions value	1.18
3	Water value	0.51
4	Construction waste (1 year)	0.03
5	Commissioning operations and maintenance value	8.47
6	Productivity and health value certified and silver	36.89
7	Productivity and health value gold and platinum	55.33
8	Less green cost premium	(4.00)
9	Total 20-year NPV (certified and silver)	48.87
10	Total 20-year NPV (gold and platinum)	67.31

buildings in which workers had productivity gains of up to 16%, including better work quality and lower absenteeism.

Energy Star is a prerequisite in LEED for existing buildings and owners and tenants place a premium on green buildings. LEED also provides financial and environmental benefits through intelligent site selection, water conversation, improved indoor air quality, waste reduction, and environmentally friendly material selections.

Healthy indoor environments can increase employee productivity according to an increasing number of case studies. Since workers are by far the largest expense for most companies (for offices, salaries are 72 times higher than energy costs, and they account for 92% of the life-cycle cost of a building), this has a tremendous effect on overall costs. (See *Green Development* by the Rocky Mountain Institute for more information.)

Project Cost Savings

Green Building projects that are well integrated and are comprehensive in scope can result in lower or neutral project development costs (Table 10.3). Rehabilitating an existing building can lower infrastructure and materials costs. Integrated design can use the payback from some strategies to pay for others. Energy-efficient building envelopes can reduce equipment needs—downsizing some equipment, such as chillers, or eliminating equipment, such as perimeter heating. Using pervious paving and other runoff prevention strategies can reduce the size and cost of storm water management structures.

Examples of Green Construction Projects

The following case studies illustrate successful applications of LEED principles in construction projects.

Case Study #1

A school built to LEED standards can accomplish significant savings.
Project size: 92,000 square feet

Project Cost: $12.93 million
Certification: LEED Silver

- Of all the waste produced by the construction, 81% was recycled; 39.5% of the construction materials were manufactured within a radius of 500 miles.
- Of the wood used in the building, 52% came from forests certified by the Forestry Stewardship Council as being sustainably managed.
- Of construction materials, 25% were made of recycled materials.
- The parking facilities had reserved parking for fuel-efficient vehicles.
- Faucets, urinals, toilets, and so on were selected to reduce water consumption by 30%.
- All paints, adhesives, sealants, and carpets contained very low levels of, or no volatile organic compounds (VOCs).

Case Study #2

The Alberici St. Louis Office Building
St. Louis, Missouri
Architects: Mackey, Mitchell Associates

The headquarters of the Alberici Corporation was completed in 2005, and received the highest possible rating from USGBC (i.e., a Platinum rating). This accomplishment was due to a highly efficient design that reduced energy usage by 60% relative to a standard design. The building featured high-efficiency HVAC, a better thermal envelope, increased daylighting and lower-lighting power densities. The building also produced renewable energy through a 65 KW wind turbine, equivalent to 18% of the building's regulated energy cost. Green strategies:

- Material and Resources: 20% by cost manufactured within 500 miles
- Over 50% of the wood specified was certified in accordance with the Forest Stewardship Council
- Over 10% of materials were salvaged or refurbished
- Construction/Demolition waste: 93% (6000 tons) diverted from landfills, pre-existing building used as structural fill
- Indoor environment: 75% of spaces naturally daylit, operable windows provide cross-ventilation. 90% of occupants have a direct outdoor view
- Underfloor air distribution
- Hydropowered, sensor-activated faucets,
- Natural linoleum flooring, Recycled rubber flooring

Note: In addition to being granted a LEED Platinum rating, this facility also received four Green Globes of a possible four.

Design Approaches for LEED

The USGBC advocates design integration as a means of creating buildings that serve the needs of users, owners, the community, and the environment. LEED should be integrated in the design at the very beginning in order to not only obtain the desired certification but to derive superior building performance as well. The design process needs to include nontraditional team members, a LEED-accredited professional (LEED® AP) and a commissioning agent/authority. In order to incorporate lean principles in a project, a professional who is knowledgeable in lean methodology such as a consultant or lean coach should be included in the design team as well.

The USGBC visualizes the design process as a collaboration of many disciplines with owners and other stakeholders to accomplish holistic design solutions that maximize building performance and a return on investment. It proposes design charrettes to bring multifaceted teams together in the predesign phase of a project to explore economic, environmental, and social issues. Very importantly, contractors and end users should be included in the collaboration, unlike the approach used in many traditional projects. This approach closely parallels lean design, in which contractors and designers resolve concerns that affect constructability and cost effectiveness. LEED charrettes include a wide range of professionals to develop a shared vision of a project, so that the extensive requirements of LEED credits can be incorporated in a design from the very outset. The attainment of LEED credits is by no means accidental; the process is highly structured and involves well-researched and deliberate actions with regard to design parameters.

The LEED standards embody such factors as urban planning, community connectivity, ecology, energy efficiency, materials, air quality, recycling, daylighting and artificial lighting. Consequently, buildings that attain high levels of LEED certification have experienced a process that integrates a very wide range of issues that not only secure the best environment for building users, but that ultimately reduce each facility's impact on the carbon footprint. Energy modeling and life-cycle cost analysis also facilitate important long-term decisions on the way each facility is designed and expected to operate in the future.

Commissioning plays a vital role in this collaboration. LEED standards require commissioning in order to ensure that the credits desired are not only programed into the design of a facility, but are also validated in actual building performance through exhaustive tests. Early use of the Commissioning Agent or Authority (CxA) clarifies the Owners' project requirements (OPR) and basis of design (BOD). The CxA also supports lean design by bringing to the team the experience gained on earlier projects that results in better constructability and in the avoidance of errors that are likely to require increased rework in the project. The CxA serves as a resource to facilitate the analysis of design choices to determine their short-term and long-term impacts.

In the earlier years following the promulgation of LEED standards, there was a misperception in the design and construction community that LEED standards resulted in higher initial costs and complicated procedures. With each passing year an increasing number of design and construction professionals have become LEED accredited, and have developed a better understanding of the procedures and processes involved in applying the standards. This has resulted in a reduction of the so-called Green Premium for LEED-based projects. Increasingly, LEED has become a part of the business model.

Lean construction benefits greatly from the foregoing collaborative approach; the design and construction team should also promote lean by avoiding the following "examples of waste related to sustainable construction"

- Ignoring the contractor's knowledge during the design
- Late recognition of green goals—when costs have escalated

- Demolition of a building that could otherwise be renovated
- Failure to reuse/recycle building elements and materials
- Lack of a "Green" design—leads to overdesign

Checklist for Environmentally Sustainable Design and Construction

- Smaller is better
- Design an energy-efficient building
- Design buildings to use renewable energy
- Optimize material use
- Design water-efficient, low-maintenance landscaping
- Make it easy for occupants to recycle waste
- Look into the feasibility of gray water and roof-top water catchment systems
- Design for future reuse
- Avoid potential health hazards: radon, EMF, pesticides
- Renovate older buildings
- Evaluate site resources
- Locate buildings to minimize environmental impact
- Pay attention to solar orientation
- Situate buildings to benefit from existing vegetation
- Minimize transportation requirements
- Avoid ozone-depleting chemicals in mechanical equipment and insulation
- Use durable products and materials
- Choose building materials with low-embodied energy
- Buy locally produced building materials
- Use building products made from recycled materials
- Use salvaged building materials when possible
- Minimize use of old-growth timber
- Avoid materials that will off gas pollutants
- Minimize use of pressure-treated lumber
- Minimize packaging waste
- Install high-efficiency heating and cooling equipment
- Install high-efficiency lights and appliances
- Install water-efficient equipment
- Protect trees and topsoil during site work
- Avoid use of pesticides and other chemicals that may leach into the groundwater
- Minimize job site waste
- Make your business operations more environmentally responsible

Challenges to Sustainable Construction

There are several challenges to sustainable construction. The building industry is currently fragmented and coordination may be difficult between architects, builders, and owners. There is a lack of government policy to encourage sustainable construction and a lack of general awareness of the benefits of green buildings. There has been development in greenfields with disconnected transportation systems rather than infill development in urban areas. Generic massive-scale developments have not utilized sustainable construction methods.

The primary barrier to the introduction of sustainable construction is the client's widely held belief that it will cost more and attract a higher risk (Landman 1999). They believe that the initial cost will increase while using new technologies, green materials, and extra design. This results, therefore, not only in higher capital cost but also in the lack of reliable accurate cost information. Most traditional construction methods only consider a short-term period, after that, all of the benefits may not be received by the builder/developer/owner. The long-term benefits from sustainable construction, taking into account life-cycle cost savings, have been less than attractive to many developers or construction companies.

Commissioning and LEED

As described in the second part of Chapter 8, commissioning is a critical component of the lean construction processes. The CxA is a third-party professional that reports to the owner directly, and acts as an extra pair of eyes and ears on the owner's behalf. Commissioning is a planned, systematic, quality control process that involves the owner, users, occupants, O&M staff, design professionals, and contractors. It is most effective when begun at the project's inception. It is essential for a project to be LEED-certified.

LEED recognizes two classes of commissioning: fundamental commissioning and enhanced commissioning. Fundamental commissioning involves having the CxA verify that a building's energy-related systems are installed, calibrated, and perform according to the OPR, BOD, and construction documents. Enhanced commissioning involves a similar task but is begun early during the design process and involves additional activities that continue after building completion. The impact of commissioning on LEED is most visible in the category of Energy and Atmosphere (EA):

EA Prerequisite 1: Fundamental Commissioning of the Building Energy Systems

EA Credit 3: Enhanced Commissioning

Goals of the commissioning (Cx) process are to:

- Improve the quality and functionality of the building and its key systems;
- Ensure consideration of facility operation and maintenance throughout design and construction;
- Enhance the quality and coordination of design and construction documents;
- Facilitate transition from construction to occupancy and use;
- Achieve higher-than-normal energy efficiency with a payback period of 7–8 years or less; and
- Support sustainability goals and LEED certification on a project-specific basis.

The Commissioning Authority serves as an objective advocate for the Owner and is responsible for:

a. Directing the commissioning team and process in the completion of the commissioning requirements,
b. Coordinating, overseeing and/or performing the commissioning testing, and
c. Reviewing the results of the system performance verification.

The Commissioning Authority shall report results, findings and recommendations directly to the Owner.

Full commissioning services are to be provided in five (5) phases:

- Predesign
- Design
- Construction
- Acceptance
- Postacceptance

Where projects involve LEED certification, the CxA shall provide both fundamental and enhanced commissioning for selected construction projects in order to meet the requirements of LEED in two places: EA Prerequisite 1, and EA Credit 3, as described in the USGBC's handbook.

CxA for Designated Building Systems—New Construction

The building systems included in the scope of the commissioning responsibility will include all or some of the following systems:

- HVAC systems
- Energy management systems (EMS)
- Lighting control systems
- Electrical power distribution systems
- Fire protection
- Fire alarm
- Smoke control
- Security alarms
- Telephone and intercommunications systems

Part B: Commissioning

The National Conference on Building Commissioning has officially defined "Total Building Commissioning" as: a "Systematic process of assuring by verification and documentation, from the design phase to a minimum of one year after construction, that all facility systems perform interactively in accordance with the design documentation and intent, and in accordance with the owner's operational needs, including preparation of operation personnel."

In effect, a building commissioning is an excellent quality assurance mechanism. For new construction projects it is a cradle-to-grave systematic process. It is optimally applied to all phases of a construction project—programing/predesign, design, construction/installation, acceptance, occupancy, and postoccupancy. It focuses on verifying and documenting that the facility and all of its systems and assemblies are planned, designed, installed, tested, operated, and maintained to meet the OPR (ASHRAE Draft guideline 0-2003). *The documentation of the owner's goals and needs; that is, project requirements (OPR) is of paramount importance. It clarifies and quantifies the needs, then records the extent to which they are met or exceeded.*

Categories of Commissioning

1. Total Building Commissioning: Verifies attainment of owners' goals and needs with regard to all building features/systems, from predesign/planning through acceptance/occupancy and postoccupancy.
2. Commissioning for equipment start-up and acceptance: The scope usually includes major MEP systems: HVAC, control systems, and electrical systems, fire suppression/alarm systems), final inspections, tests, and basic operating performance.
3. Commissioning for certification (LEED, Energy Star, Green Globes, etc.): The scope includes design reviews and construction inspections to certify performance goal attainment.
4. Commissioning for functional performance testing: Testing of selected building systems including the building envelope to verify that the respective performance expectations have been met.

For existing buildings, there are two main approaches: (1) Where a building/facility was provided with building commissioning at the time of construction, recommissioning ensures that the established performance levels are sustained. (2) Where building systems were never formally commissioned, retro commissioning describes the application of commissioning procedures to existing building systems.

Importance of Commissioning

The construction process is fraught with many difficulties and challenges. Buildings consist of many components and systems, many of high complexity based on the latest digital technology. Since the 1980s, this complexity has been increasing in U.S. construction while designers, suppliers, and technicians have become more specialized. This situation would imply a need for greater collaboration between the parties, but it has not materialized as they have protected their self-interest in order to reduce their operating costs and potential liabilities. The resulting gaps in responsibility often lead to inadequate design detail as designers pass on liability for system performance to contractors; the contractors in turn are expected to resolve installation details through shop drawings that often have critical details missing or incomplete. Equipment is often located with inadequate clearances for proper maintenance. Given the many interactions between subcontractors and trades that assemble building systems under difficult field conditions, it is not unusual for some systems to fail or simply not work as they were originally specified.

This shortcoming can work in direct opposition to lean thinking. A facility that has been designed for lean operations, and configured so that the construction process is based on lean concepts, can fail to meet expectations if critical systems do not function as intended. The commissioning process ensures that professionals are focused on ensuring that all systems work, while everyone is engaged in frantic activity to erect a facility.

Commissioning and Lean Construction

Lean construction requires a collaboration of all design and construction stakeholders. There is evidence that the early involvement of contractors in the design process, for example, leads to more constructible designs. By the same token, commissioning agents improve both the design and construction processes by enabling designs that use resources cost effectively, and that benefit from the quality assurance provided by the commissioning process. This approach minimizes waste, not only in the design of facilities, but also in their subsequent operation.

In renovation and addition projects it is important to implement recommissioning procedures. Today's and future construction must be based on energy-efficient standards; while an addition might be based on up-to-date energy standards, it would be counterproductive not to bring the preexisting construction up to similar performance levels.

The engagement of commissioning services needs to be dovetailed with the requirements for sustainability classifications. In LEED standards, additional points are obtained when enhanced commissioning is implemented. Enhanced commissioning has the Cx involved to a greater extent in the predesign/planning phase as well as in the acceptance and postacceptance phases.

An important benefit of the commissioning process is the quality assurance provided by ensuring that installation work is faithful to the plans and specifications. Problems and conflicts are detected early when the CxA asks installers to verify installation details at a time when corrective work can be done cost-effectively if it is needed. In the case of HVAC systems, for example, this can be very beneficial; changes to ductwork configuration may avoid conflicts, but airflow quantities may be very difficult to compensate for once ceiling spaces are filled with various plumbing, sprinkler systems, and electrical raceways. The Cx review process ensures that user/client satisfaction is secured with the final product, in keeping with lean principles.

Building performance can be validated by using the commissioning process for quality assurance beyond the requirements of LEED standards. For example, CxA review of the building envelope is not required by LEED, but obligates the contractor to demonstrate that there are no leaks or early points of failure.

Commissioning Service Providers

As previously stated, the CxA is an objective, independent advocate for the building owner. That professional has significant design and hands-on experience with building mechanical and electrical systems. The CxA is hired or assigned by the owner, contractually independent of the design firm, the construction firm, subcontractors, or equipment suppliers.

Rationale for Commissioning

Commissioning reduces a building owner's total cost of ownership. Energy costs are a major component of a building's operating costs. It is likely that these costs will increase in the future as utilities respond to market uncertainties. Even though a facility may have been designed to be energy efficient, contractors focus on getting equipment installed in place to meet project deadlines. It is easy for errors to occur that prevent equipment from performing as specified. Commissioning is a formal requirement for facilities to qualify for LEED certification. It provides assurance that a facility's systems are functioning as designed, and that LEED or other certification is justified. With regard to scope, commissioning may be applied to a complete facility or to individual building systems such as HVAC.

Other Benefits of Commissioning

- It clarifies owners' project requirements (OPR). These up-front requirements are an important foundation for project design. The CxA's input on the consequences of these requirements helps to improve the overall probability of desired system performance.
- Project designs have fewer errors/imperfections.
- Project cost is reduced, as change orders are avoided.
- Delays, waste, and rework are minimized.
- Punch lists are reduced.
- Decision-making and construction are speeded up.
- The operation and maintenance of facilities are improved, especially with regard to electrical, mechanical, and plumbing. Operating costs (including energy) are reduced because of optimum performance.
- There is trouble-free turn over—systems are fully functional.
- There are fewer warranty calls to contractors.
- Equipment life is extended, as it works within normal limits.
- Air quality is improved and occupants experience fewer illnesses.

Commissioning versus Testing, Adjusting, and Balancing

Testing, adjusting, and balancing (TAB) are routinely performed on newly installed HVAC systems, but this action does not replace the need for the formal commissioning of these systems. TAB specialists check water and air flows under normal working conditions, but are not required to test these systems under the extreme conditions that may occasionally occur. Their scope of work does not include validating that the OPR have been met; rather, they help to adjust systems such as the HVAC so that they function satisfactorily.

Full building commissioning includes a much wider variety of building systems than does the test and balance process; TAB focuses on HVAC and electrical, and mechanical systems. On the other hand, full building commissioning includes the building envelope—roofing, doors, windows, communications systems, and life-safety systems. The qualifications of the CxA exceed the requirements of TAB specialists. In a typical new construction project, the CxA is brought on board as early as the predesign stage. A competent CxA is expected to interact with the owner and designers to define project requirements for the owner that also promote the best environmental practices.

Commissioning Cost/Benefit Analysis

Estimates of commissioning cost vary among different sources; however, there is evidence that these costs are recovered within a relatively short period of time from the savings in operating costs. Guidelines published by the Commissioning Group (ACG) of the Associated Air Balance Council (AABC) identify the cost of commissioning HVAC systems as between 2 and 5% of HVAC construction cost. Other estimates place costs between 2–3% of total mechanical cost for commissioning HVAC and control systems and 1 to 2% of total electrical cost for commissioning electrical systems. Overall, commissioning HVAC and electrical systems in typical buildings costs 0.5–1.5% of total construction cost (Table 10.4).

The University of Wisconsin's (UW) Commissioning training program points out that the commissioning process should clearly document that its total cost through 1 year of occupancy has been recovered by the end of design or early construction. UW indicates that if this is not the case, then the process is not being properly implemented. For each $80,000 paid for commissioning, returns have been as high as $500,000. UW points to many design-based savings in addition to a smoother construction process such as

- Needs identification early in the process
- Proper system/component selection
- Reduction of requests for information (RFIs)
- Problems are discovered earlier when easier to correct

Overall, the commissioning process benefits all construction stakeholders: owner, occupants, designers, contractors, subcontractors, testing agents, manufacturers/vendors, bonding companies/insurers, and LEED professionals.

Summary of the Commissioning Process

The following factors ensure a successful commissioning process:

- Commitment by the owner/client to the process, and ongoing support. Many contractors do not attach great importance to the commissioning process and may not respond to communications in a timely manner.
- Early clarification of the CxA's role and requirements in the contract documents as well as in preconstruction meetings.

TABLE 10.4

Estimated Commissioning Costs

	Commissioning Scope	Estimated Cost Range
1	Whole building: controls, electrical, mechanical; predesign through warranty	0.5–3% of total construction cost
2	HVAC and automated controls systems only	1.5–2.5% of the mechanical contract cost
3	Electrical system only	1–1.5% of the electrical contract cost
4	Energy-efficiency measures	$0.23–0.28/square feet ($2.48–$3.01/square meter

- An experienced project manager for commissioning that represents the authority.
- A clearly defined commissioning plan that all the parties understand.
- A cooperative attitude by all the parties to the project.

To be fully effective, commissioning should begin with the predesign stage of the project and continue beyond the warranty phase. As stated previously, the commissioning authority can help the owner/client to define the OPR to take advantage of technological requirements as well as cost-effective construction methods. This involvement enables designers to adopt a holistic design that incorporates energy efficiency and sustainability, minimizing environmental pollution, and maximizing life-cycle cost effectiveness.

The building commissioning scope of work should be interwoven into the project's goals for performance, quality, safety, and innovation. The CxA should be the objective "eyes and ears" of the owner/client throughout the project to ensure that construction delivery decisions do not diminish the functional performance of the equipment specified for the project. At the substantial completion stage, the CxA oversees functional performance tests to ensure that the facilities' systems work as promised. The CxA ensures that the O&M staff are properly trained by the contractor in all the procedures and equipment needed for reliable operation. In the post-warranty stage the CxA may recommission a facility to ensure that systems continue to work cost effectively.

Qualifications

The commissioning authority (CxA) requires a broad and deep set of skills to be effective. Technical expertise is needed in building design, operations, and diagnostics. This is especially important with the ongoing emergence of sophisticated electronic control systems for communications, security, alarms, energy management, building automation systems, and so on. In particular, today's CxA should be LEED accredited in order to meet the qualifications for LEED-certified projects.

Cutting-edge approaches involve the use of computer modeling to visualize the expected performance of a system, then verifying that performance in the field. Management and communications expertise is also needed to enable the CxA to work effectively with others and maintain cooperation and collaboration. Expertise is also essential in documenting procedures as well as providing training to facility O&M staff and users. The following factors should be considered in selecting a commissioning provider:

1. **Related experience:** At a minimum, the company's qualifications and experience should include the following:
   A. Preferably at least 10 years experience with the types of building, HVAC, and control systems included in the subject project.
   B. Experience in commissioning at least two projects of similar size and of similar equipment to the current project. This experience should include writing functional performance test plans and assembling a complete commissioning plan.
   C. Experience in the testing, design, specification, or installation of commercial building mechanical and control systems. and other systems being commissioned.

    D. Experience working with project teams, project management and conducting scoping meetings; good team building skills, strong communication skills, especially documentation.

    E. Experience in planning and delivering O&M training.

    F. Direct responsibility project management of at least two (2) commercial construction or installation projects with mechanical costs greater than or equal to current project costs.

    G. Experience in design installation and/or troubleshooting of direct digital controls and energy management systems (EMS) if applicable.

    H. Demonstrated familiarity with metering and monitoring procedures. Knowledge and familiarity with air and water testing and balancing.

2. **Professional/technical staff:** Typical requirements, depending on project size would include the CxA's organization having a staff of no fewer than five (5) field technicians. Personnel performing the work should preferably have a mechanical engineering degree from an accredited college or university or a state-issued mechanical contractor's license and/or certified by either the Building Commissioning Certification Board (BCCB), AABC, or the National Environmental Balancing Bureau (NEBB).

Certification Organizations

It is important to ensure that a commissioning agent/authority has taken an appropriate course of study, and has the experience necessary to provide the quality of expertise needed to justify the cost of commissioning services. In 2009, there were five organizations that certify commissioning providers. Each organization had its own set of requirements and a different title for the providers it certified. The organizations and certifications are listed below; requirements for several of these organizations are listed below also:

1. Building Commissioning Association (BCA), Certified Commissioning Provider
2. National Environmental Balancing Bureau (NEBB), Systems Commissioning Administrator
3. Associated Air Balancing Council Commissioning Group (ACG), Systems Commissioning Administrator
4. University of Wisconsin–Madison (UWM), Accredited Commissioning Process Provider
5. Association of Energy Engineers (AEE), Certified Building Commissioning Professional (CBCP®)

Building Commissioning Certification Board (BCCB)

The BCCB confers the Certified Commissioning Professional (CCP) designation. The CCP must have at least 3 years of experience as a commissioning provider in a lead role, must have been continuously employed as a commissioning provider in at least three full years during the previous five years. The CCP must have completed—as lead commissioning provider from project design through the end of construction—three or more commissioning projects for new construction or major renovations of buildings totaling at least 150,000 square feet and at least $30 million, or two or more projects of at least 100,000 square feet and $15 million, plus the BCA training.

National Environmental Balancing Bureau (NEBB)

The NEBB certification covers HVAC systems commissioning, plumbing systems, and fire protection systems (as of 2007). Firms must have operated for a minimum of 12 months continuously with full-time employees as an installing piping contractor, sheet metal contractor, mechanical contractor, test and balance contractor, or engaged in building systems commissioning work. Supervisors must have a minimum of 10 years experience including four years of HVAC-related experience, two years of TAB experience, and four years of supervisory experience in TAB or Building Systems Commissioning.

Associated Air Balance Council (AABC)

The AABC confers the Certified Commissioning Authority (CxA) designation to professionals and their employers. The applicant company must employ at least one registered architect and professional engineer, or certified test and balance engineer. One of these positions must be the company's designated representative who is responsible for overseeing all commissioning activities. Successful completion of the CxA examination results in both the individual and the company receiving the CxA designation. The CxA must have experience commissioning at least two projects of similar size and of similar equipment to the current project. This experience should include:

- Writing functional performance test plans and assembling a complete commissioning plan.
- Direct responsibility project management of at least two commercial construction or installation projects with mechanical costs greater than or equal to current project costs.

Commissioning Requirements in Construction Documents

(Reproduced with permission from the USGBC *New Construction and Major Renovation Reference Guide* 3rd ed., October 2007.)

- Commissioning team involvement
- Contractors' responsibilities
- Submittals and submittal review procedures
- O&M documentation/manuals
- Meetings
- Construction verification procedures
- Start-up plan development and implementation
- Functional performance testing
- Acceptance and closeout

- Training
- Warranty review site visit

Sample of Commissioning Contract Requirements

A. General Scope of Services:

1. Serve as the Commissioning Authority for Building Commissioning (CxA) for a specified variety of major building systems, including HVAC systems, in construction projects.

2. Be responsible for
   - directing the commissioning team and process in the completion of the commissioning requirements
   - coordinating, overseeing, and/or performing the commissioning testing, and
   - reviewing the results of the system performance verification. The Commissioning Authority (CxA) shall report results, findings, and recommendations directly to the Owner.

3. Full commissioning services are to be provided in five (5) phases: predesign, design, construction, acceptance, and postacceptance. One or more of these phases may be excluded on a project-specific basis.

4. Where projects involve LEED certification, the CxA shall provide both fundamental and enhanced commissioning in keeping with the requirements of the USGBC.

5. The building systems included in the scope of the commissioning responsibility will include all or some of the following systems where applicable:
   - HVAC systems
   - EMS
   - Lighting control systems
   - Electrical power distribution systems
   - Fire protection
   - Fire alarm
   - Smoke control
   - Security alarms
   - Telephone and intercommunications systems
   - Cable TV and CCTV systems
   - Elevator controls
   - Building envelope

Commissioning Agent Responsibilities (Figure 10.2, Table 10.5):

1. **Predesign phase:** Assist Owner and project team to develop OPR and BOD
   a. The OPR should detail the functional requirements of a project and expectations with regard to the systems that will be commissioned.

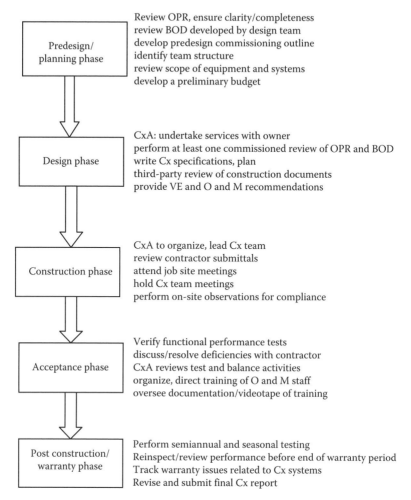

FIGURE 10.2
Process flow for commissioning.

b. The BOD provides a narrative describing the design of the systems to be commissioned, including design assumptions. Review scope of equipment and systems.

Prepare a predesign commissioning outline. It shall contain:

- A description of the commissioning process
- Identification of the commissioning team's structure, with a summary of the roles and responsibilities of each team member
- Identification of the systems to be commissioned
- Preliminary commissioning time requirements for each project phase
- Commissioning cost estimates

TABLE 10.5

Comparison of Fundamental and Enhanced Commissioning

	Tasks	Fundamental Cx	Fundamental and Enhanced Cx
1	Develop basis of design	Design team	Design team: CxA may assist
2	Incorporate commissioning requirements in construction documents	Project team or CxA	Project team or CxA
3	Conduct Cx design review before midconstruction docs	N/A	CxA
4	Develop and implement a Cx plan	Project team or CxA	Project team or CxA
5	Review Cx-related contractor submittals	N/A	CxA
6	Verify installation and performance	CxA	CxA
7	Develop a systems manual for Cx items	N/A	CxA
8	Verify training requirements met	N/A	Project team or CxA
9	Complete a summary Cx report	CxA	CxA
10	Review building operation within 10 months after completion	N/A	CxA

2. **Design phase**

   - Perform, at a minimum, one commissioning design review of the OPR, BOD
   - Review design to ensure it accommodates commissioning
   - Write commissioning specifications defining contractor responsibilities
   - Write the commissioning plan
   - Establish the project schedule for commissioning
   - Perform plan reviews during the design phases of the project, identifying items that cause problems in obtaining the final balance conditions required by the design. Upon completion of the review, a written report of recommendations shall be submitted.
   - Conduct a third-party review of the construction documents prior to bidding

3. **Construction phase**

   - Organize and lead the commissioning team
   - Review contractor submittals applicable to the systems being commissioned for compliance with the OPR and BOD. This review shall be concurrent with A/E reviews and submitted to the design team and the owner.
   - Update the commissioning plan to reflect equipment and controls data from the submittals.
   - Perform on-site observations during construction for compliance with the specifications and prevailing industry standards. Discuss deficiencies with the appropriate contractor's representative and report findings to the owner's representative.
   - Maintain an issues log, tracking all issues that arise, to ensure they are resolved.
   - Carry out and document system verification checks

- Carry out and document system start-ups
- Witness equipment and system start-ups as deemed necessary; ensure documentation is complete.
- Review test and balance reports

4. **Acceptance**

- Review test and balance (TAB) activities as specified for the project
  - Carry out functional performance tests (FPT) on all designated systems
  - Collaborate with the contractor to retest repaired items
  - Organize and direct the training of the O&M staff for effective, on-going O&M of all systems
  - Videotape O&M staff training sessions
  - Provide full documentation

5. **Postacceptance**

- Perform and document any required "off season" functional performance tests
- Review building operations within 10 months after substantial completion with O&M staff and occupants. Include a plan for resolution of outstanding commissioning—related issues.
- Revise and submit final commissioning report
- Follow-up with post-warranty validation and/or recommissioning.

Special Requirements for Retro-Commissioning

Retro-commissioning seeks to ensure the functionality of equipment and systems and also to optimize how they operate together in order to reduce energy waste and improve building operation and comfort. Thus, the goal of ensuring comfort and productivity of the building occupants accompanies the goal of cost savings.

Retro-commissioning occurs after construction, as an independent process, and its focus is usually on energy-using equipment such as mechanical equipment, lighting, and related controls. It may or may not emphasize bringing the building back to its original intended design.

Retro-commissioning begins with the planning phase, which consists of identifying project objectives, targeting systems for improvement, and defining tasks and responsibilities. A plan for conducting the work should then be prepared. An investigation phase follows, in which on-site assessment and testing are conducted. This phase allows deficiencies to be found and the scope of work to be refined. Once the scope is finalized, the improvements are then carried out in an implementation phase, and their success is validated. Finally, the completed improvements are "handed-off" to the owner along with information and knowledge gained during the process to help ensure long-term performance for the owner.

Retro-commissioning (EBCx) steps: The CxA will carry out the following duties:

1. Planning phase

   Develop retro-commissioning objectives for approval by the Board

- Review available documentation and obtain historical utility data
- Develop the retro-commissioning plan
- Prepare commissioning cost estimates for the specific project including the (a) cost of services to be provided, and (b) payback from energy and other savings

2. There is no design phase.
3. Investigation phase
   - Perform site assessment
   - Obtain or develop missing documentation
   - Develop and execute functional test plans
   - Analyze results
   - Develop master list of deficiencies and improvements
   - Recommend cost-effective improvements for implementation
   - Identify the scope of major repairs/installation to be performed by others

4. Implementation phase
   - Implement minor repairs and improvements /or conduct field visits, inspections, and oversight of tests (by others) to review work performed by others
   - Review Pre-Functional and Functional Testing and the interoperability of systems and components (Performance Testing) to confirm compliance with the retro-commissioning plan
   - Retest and re-monitor for results
   - Fine tune improvements if needed
   - Revise estimated energy savings calculations

5. Project hand-off and integration phase
   - Prepare and submit final report
   - Compile or update an indexed systems manual, which contains important building documentation such as construction record documents, specifications, submittals, narrative description of the sequence of operations, training materials, ongoing optimization guidance
   - Perform deferred tests (if needed)
   - Develop and provide an O&M plan for the facility, including guidelines for implementing a new preventive maintenance plan
   - Develop a comprehensive training plan for O&M staff. Train staff on system operation and maintenance. Prepare a video recording of this training for future reference.
   - Develop a recommissioning plan/schedule (3-year schedule), a plan for operational sustainability and ongoing commissioning

6. Persistence phase
   - Implementation of systems and tools to support both the persistence of benefits and continuous performance improvement over time.

- Monitoring and tracking of energy use
- Trend key system parameters to detect problems early and assess system performance

Building Systems to Be Retro-Commissioned (EBCx) Existing Construction

The building systems included in the scope of the recommissioning responsibility will include all or some of the following systems:

- HVAC systems
- EMS
- Electrical power distribution
- Lighting control systems
- Emergency power generators and automatic transfer switching
- Uninterruptible power supply systems
- Fire protection
- Fire alarm
- Smoke control
- Plumbing systems, water distribution, sanitary
- Telephone and intercommunications systems
- Cable TV and CCTV systems
- Elevator controls

Questions for Discussion

1. Describe the benefits of sustainable buildings
2. What is the purpose of having a LEED-Accredited Professional (LEED® AP) in the design team for a lean project?
3. Some members of the construction industry believe that implementing both LEED and lean in a project may inflate costs. Discuss.
4. Why should an owner pay for commissioning, when fees are already being paid to architects and engineers?
5. What specific services would you expect from the CxA at each phase of design and construction?
6. What are the minimum qualifications you would expect the CxA to have?
7. If an owner is not seeking LEED certification or another environmental certification, is the cost of having a commissioning authority (CxA) justified?
8. How does enhanced commissioning differ from fundamental commissioning?
9. In what way is sustainability linked with lean construction?
10. In what way does commissioning support lean construction?

Appendix: Certification (Certifying Organizations, Designations, and Web Sites)

Associated Air Balancing Council Commissioning Group (ACG), Certified Commissioning Provider

- http://www.commissioning.org

Association of Energy Engineers (AEE), Certified Building Commissioning Professional (CBCP®)

- http://www.aeecenter.org

Building Commissioning Association (BCA), Certified Commissioning Professional (CCP)

- http://www.bcxa.org

National Environmental Balancing Bureau (NEBB), Systems Commissioning Administrator

- http://www.nebb.org

University of Wisconsin–Madison (UWM), Accredited Commissioning Process Provider

- http://epdweb.engr.wisc.edu

Web Sites for Locating Commissioning Information

Building Commissioning Association

- http://www.bcxa.org/

Commissioning for Better Buildings in Oregon

- http://www.energy.state.or.us/bus/comm/bldgcx.htm

GSA Project Management Center of Expertise

- http://pmcoe.gsa.gov/What_We_Offer/Links/links.asp

Total Building Commissioning

- http://sustainable.state.fl.us/fdi/edesign/resource/totalbcx/

U.S. Green Building Council

- http://www.usgbc.org

Whole Building Design Guide

- http://www.wbdg.org/

Commissioning: Related Organizations with Web References

Air Conditioning and Refrigeration Institute

- http://www.ari.org/

Air Movement and Control Association

- http://www.amca.org/

American National Standards Institute

- http://www.ansi.org/

ASHRAE

- http://www.ashrae.org/

ASHRAE Puget Sound Chapter

- http://www.pugetsoundashrae.org/

Chartered Institution of Building Services Engineers (UK)

- http://www.cibse.org/

Commissioning Specialist's Association (UK)

- http://www.csa.org.uk/

DOE FEMP/GSA Building Commissioning Guidelines

- http://www.eren.doe.gov/femp/techassist/bldguide.pdf

e-Design online

- http://fcn.state.fl.us/fdi/

Federal Energy Management Program

- http://www.eere.energy.gov/femp/techassist/bldgcomgd.html

National Conference on Building Commissioning and Portland Energy Conservation Inc. (PECI)

- http://www.peci.org/

National Fire Protection Association

- http://www.nfpa.org/

Oregon Office of Energy

- http://www.energy.state.or.us/bus/comm/bldgcx.htm

Portland Energy Conservation, Inc.

- http://www.peci.org/cx/

References

Kibert, C. J. 1994. *Proceedings of the First International Conference on Sustainable Construction.* Center for Construction, FL: University of Florida.

Landman, M. 1999. *Breaking through the barriers to sustainable building: insights from building professionals on government initiatives to promote environmentally sound practices.* Medford, MA: Department of Urban and Environmental Policy, Tufts University.

U.S. Green Building Council (USGBC). 2007. *New construction and major renovation reference guide, version 2.2,* 3rd ed. Washington, DC: USGBC.

Bibliography

Heschong. 1999. *Skylighting and retail sales: An investigation into the relationship between daylighting and human performance*. Gold River, CA: Heschong Mahone Group. Available at http://www.h-m-g.com/Daylighting.

Odom, J. D., and G. DuBose. 1999. *Commissioning buildings in hot humid climates*. Meridian, CO: CH2MHILL.

Roodman, D. M., and N. Lessen. 1995. *Worldwatch paper 124: A building revolution*. Washington, DC: Worldwatch Institute.

11

Selected Performance Improvement Tools and Techniques

Performance Improvement in Construction

Lean construction seeks to improve construction performance primarily by integrating the actions of the stakeholders in the supply chain. This approach can be made even more effective by applying a number of analytical techniques that have been adapted to the construction environment. In a traditional environment, these techniques can provide significant benefits at the activity level. In a lean project environment, they can be even more beneficial by supporting project-wide optimization.

Performance improvement requires a frame of reference in order to be effective. It provides that reference point by determining the current state of key processes in order to measure the impact of subsequent improvement strategies. It provides the information needed for process control and makes it possible to establish challenging yet practical goals. These actions support the implementation of competitive business strategies. Although some construction managers recognize the importance of performance measurement, it has not been widely implemented in a formal sense. These managers often base their decisions on intuition and common sense, as well as on a few broad financial measures that are no longer adequate in today's competitive environment.

The above mentioned techniques, therefore, need to be applied in the context of performance measurement. Unless the current state of a process is measured it cannot be fully understood. If it is not fully understood, then the impact of performance improvement techniques cannot be properly evaluated.

The following techniques may be used to increase construction performance.

- Work measurement
- The learning curve
- Cycle time analysis
- Simulation
- Quality Function Deployment (QFD)

Work measurement techniques have been traditionally used by industrial engineers in the manufacturing and service industries. They include:

- Work sampling
- Stop watch time study
- MTM

Construction professionals can also use work measurement techniques to increase construction productivity. Whereas standard work times are often used by the manufacturing and service industries, these standards need to be reviewed and updated for the construction industry. These standards can be tailored further to specific projects to reflect the logistics of the work site and to adjust the standards to represent methods improvement. The more accurate the information that is available on work standards, the better construction managers can conduct the preplanning of projects and exert control over the costs and schedules of these projects. Many construction standards need to be reengineered to reflect the use of technology in work processes.

Work Sampling

Work sampling is a useful method for conducting work measurement studies. It can estimate the proportions of the total time spent on a task in terms of its various components. It is particularly useful in distinguishing between direct and indirect work in situations where stopwatch time study is not practical or appropriate. Work sampling involves taking a small portion or sample of the occurrences in an activity being studied in order to make inferences about the frequency of occurrences in the overall activity. In construction, for example, certain tasks have high variability—installing drywall panels may involve taking an increasing number of steps to take them from a stack to the point of installation. Also, the panels are often not installed as whole sheets—workers may need to cut holes in the panels for electrical outlets and other devices. On the other hand, it is useful to know what percentage of the time workers are actively working instead of being idle, and work sampling can provide that information.

Work sampling's origins lie in its application as ratio delay studies by L. H. C. Tippett in the British textile industry (Tippett and Vincent 1953). Ralph Barnes (1980, 2009) also documented the work sampling technique. It is also known as activity or occurrence sampling.

A process is observed at intervals such that each activity has an equal chance of being observed; the probability of an occurrence is based on the binomial distribution. For convenience, the normal distribution is used, especially when the number of observations is large.

Typically the variable p is used to represent the proportion of an occurrence; that is, the proportion of a sample that will have a particular feature or characteristic.

Proportion $p = x/n$ where x is the number of samples that exhibit a particular characteristic, and n is the size of the sample.

Accuracy is expressed by the limit of error. It refers to the range in which the value of p will lie. It is the percentage of variation above and below the sampling estimate where the true value will lie, for a given confidence interval.

Absolute accuracy (A) = p · r where p is the proportion and r is the relative accuracy.

If the proportion p = 40%, and relative accuracy r is 5%, then the desired absolute accuracy

$$A = 0.4 \times 0.05 = 0.02 \text{ i.e., (2\%).}$$

The true value of p will lie in the range

$$40\% \pm (40 \times 0.05) = 40\% + 2\%, \quad \$0\% - 2\% \to 42\% \text{ to } 38\%$$

Confidence Interval

The confidence interval refers to the dependability of an estimation based on a sample. It denotes the long-term probability that the ratio p will lie within the limits of accuracy. A 95% confidence interval implies that 95 of 100 samples would contain the true value of p. The higher the value of the confidence interval, the greater the number of samples that are required for a given level of accuracy.

In most construction situations, a confidence interval of 95% and a limit of error or accuracy of 5% are generally adequate. The percentage of occurrences is also usually between 40 and 60%.

With:

$$\text{Confidence interval estimate: } P \pm Z \cdot \sqrt{\frac{p(1-p)}{n}}$$

Table 11.1 shows the number of standard deviations above and below the mean of a normal distribution that represents the population sample. A confidence level of 95% corresponds to 1.96 standard deviations (approximately two) above and below the mean.

Similarly, a confidence interval of 68% corresponds to one standard deviation on either side of the mean.

Number of Samples Required

In order to conduct a sampling study, an observer needs to predetermine the confidence interval and desired limit of error. The number of samples required may be determined from the equation below. n = number of samples, z = number of standard deviations, s = limit of error, and p = category proportion.

$$n = Z^2 \cdot \frac{p(1-p)}{S^2}$$

If a number of observations n is selected first, then the resulting limit of error S may be determined from the equation

$$S = Z \cdot \sqrt{\frac{p(1-p)}{n}}$$

An example is shown here of a preliminary estimate from a work sampling study with a percentage occurrence p of 40%.

Confidence level desired = 95%
Accuracy desired = 5%.

TABLE 11.1

Confidence Levels and Corresponding "Z" Values of Standard Deviation

Confidence Level	Z Values (Number of Standard Deviations)
68%	+−1.00
90%	+−1.64
95%	+−1.96
99.73	+−3.0

Find the number of samples needed to meet those requirements (see Table 11.1).
Number of standard deviations z = 1.96 from Table 11.1

$$p = 0.4, \quad s = 0.05$$

Substituting: n = $(1.96)^2 \cdot 0.4(1-0.4)/(0.05)^2 = 3.842 \cdot 0.24/0.0025 = 369$ samples
Check your understanding:

1. How many samples would be needed for a 90% confidence interval?
2. At the 90% confidence interval, what accuracy would be obtained with 721 samples?

Work Sampling Procedure

- Establish the purpose of the sampling study
- Clarify the categories of the activities to be measured
- Estimate percentage (p) with a trial study
- Select the confidence intervals and the accuracy desired
- Calculate the number of observations needed
- Schedule observations based on randomized observation times
- Develop data collection forms for sampling
- Collect the data
- Summarize the results

Duration

The duration of the study will be determined based on the total number of observations needed and the number that can reasonably be taken each day. If 384 observations are needed and 40 can be taken per day, the number of days required = 384/40 = 9.6 days. For planning purposes, 10 days would be required for collecting the required number of samples.

Randomization

The validity of work sampling is based on the assumption that each component of an activity has an equal chance of being observed. The observations are required to be random, unbiased, and independent. The use of a random number table provides a convenient method for structuring a study so that the required number of observations can be planned for, within the time frame allotted. Each observation should be made at a time determined by chance such that no bias is introduced by having the subjects anticipate the observations. The observation times can be selected from random number tables or from electronic random number generators.

Stopwatch Time Study

The stopwatch time study is based on the successive observation of a repetitious task to measure the time taken for each of its components. The purpose of such a study is to determine a reasonable and predictable time for the performance of a task. This is termed "normal time." In turn, this time is adjusted to create a "standard time," and serves as a mechanism for determining the staffing required to accomplish a given task or set of tasks.

The greatest benefit of stopwatch time is derived when it is based on a "best method" or "best practice." It would be pointless to dedicate extensive effort to developing a standard based on a method that is not the most effective one. Before embarking on the time study, considerable attention should be given to clarifying a best method that is agreed upon by all concerned. If workers "buy in" to the method, then the time standards that are based on it are much more likely to be accepted.

Allowances

Workers develop fatigue by performing a task repeatedly. They also need to take breaks for personal time, delays, and other routine activities that are not a part of the task itself. Standard time is determined by adjusting normal time to include allowances. A typical stopwatch time study is structured as follows:

1. Document the method used for the activity in detail. The work cycle should be subdivided into discrete steps or elements.
2. Measure and record elemental times. Observe multiple cycles of the same activity to meet accuracy requirements, based on the variability of the elemental times.
3. Rate the effort exerted by the subject in order to establish a fair base for the elemental times. The observer must have the experience and expertise necessary to determine if a subject is working at a faster or slower than normal pace. The standard should be based on a so-called average worker, working at a normal pace that can be sustained. The observer may rate individual elements of the activity separately, but should provide an overall effort rating for the entire activity.
4. Calculate normal time. This is done by multiplying the observed time by the performance rating. For example: mean observed time = 2.0 minutes per piece; effort rating = 115%; and normal time = $2.0 \times 1.15 = 2.3$ minutes per piece.
5. Calculate allowances. For example, the allowance for personal time, delays, and other activities may be 25%.
6. Calculate standard time. Standard time = normal time $\times$ allowance factor = 2.3 minutes $\times 1.25 = 2.875$ minutes per piece.

Methods Time Measurement

Methods time measurement (MTM) can be used to develop engineered standards. The MTM is based on the concept that a method must first be developed, elemental steps defined, and standard times developed. The standard must be based on the average times necessary for trained experienced workers to perform tasks of prescribed quality levels, based on acceptable trade practices. This approach is most practical with repetitious tasks of a relatively short duration.

In the MTM system, operations are subdivided into tasks; tasks are further reduced to individual body movements such as reaching, grasping, applying pressure, positioning, turning, and disengaging. Other movements include eye travel; focus; and body, leg, or foot motions. Each body movement is subdivided into individual actions, such as reaching

TABLE 11.2

Examples of Methods-Time Measurement
Application Data

Activity	TMU	
Reach 2″	4	
Grasp (simple)	2	
Turn	6	1 TMU = 0.00001 hour
Regrasp	6	= 0.0006 minute
Look (eye time)	10	= 0.036 second
Leg motion	10	
Kneel on one knee	35	
Arise	35	
Total	108	

Note: All times include a 15% allowance
Normal time = 108 TMU × 0.036 seconds/TMU = 3.89
  seconds.
In calculating the standard time, provision must be made
  for breaks.
If personal allowances are 25% then standard
  time = 3.89 × 1.25 = 4.86 seconds.

two inches, grasp, apply pressure, turn, and so on. Each action is assigned a standard time stated in time measurement units (TMU; Table 11.2).

In applying the MTM system (or any other standardized measurement system) it cannot be overemphasized that an appropriate method must first be established that can be successfully applied by the average, trained worker at definitive quality levels. The effect of the learning curve should also be considered when establishing work standards to ensure that repetition does not render the task times excessively long. Further details on MTM can be obtained from the current edition of the *Handbook of Industrial Engineering*, J. Salvendy, editor.

Learning Curve

A learning curve is the phenomenon demonstrated by the progressive reduction in the time taken by an individual, or by a team to perform a task or a set of tasks on a repetitious basis. The individuals performing the task or project become more proficient with each repetition; the observed improvement serves as a motivator and a learning tool resulting in successively shorter performance times. The learning curve is represented by an equation of the form

$$T_n = T_1 \cdot n^{(-a)}$$

where Tn is the time for the nth cycle, T1 is the time for the first cycle, N is the number of cycles and a is the constant representing the learning rate. This equation produces a hyperbolic curve.

In order to determine the learning rate of a given activity, the time study may be applied to a worker who is performing the task. For example, masons installing concrete blocks to form a wall would be timed as they perform successive iterations of the process. The learning curve can be applied to construction projects. It can be highly relevant in repetitive projects such as housing construction, but the success of this application requires the IE to understand that interruptions to the construction process limit its use. Examples of such interruptions include prolonged shutdowns or over holidays.

Construction tasks are often varied and nonrepetitive, so the user has to apply the concept very judiciously. On-site managers who understand the learning curve rates for different types of tasks can improve work performance by selecting alternative work methods, especially with less experienced crafts persons. There are three distinct phases:

1. When construction crews are familiarizing themselves with a process.
2. When a routine is learned so that coordination is improved.
3. A deliberate and continuing effort to improve with successive iterations of the process.

Learning curves for construction typically fall in the 70–90% range.

The curve in Figure 11.1 represents a project involving the installation of a number of generator units. The expert's estimate for carrying out this work was 11,000 man hours. The contractor's bid was lower—7200 hours per unit. It is unlikely that a bid based on the expert's estimate would have been successful. The use of the learning curve allowed the contractor to complete the project at an even lower level of man-hours—5900 hours per unit. By using the benefit of the learning curve, the contractor was able to reduce the labor hours for the project by $1200 \times 8 = 9600$ hours over eight installations. This savings could translate directly to an increased profit margin (Figure 11.1).

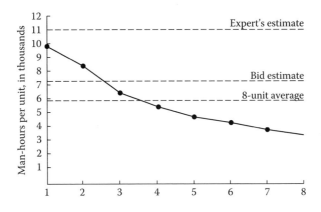

FIGURE 11.1
An arithmetic plot of worker hours expended for installing successive generator units. (With permission of G.A. Howell).

Example: Learning Curve Calculations

A construction crew is carrying out a repetitive task. The first cycle takes the crew 5 hours to complete. The third cycle takes the crew 4 hours. Calculate the learning rate.

$$i = 1 \text{ (first cycle)}$$
$$j = 3 \text{ (third cycle)}$$
$$r = 5 \text{ hours}$$
$$s = 4 \text{ hours}$$
$$r/s = (j/i)^{\wedge n}$$
$$5/4 = (3/1)^{\wedge n}$$
$$1.25 = 3^{\wedge n}$$
$$n\log 3 = \log 1.25$$
$$n = \log 1.25/\log 3 = = 0.203$$
$$\text{learning rate} = 2^{\wedge n} = 1/2^{\wedge n} = 1/1.151 = 0.868 \text{ or } 86.8\%.$$

How long should it take to complete the fourth cycle of the task?

$$i = 1$$
$$j = 4$$
$$r = 5$$
$$s = ?$$

From above:

$$\text{learning rate} = 86.8\%, n = 0.203$$
$$r/s = (j/i)^{\wedge n}$$
$$5/s = (4)^{\wedge 0.203} = 1.3205$$
$$s = 5/1.3205 = 3.79 \text{ hours}$$

The fourth cycle takes 3.79 hours. Over a period of time, the use of the tool can be improved by adding automation to the process.

Cycle Time Analysis[*]

Cycle time is the elapsed time from the start of a process until it is completed. It is usually desirable for cycle times to be reduced as this enables construction equipment and resources to be used to provide increased outputs cost effectively. The application of process improvement methods can identify best methods for carrying out repetitious work activities that are a major component of construction projects. By optimizing such work cycles, construction costs (and durations) can be reduced; profitability is directly enhanced by the resulting improvements in resource utilization.

According to Griffis, Farr, and Morris (2000), cycle time analysis is a deterministic procedure that analyzes each action associated with a cyclical activity, develops the time to perform it, and sums all such activities to establish a total cycle time. The total cycle time of a system

[*] Reference publication by McGraw-Hill Companies, Inc: Title: Construction Planning for Engineers, I/E (C) 2000. Authors: FH Griffis and John Farr. Pages 196, 197, and part of 198.

$$t_s = \sum_{j=1}^{n_{(s)}} t_{sj},$$

where T_{sj} = service time for server s performing action j and n(s) = total number of actions or segments in activity s.

Time to service one customer

$$T_s = [C_t / C_s] \times t_s,$$

where C_s = capacity of the servicing activity and C_t = capacity of the item being serviced.

The following example models an excavation project in which a backhoe is used to excavate a trench for the installation of a water line (Griffis, Farr, and Morris 2000*, Construction Planning for Engineers, copyright McGraw-Hill Companies, Inc.). Dump trucks are used to haul the excavated material 2 miles from the site. The backhoe has a capacity of 1 cubic yard, the dump truck 10 cubic yards, and the trench is 6 feet deep and 4 feet wide. The cycle time for loading one truck can be determined as follows:

Excavation cycle:

Dig	30 seconds
Swing	15
Dump	15
Swing	15
Total	75

$T_s = 75$ seconds per cycle

Swell factor of soil = 25%

Ct = 10 loose yards/1.25 = 8 bank cubic yard

Loader capacity = Cs = 1 loose yard/1.25 = 0.8 bank yard

1. The time to service the truck = $T_s = [C_t / C_s]$. $t_s = 8$ bank yard/0.8 bank yard (75 seconds per cycle) = 750 seconds per cycle
2. The service rate μ is the reciprocal of the service time = $60/T_s$
3. The production rate Ns = C_t, μ

If the activity involves servicing a customer, the cycle time of the serviced activity:

$$T_{sj} = \sum_{J=1}^{n_{(s)}} Et_{sj}.$$

The arrival rate of the serviced activity/customer $\lambda = 60/T_t$ (where T is in minutes). The productive time per hour is calculated with the factor θ. The production rate Nt = θ. Ct, λ. Example (continued) to calculate backhoe output and completion time.

$$\mu = 1/T_s = 60s/min/750 \text{ s per truck} = 0.08 \text{ truck per minute.}$$

---

* Reference publication by McGraw-Hill Companies, Inc: Title: Construction Planning for Engineers, I/E (C) 2000. Authors: FH Griffis and John Farr. Pages 196, 197, and part of 198.

Production rate Ns = ct. μ = (60 min/h) (8 bank cubic yd per truck) (0.8 truck per minute) = 38.4 bank cubic yard. Quantity of material to be removed Q = 6 × 4 × 6000/27 = 5334 bank cubic yard

$$T_s = Q/Ns = 5334/38.4 = 139 \text{ hours}$$

Calculation of truck output based on 50 productive minutes/hour Assume average speed of 20 mph:

Hauling (2 miles):	6 min
Return trip	6 min
Dumping, turning, acceleration	2 min
Loading	12.5 min
T_t	26.5 min per trip

λ = 60/26.5   = 2.264 trips/hour
N_t = θ.ct. λ = 0.833 × 8 × 2.264 = 15.1 bank cubic yard/hour

Total time required (T_t) = Q/N_t = 5334/15.1 = 353 hours or 51 7-hour days. This time frame is the optimal time calculated for this activity*.

Simulation

Simulation may be a physical or numerical representation of a system; that is, a group of entities that interact in accordance with defined rules of behavior. Although physical simulation has been well utilized to test engineering systems such as using wind tunnels to represent wind load conditions, numerical simulation has not been used as extensively with the construction process itself. Given the high cost of labor and related lost time, numerical simulation can effect significant economies in the construction environment, especially as an aid to preplanning. Traditionally, experienced estimators visualize how each stage of a project will be executed, and plan the demand for work crews, materials, and equipment; completion forecasts, and costs. This deterministic approach has severe limitations, as many variables interact in real time. Simulation, on the other hand, includes the randomness that is likely to occur in each of many activities in a complex project. A major factor in the success of simulation is the representation of activity times with appropriate mathematical distributions.

Simulation models can be broadly grouped into three types: iconic, analog, and analytical. Iconic models are physical replicas of the real systems on a reduced scale. This type of model is common in engineering. In aircraft design, wind tunnels are used to simulate the environment around an aircraft in flight. In the design of large engineering structures, such as skyscrapers, dams, bridges, and airports, three-dimensional architectural models are often prepared to provide a realistic view of the design.

A simulation model, in which the real system is modeled through a completely different physical media, is called an analog model. In studying the response of engineering structures to various intensities of earthquakes, it is impossible to build a small model of the

* Reference publication by McGraw-Hill Companies, Inc: Title: Construction Planning for Engineers, I/E (C) 2000. Authors: FH Griffis and John Farr. Pages 196, 197, and part of 198.

earthquake zone using rocks and soils and to generate earthquakes at the command of the experimenter. However, if the dynamic property of quake waves is known, an instrument may be constructed to generate a similar type of force motion.

For problems in which the characteristics of the system components and system structure can be mathematically defined, then an analytical model constitutes a powerful simulation tool. It may be composed with systems of equations, boundary constraints, and heuristic rules, as well as numerical data.

Simulations may be either static or dynamic; static simulation represents the status of a system at any point in time, whereas dynamic simulation shows changes over time. Monte Carlo simulation is the former type; it is based on input variables represented by distributions. These distributions, in turn, are linked by decision rules and are used to determine the distributions of outcome variables of interest. In Monte Carlo simulation each activity is assumed to be a random variable that can be represented by a known probability distribution. The duration of the activity is expected to follow the defined probability distribution instead of being a point estimate value. The simulation process uses random numbers to assign this duration during each iteration. This section focuses only on the method of simulation called the Monte Carlo Simulation, an analytical model.

What Is Monte Carlo Simulation?

Monte Carlo simulation is an iterative procedure and may be described as an input–output study with feedback provided to guide the changes in the input parameters. It can imitate a real-life system, especially when other analyses are too mathematically complex or too difficult to reproduce. The inputs define the set of events and conditions to which the system can be subjected in the real world, and the outputs predict the system response. By studying the outputs at the end of each simulation run, one can learn more about the system behavior and may adjust the inputs accordingly.

At the core of Monte Carlo simulation is random number generation. The computer generates a sequence of numbers, called random numbers or pseudorandom numbers. The numbers generated are considered to be absolutely random and without a pattern. Because of the element of chance, we often call it a Monte Carlo simulation. A Monte Carlo simulation is therefore a probabilistic model involving an element of chance and, hence, it is not deterministic.

The random behavior in games of chance is similar to how Monte Carlo simulation selects variable values at random to simulate a model. When we roll a die, we know that either a 1, 2, 3, 4, 5, or 6 will come up, but we do not know what will come up for any particular roll. The same situation applies to the variables that have a known range of values but an uncertain value for any particular time or event (e.g., interest rates, staffing needs, stock prices, inventory, phone calls per minute). An analogy to the above example of rolling a die is the generation of random numbers. The examples provided in this chapter illustrate simulations based on generating random numbers. However, it is important to understand the limitations of the Monte Carlo simulation technique.

Limitations

Despite the many applications and advantages, the following are some limitations of computer simulations (Figure 11.2):

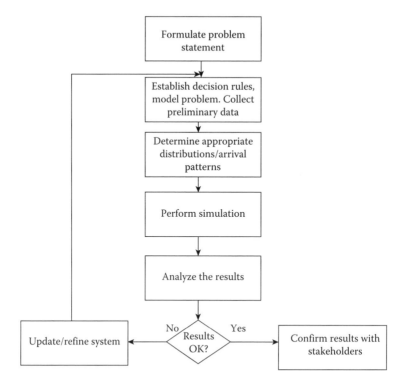

FIGURE 11.2
The simulation process.

- The simulation may be expensive in time or money to develop.
- Simulation by itself is *not* an optimization technique. We have to perform a number of simulations and then choose an optimal solution from their results. Because it is impossible to test all alternatives, we can sometimes only provide good solutions but not an optimum solution.
- Because a simulation is probabilistic involving an element of chance, we should be careful of our conclusions.
- The results may be difficult to verify because often we do not have real-world data.
- Advantages: The most important concepts can be represented in the simulation model. Changes in the input variables can be tested very quickly and efficiently.
- Disadvantages: The development of the model can be time consuming and costly. It may be difficult to verify and validate the model. Data collection and analysis can at best provide suitable approximations to these distributions.

Example 11.1

Dump trucks transport aggregates from an original source to stockpiles. When trucks arrive at the stockpiles they must turn around and then dump the aggregates into the stockpiles from the back. Truck interarrival times are distributed according to Table 11.3.

TABLE 11.3

Truck Interarrival Time and the Corresponding Probabilities

Interarrival time (minutes)	4	5	6	7	8	9
Probability	0.10	0.25	0.30	0.20	0.10	0.05

The times for turnaround and dumping are distributed as shown in Table 11.4:

TABLE 11.4

Turnaround and Dumping Time and the Corresponding Probabilities

Turnaround and dumping time (minutes)	4.0	4.5	5.0	5.5	6.0	7
Probability	0.10	0.15	0.35	0.25	0.10	0.05

Use Monte Carlo simulation for the arrival and departure of 10 trucks starting with the arrival of the first truck. Estimate the mean waiting time of the trucks. The following sequence of random numbers have been generated (Table 11.5).

TABLE 11.5

Generated Random Numbers

For interval times of trucks	5	65	78	51	23	89	11	63	56	
For turnaround and dumping	69	34	41	10	14	16	8	2	51	7

Note: Waiting time is the time in the queue and does not include the time taken for turnaround and dumping.

Solution
Step 1:

TABLE 11.6

Assign Random Numbers for Interarrival Times

Interarrival Times (minutes)	Probability	Cumulative Probability	R.N
4	0.10	0.10	00–09
5	0.25	0.35	10–34
6	0.30	0.65	35–64
7	0.20	0.85	65–84
8	0.10	0.95	85–94
9	0.05	1.00	95–99

TABLE 11.7

The Interarrival Times from the Random Numbers Generated

Random number	5	65	78	51	23	89	11	63	56
Interarrival times (minutes)	4	7	7	6	5	8	5	6	6

Step 2: Assign random numbers for turnaround and dumping times.

TABLE 11.8

Random Numbers

Turnaround and Dumping Times (minutes)	Probability	Cumulative Probability	R.N.
4	0.10	0.10	00–09
4.5	0.15	0.25	10–24
5	0.35	0.60	25–59
5.5	0.25	0.85	60–84
6	0.10	0.95	85–94
7	0.05	1.00	95–99

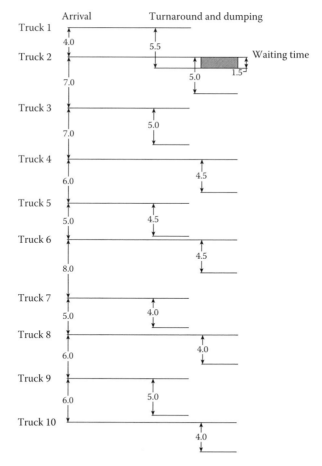

FIGURE 11.3
Simulation result for 10 trucks.

TABLE 11.9

The Turnaround and Dumping Times from the Random Numbers

Random number	69	34	41	10	14	16	8	2	51	7
Turnaround and dumping times (minutes)	5.5	5	5	4.5	4.5	4.5	4	4	5	4

Step 3: Draw Figure 11.3, the simulation result for 10 trucks. From the simulation, only the second truck has to wait for 1.5 minutes. The total waiting time is therefore 1.5 minutes. Hence, the average waiting time is 1.5/10 = 0.15.

Other Applications of Monte Carlo Simulation in Construction

Monte Carlo simulation is a very versatile method of risk analysis that can be applied to many diverse applications. In the construction industry, Monte Carlo Simulation can be applied to the risk analysis of CPM schedules and range estimating. Applications exist that will perform Monte Carlo within Primavera Project Planner and on their own. Complex Monte Carlo simulations can also be set up in a spreadsheet application such as Microsoft Excel.

Exercise Question

A concrete company supplies building contractors A, B, C, and D with units of ready-mix concrete. Contractor A orders seven units every 4 days; B orders every 5 days but the size of its order varies according to the Table; C orders five units whenever it orders, but the interval between successive orders varies according to Table 11.10:

TABLE 11.10

Summary of Orders: Contractor A

Size of B's Order (Units)	P	Interval of C's Order (Days)	P
5	0.2	3	0.1
7	0.3	4	0.4
9	0.3	5	0.3
11	0.2	6	0.2

Contractor D orders concrete with varying size as well as varying interval, according to Table 11.11:

TABLE 11.11

Summary of Orders: Contractor D

Size of D's Order (Units)	P	Interval of D's Order (Days)	P
3	0.1	2	0.2
6	0.2	3	0.3
9	0.4	4	0.3
12	0.3	5	0.2

Simulate 25 days demand for concrete starting from an initial day when all four contractors order concrete. Obtain, from the simulation, an estimate of the mean demand per day for concrete.

Quality Function Deployment

Quality function deployment (QFD) is defined as "a technique to deploy customer requirements into design characteristics and deploy them into subsystems, components, materials, and production processes." It is impossible to accurately measure the quality performance of the construction industry in general because every project has its unique features. The quality performance depends on the segment such as general contracting, design/build (D/B), design, and so on where the firm is active. This topic concentrates on only one segment of the construction industry, namely the D/B sector because of its extensive use in the current scenario. The D/B is defined as the owner-driven project delivery system in which one integrated entity forges a single contract with the owner. The single entity assumes single source risk and responsibility. The firm handles all phases of a project from planning, conceptual and preliminary design, detailed design, and procurement through construction to operation with sole responsibility within their organization. The responsibilities are delegated to diverse functions such as contract, design, procurement, construction, and servicing as the conceptual design shifts to detailed design and construction to final product.

Many professionals involved in D/B insist that the advantages of the D/B method greatly outweigh its disadvantages. The advantages and disadvantages to project participants are well documented in the literature. The advantages of a D/B system include its efficiency, a minimal possibility of claims and changes, budget integrity, and a strong working relationship among the parties. Value engineering is integrated into the design process much sooner than in traditional methods. Constructability is improved, and this generates savings in construction overhead costs and project-financing costs by reducing the time required to deliver a project. These advantages are mostly due to the fact that the D/B method provides a single source of responsibility.

These advantages are valuable only when the D/B project is performed to an acceptable level of quality. In other types of project delivery the owner can have a designer or other professional manage or monitor the quality performance of a contractor, but the owner does not retain the authority of these checks and balances in a D/B project. For that very reason, the management of quality is an important issue in the delivery of D/B projects, requiring a more elaborate measurement tool of quality performance than any other project delivery system.

QFD can serve to enhance perceived quality in construction projects by identifying the owner's most important needs and incorporating them in the design and planning phases. As described by Bossert (1991), QFD provides a systematic method of quantifying users' needs and reflecting these needs in the features of the respective products/services. In this case, the built facilities should more closely meet users' needs than is derived through current design practices that rely mostly on practitioners' experience.

A QFD model may be used to represent six basic project management areas: project scope (functional requirement), budget costing, scheduling, land requirements, technical and safety requirements, and statutory and environmental requirements. Data from two projects of different types, nature and scale can be entered into the QFD model for testing purposes. QFD can enhance the project planning process in the following ways:

1. QFD serves as a road map for navigating the planning process and always keeps track of customer requirements and satisfaction. This actually helps eliminate human inefficiency.

2. The process of building a QFD matrix can be a good communication facilitator that helps break through the communication barriers between client and the designer and among members of the design team.

3. QFD can be an excellent tool for evaluating project alternatives, balancing conflicting project requirements, and establishing measurable project performance targets.

4. QFD can be used as a quick sensitivity test when project requirements change.

Used as a project planning tool, QFD can bring benefits and enhancements to civil engineering capital project planning. Some research topics suggested for further study are streamlining the QFD process, computer-aided QFD applications, evaluation of the cost and benefits of using QFD, use of QFD in detailed design, and how to integrate QFD with total project quality management systems.

Case Study

This case represents a study that was carried out with the Hong Kong Public Works Program (Ahmed, Sang, and Torbica 2003). The objective of this study was twofold: (a) to explore the suitability of QFD in the planning and design of capital projects, and (b) to propose a QFD application model that could be readily used in the planning and design process. To accomplish this objective, a four-stage research method was adopted.

Civil engineering capital projects, irrespective of type, size, and complexity, go through a typical development life cycle that can be divided into two stages: planning and design and construction and implementation. During the planning and design stage, the owner's concept of a project is converted into a finished design. In this study we use the Hong Kong government's procedures and practices, called the Public Works Program system, as a generic model to illustrate the project planning process. The project development process is similar in other countries, both in the private and public sectors.

The objective of this case study is to demonstrate how QFD can be used in capital project planning and to test the validity of the proposed model. Two levels of the House of Quality (HOQ) will be developed using the project data extracted from the various phase deliverables of the project for back-analysis (Figure 11.4).

Two projects are reviewed: The first, Project A, is an upgrade of an existing sewage treatment facility (STF) currently in the preliminary design phase. The project undergoes the traditional method of project development, and the project requirements, feasibility study, and preliminary design are all conducted internally by an engineering team in the relevant works department of the client (Hong Kong government; Figure 11.5). The existing STF is the only one in the area and provides sewage treatment for the vicinity. Ever since its first completion in the late 1980s, the facility has undergone various stages of upgrading and expansion; the last upgrade was completed in 1997.

In 1992, the Hong Kong government concluded, after conducting a number of studies, that further expansion would be required as a result of anticipated population increases. It was anticipated that the present flow capacity would be reached by 2004. Two new effluent quality requirements regarding bacteria and ammonia removal were recently imposed by the Environmental Protection Department. A complete review of the existing treatment processes was proposed to determine the most practical and economical way to provide disinfection and ammonia removal facilities.

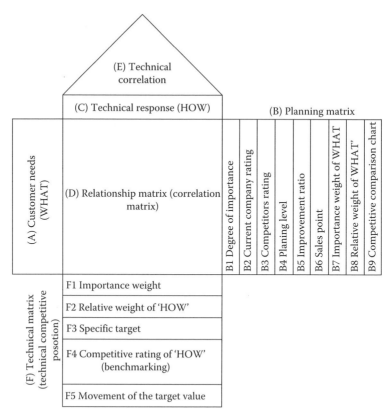

FIGURE 11.4
Sections of house of quality.

House of Quality (HOQ) Level 1

The QFD model and process flows are depicted in Figure 11.6. The project requirements are abstracted from the client project brief (CPB), which is the phase deliverable of the project requirement phase. The process begins by collecting the initial client's objectives for the project to build up the "what" section of the HOQ. Calculation of the importance weights (technical importance) and the relative weights (rounded off to whole numbers). The "how" was calculated by the following formulas:

$$\text{technical importance}(i) = \sum_1 \text{importance rate}(i) \times \text{corrlation}(i), \tag{11.1}$$

$$\text{relative weight}(i) = \frac{\text{technical importance}(i) \times 5}{\text{Max[technical importance}(i)]}. \tag{11.2}$$

The "roof" section represents any interactions or conflicts among the elements in the "how" section. If two elements of "how" are in conflict with each other, then improvement of one element will mean worsening the other. The roof section helps the QFD team recognize the conflicting situation early to avoid waste of resources. Deciding what element should be improved more and what could be traded off a little depends on the degree of

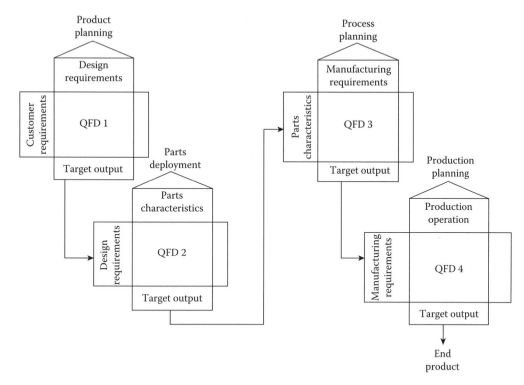

FIGURE 11.5
Chain of QFD matrix.

importance of the element. Some of the major items in the technical correlation section are discussed below (Figures 11.7 and 11.8).

1. Scopes (1) and (2) are mutually exclusive. The total capacity required for the additional flow is the sum of two. If the capacity provided by upgrading existing facilities increases, then the capacity required from new units will decrease and vice versa.

2. Costs (1), (2), and (3) correspond to scopes (1), (2), and (3), respectively, and will follow the same correlations.

3. All scope items tie positively with the corresponding cost items and adversely with schedule items (time for completion).

4. Costs will go up if the design treatment capacity increases (technical items).

5. Scope (1) will adversely affect the choice of sitting because the existing sewage facility may not have sufficient space to locate all new units.

6. Project contingencies are a percentage of the cost items.

7. Environment (1) is a positive function of cost (4).

House of Quality (HOQ) Level 2

For building up the second level of the HOQ, the target values and relative weights of engineering solutions in HOQ-1 are transferred to the "what" section in HOQ-2. Following

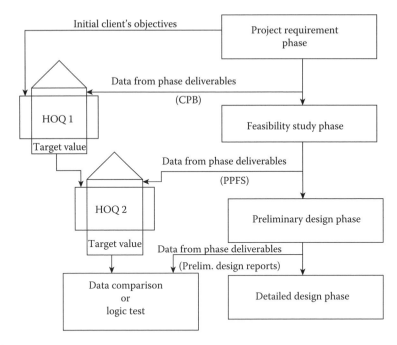

FIGURE 11.6
Process flow of QFD modeling for the case study.

that, the engineering proposals explored in the preliminary project feasibility study are reproduced as the "how" section in HOQ level 2. Other sections of HOQ-2 will repeat the same processes and analysis as in HOQ-1.

Observations from Quality Deployment

After going through two levels of the HOQ, the project profile gradually takes shape. More than 30 detailed project requirements are generated from the nine basic project requirements. These project details are the refinement of the initial project objectives. Throughout the QFD process, the utmost in customer requirements are always maintained. For instance, the design flow is 130,000 cm/day because the new sewage treatment facilities must be able to handle the flow generated from the population in year 2001 plus a 30% reserved capacity. Interconnection facilities (scope 5) are required to connect the new units with the existing facilities so that together they can accommodate the future flow and the new effluent standard. A quick look at the schedule section in Figure 11.9 can demonstrate how decisions and tradeoffs can be made using the QFD structure. The relative weights for "upgraded to category A by December 2002" (schedule 4) and "contracts award by June 2003" (schedule 5) are 0 and 1, respectively, indicating low importance. As a result, the project delivery program delays beyond the end of 2004 to July 2006 (schedules 6 and 7).

A quick sensitivity test can be easily conducted if any one of the requirements changes in any phase. For instance, if the space in the existing STF is not sufficient to site the new sewage treatment facilities, then the project needs to site at the adjacent landfill. The impact on the scope, cost, schedule, and technical aspects as well as the environment could quickly be visualized from the HOQ. By demonstrating that each of the initial objectives is fully

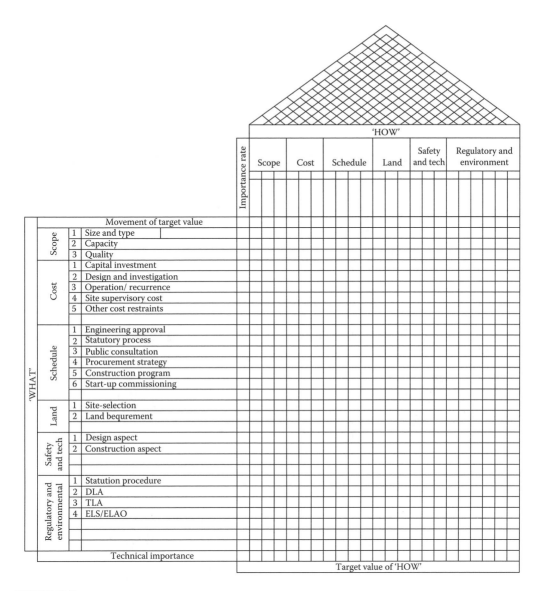

FIGURE 11.7
Generic house of quality (HOQ) template.

addressed, all outputs are traceable back to their origin, and each target value is logical, one can conclude that the QFD model has passed the logic test criteria.

Comments on Output Target Value

Figure 11.9 compares customer needs designated "What" with the technical response, designated "How." The columns under the roof of the HOQ are subdivided into Scope, Cost, Schedule, Land, Safety, and Technology and Regulatory and Environmental. The Cost

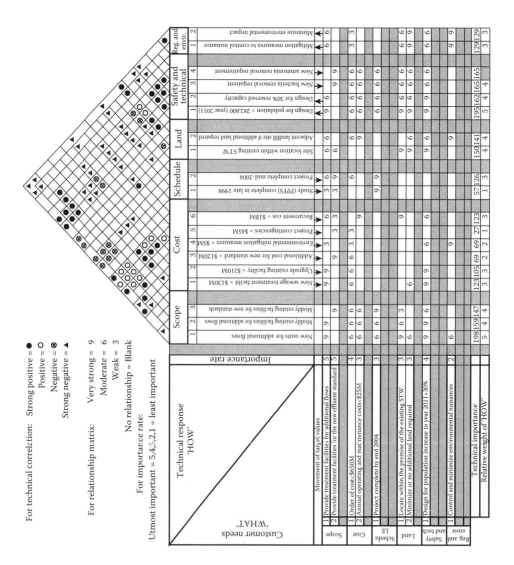

FIGURE 11.8
House of quality level 1.

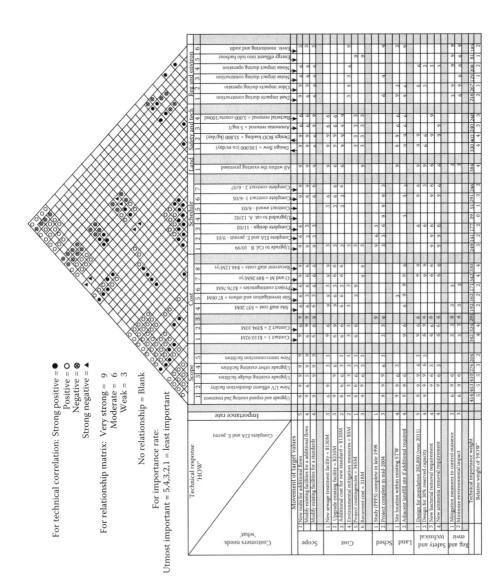

FIGURE 11.9
House of quality No. 2.

section has eight columns with different cost scenarios. By the same token, the Schedule section has seven columns with different time frames.

The target values of cost and schedule do not comply with the initial client's objectives. The total project estimate and annual recurrent expense exceed the initial budgets. These values were $681.78 million comprised of cost 1 + cost 2 + cost 6, as compared with $650 million and $93.38 million = cost 7 + cost 8 versus $25 million). The delivery program was also delayed beyond the end of 2004 (schedules 6 and 7). This was due to a number of small changes made by the client during the process of project planning and design.

In traditional project planning it is very difficult to identify the impact of one small change on the whole project due to the complexity of a civil engineering project and the interrelationship among various project elements. QFD provides a road map for navigating the process as a whole. All project deficiencies encountered on project A could have been identified had QFD been used in the planning process. Not only can signs of deficiencies be identified early, but the project deficiencies can also be traced back to their origin.

As a result, the client and the designer would have been in a position to act early to mitigate the potential loss to the project. The quality matrix developed is usually of substantial size and complexity, which is a great burden to the designer and demands substantial initial staffing resources. However, the matrices produced most likely cannot be reused for other projects. It is also noted that some elements in the matrices are quite independent and do not involve much interrelationship with the other elements. If it were feasible to devise standard QFD matrices for such subsystems, the size of the HOQ as well as the necessary staff resources could be greatly reduced. Such reduction in matrix size would not only cut costs, but would also reduce the chances of overlooking some important issues. Further investigation and research in this direction are worthwhile and should be encouraged.

Summary

The case study explores the applicability of QFD in civil engineering capital project planning. A QFD model is proposed that concentrates on the six basic project management areas: project scope (functional requirements), budget costing, scheduling, land requirements, technical and safety requirements, and statutory (regulatory) and environmental requirements. Data from two projects of different type, scale, and nature are fed into the model for validity testing. Verification has given encouraging results, suggesting the validity of the QFD model. It is found that the use of QFD can enhance the project planning process in the following ways:

1. QFD serves as a road map for navigating the planning process and always keeping track of customer requirements and satisfaction. This actually helps eliminate human inefficiency.

2. The process of building a QFD matrix can be a good communication facilitator that helps break through the communication barriers between the client and the designer and among members of the design team.

3. QFD can be an excellent tool for evaluating project alternatives, balancing conflicting project requirements, and establishing measurable project performance targets.

4. QFD can be used as a quick sensitivity test when project requirements change.

To ensure the best utilization of QFD in the project planning process, the following points need to be taken into consideration:

1. While there are many different ways to apply QFD, it must be applied as early as possible in the planning process, starting with the original customer's requirements. QFD has no magic to get a quick result, nor is it a remedial measure to save a project when its original customer's requirements are already sidetracked.

2. Empowerment from the client side is necessary. A client who is just there to provide the necessary funding and land for the project is not enough. To ensure success, a client must actively participate in the QFD process and work with the designers to provide policy directives, refine the project objectives, set project priorities, and make decisions when conflicts occur.

3. Total involvement of team members is necessary. All team members, who represent different fields, need to work together to share the common goal of the project and make their valuable contributions supplement each other.

There are many barriers to the successful use of QFD in civil engineering capital project planning. The concept of QFD is still new to civil engineering professionals, who require extensive training to become familiar with the QFD tools. Different enabling departments within the client organization will provide consultative input to the project in various project stages. Sometimes individual departments' objectives may be in conflict with those of others or with a project's common objectives. To overcome this barrier, the enabling departments should be part of the QFD team so that the project vision and project objectives are clearly understood and maintained throughout the client's organization. Also, the use of QFD will require an initial up-front investment. The uniqueness of every civil engineering capital project makes it difficult to reuse a QFD matrix on more than one project.

Questions for Discussion

1. How can the development of standard times for repetitious tasks benefit the construction process?

2. A repetitious task takes 3 hours for the second cycle, and 4 hours for the fourth cycle. Calculate the percentage learning rate.

3. The learning curve: The first unit takes 5 hours. The third unit takes 4 hours. Determine the learning curve expressed as a percentage. If the learning curve remains the same, how long would the fifth unit take? If the learning curve were 80%, how long would the third unit take?

4. What are the components of the "House of Quality"?

5. What is gained by using the structure of the QFD process as opposed to simply having designers rely on their experience?

6. Explain how the QFD process could be used in target value design for lean construction.

References

Ahmed, S. M., L. P. Sang, and Z. M. Torbica. 2003. Use of quality function deployment in civil engineering capital project planning. *Journal of Construction Engineering and Management* 129(4): 358–68.

Barnes, R. 1980, 2009. *Motion and time study design and measurement of work*, 7th ed. New York: Wiley.

Griffis, F. H., J. V. Farr, and M. D. Morris. 2000. *Construction planning for engineers.* New York: McGraw-Hill.

Oglesby, C., H. W. Parker, and G. A. Howell. 1989. *Productivity improvement in construction.* New York: McGraw-Hill Inc.

Tippett, L. H. C., and P. D. Vincent. 1953. Statistical investigations of labour productivity in cotton spinning. *Journal of the Royal Statistical Society* 116: 256–71.

Bibliography

Al-Sudairi, A. A., J. E. Diekmann, A. D. Songer, and H. M. Brown. 1999. *Simulation of construction processes: Traditional practices versus lean principles.* Berkeley, CA: University of California.

Bossert, J. L. 1991. *Quality function deployment: A practitioner's approach.* Milwaukee, WI: ASQC Quality Press.

Burati, Jr., J. L., and T. H. Oswald. 1993. Implementing TQM in engineering and construction. *Journal of Management in Engineering, ASCE* 9(4): 456–70.

Kubal, M. T. 1994. *Engineered quality in construction.* New York: McGraw-Hill.

Lantelme, E., and C. Formoso. 2000. Improving performance through measurement: The application of lean production and organizational learning principles. *8th International Conference of the International Group for Lean Construction*, July 17–19. University of Sussex, Brighton.

Salem, O., J. Solomon, A. Genaidy, and I. Minkarah. 2006. Lean construction: From theory to implementation. *Journal of Management In Engineering, ASCE* 22(4): 168–75.

12

Safety Management

Introduction

The construction industry has a reputation for being unsafe, due to its historically high rate of occupational accidents (Bureau of Labor Statistics 2006). In the year 2006 alone, construction accounted for 21% of all deaths and 11% of all disabling injuries/illnesses in private industry in the United States. But no industry has a good record, some are just worse than others. It is often claimed that the nature of construction work makes accidents inevitable, but that is an unacceptable position to take.

In the context of lean construction, injuries and fatalities have consequences that work in opposition to a lean philosophy. While lean seeks to reduce or eliminate waste and to deliver more value to the customer, accidents have many negative consequences including lost work hours. Minor injuries may cause a temporary distraction to treat the affected individuals. Not only does an injured worker reduce the capacity of the work force, but other work crews generally lose additional time in the treatment of that individual. Serious injuries and fatalities inevitably cause major work stoppages and dampen workers' morale as well as their productivity.

Safety management is a relatively new method of controlling safety policies, procedures, and practices within a company. It is currently being implemented by many construction companies to limit their liabilities and costs, thereby making them more competitive in the construction marketplace. Safety management is a dynamic process operating in a constant state of change. Therefore, the process must be constantly monitored and adjusted to achieve the desired goals.

Most of the accidents that occur on a job site could be prevented if contractors treated safety as a priority. Safety programs can help to prevent accidents with injuries and deaths at the job site, and these programs provide an attractive return on investment through higher-production rates, lower workers' compensation premiums, and so on. They also improve a company's reputation and provide motivation for the labor force. The decision to implement and run a good safety program rests solely in the hands of a company's top management. They are the only ones responsible for designing and implementing procedures to reduce accident rates.

The total cost of occupational fatal and non-fatal accidents in 2004 was $49.7 billion representing 10.72% of the total turnover of the construction industry. The total cost of occupational injuries/illnesses and fatality incidents can threaten the survival of a construction company in a highly competitive environment. For example, a company operating at a 4% profit margin would have to increase contract prices by $400,000 to pay for a $16,000 injury, such as the amputation of a finger.

Safety improvement begins with a systems-thinking approach, using a methodology that is similar to the approach of TQM. In fact, existing safety management systems can be significantly improved with the adoption of core TQM principles and procedures. Weinstein (1997), recommends a number of applications of TQM principles that are helpful to occupational or work safety. These can be referenced in Chapter 9. Ergonomics also has a significant impact on worker safety in material handling tasks, this can be referenced in Chapter 13.

Construction site safety is an area of concern for employers of construction workers, this concern for safety is extended to the employees of subcontractors as well. The concern for safety has intensified in the past two decades, primarily due to the escalating costs of workers' compensation insurance (WCI), the rise in the number of liability suits, the intensification of safety regulations, and the mandate by owners to address construction worker injuries. As a result, there has been a relatively steady decline in the incidence of fatalities and disabling injuries. An issue of particular concern is the attitude that construction work is inherently unsafe and a project's design has no effect on safety. This attitude must be dispelled.

How Safe Is Construction?

According to the U.S. Department of Labor, Bureau of Labor Statistics (BLS) during 1995 through 2004 there were 1,954,223 occupational injuries/illnesses among construction workers, representing approximately 10.5% of the total injuries/illnesses involving days away from work in the private sector as shown in Table 12.1.

Dividing the construction industry into three major divisions as per the Standard Industry Classification code (SIC code), Table 12.2 represents the statistics for non-fatal incidents and Table 12.3 shows the statistics for fatal incidents recorded for General Building Contractors (SIC 15), Heavy Construction Contractors except Building Contractors (SIC 16), and Specialty Trade Contractors (SIC 17). The data indicate that specialty contractors accounted for the highest average incident rates; that is, 67.5% of all injuries/illnesses and 58% of all fatalities.

The BLS has reported that the average number of days away from work for construction workers caused by non-fatal occupational injuries was approximately 4.5 in 2004. The average number of injuries and illnesses per 100 full-time workers can be calculated using Equation 12.1 and is termed the *Incidence Rate*. In construction from 1995–2004, the average incidence rate is found to be 4.1. This indicates that about 1 of every 25 construction workers had an injury/illness during this period. For the entire private sector, the average incidence rate was approximately 3.0 for the same time period, supporting the earlier finding that the accident rate in construction is higher than in other industries.

$$IR = \frac{N}{EH} \times 200,000, \tag{12.1}$$

where IR = incidence rate, N = number of injuries and illnesses, and EH = total hours worked by all employees during the calendar year. The base for 100 equivalent full-time workers (working 40 hours per week, 50 weeks per year) is 200,000.

TABLE 12.1

Fatal and Non-Fatal Occupational Accidents (Bureau of Labor Statistics, 2006)

| Year (1) | Non-Fatal Cases Involving Days away from Work | | | Fatalities | | |
	All Private Industries (2)	Construction (3)	Percentage (%) (4)	All Private Industries (5)	Construction (6)	Percentage (%) (7)
1995	2040929	190591	9.34	5497	1055	19.19
1996	1880525	182334	9.70	5597	1047	18.71
1997	1883380	189839	10.08	5616	1107	19.71
1998	1730534	178341	10.31	5457	1174	21.51
1999	1702470	193765	11.38	5488	1191	21.70
2000	1664000	194400	11.68	5347	1155	21.60
2001	2031098	222811	10.97	5281	1226	23.22
2002	1952591	204436	10.47	4978	1125	22.60
2003	1936639	197731	10.21	5043	1131	22.43
2004	1836485	199975	10.89	5177	1224	23.64
Total	18658651	1954223	–	53,481	11,435	–
Average	18,658,65	19,542	10.50	5348	1144	21.43

Source: Bureau of Labour Statistics. *Fatal and non-fatal Occupational Statistics from 1992–2004.* Online at http://www.bls.gov. Accessed on March 1, 2006.

Equation 12.1 can also be used to estimate the incidence rate of fatalities with N representing the number of fatalities. The analysis shows that during a 10-year period from 1995 to 2004 approximately one of every 3900 construction workers died on the job. It is often claimed that the nature of construction work makes accidents inevitable. However, many companies have improved their safety plans, including developing "Zero Accidents Programs," and a great improvement can be seen in their current safety records (Bureau of Labor Statistics 2006).

Cost of Occupational Injury/Illness Accidents

The costs associated with fatal and non-fatal occupational accidents can be classified as economic costs and non-economic costs (Dorman 2000). Non-economic costs are associated with the deep human emotions that arise when life is unnecessarily shortened or impaired. It is hard to quantify these costs in numbers.

Many studies have been conducted to estimate the real cost of accidents for the construction industry. In 1979 the Department of Civil Engineering at Stanford University, under contract with the Business Roundtable (BR), found that accidents cost the construction industry $8.6 billion or about 6.5% of the $137 billion spent on industrial, utility, and commercial construction (Fullman 1984). The researchers estimated that adequate safety programs could reduce annual accident costs by approximately $2.75 billion. The probable cost of such programs would be about $86 billion, making them cost effective by a ratio of 3.2–1. Everett and Frank (1996) estimated that the cost of occupational accidents to the construction industry in 1994 was $31.94 billion, representing 11.9% of the total turnover of the construction industry in that year.

TABLE 12.2

Non-Fatal Occupational Accidents Involving Days Away from Work in Private Construction Classified by the SIC Code (Bureau of Labor Statistics, 2006)

Year (1)	General Building Contractors		Heavy Construction Contractors Except Building		Specialty Trade Contractors	
	Total (2)	Percentage (%) (3)	Total (4)	Percentage (%) (5)	Total (6)	Percentage (%) (7)
1995	40,315	21.15	27,857	14.62	122,421	64.23
1996	37,161	20.38	24,778	13.59	120,395	66.03
1997	36,947	19.46	26,132	13.77	126,760	66.77
1998	38,551	21.62	25,638	14.38	114,152	64.01
1999	38,968	20.11	23,379	12.07	131,418	67.82
2000	36,992	19.03	24,244	12.47	133,164	68.50
2001	35,873	16.10	32,308	14.50	154,630	69.40
2002	32,845	16.07	30,735	15.03	140,856	68.90
2003	31,125	15.74	28,764	14.55	137,842	69.71
2004	30,405	15.20	30,613	15.31	138,957	69.49
Total	359,182	–	274,448	–	1,320,595	–
Average	35,918	18.49	27,445	14.03	132,060	67.49

Source: Bureau of Labour Statistics. *Fatal and non-fatal Occupational Statistics from 1992–2004.* Online at http://www.bls.gov. Accessed on March 1, 2006.

TABLE 12.3

Fatal Occupational Accidents in Private Construction Classified by the SIC Code (Bureau of Labor Statistics, 2006)

Year (1)	General Building Contractors		Heavy Construction Contractors Except Building		Specialty Trade Contractors	
	Total (2)	Percentage (%) (3)	Total (4)	Percentage (%) (5)	Total (6)	Percentage (%) (7)
1995	176	16.68	246	23.32	618	58.58
1996	184	17.57	248	23.69	606	57.88
1997	194	17.52	252	22.76	648	58.54
1998	213	18.14	272	23.17	680	57.92
1999	183	15.37	280	23.51	710	59.61
2000	188	16.28	282	24.42	685	59.31
2001	199	16.23	302	24.63	725	59.14
2002	164	14.58	288	25.60	673	59.82
2003	171	14.50	290	25.64	670	59.24
2004	190	13.97	307	25.08	727	59.40
Total	1862	–	2767	–	6742	–
Average	186	16.08	277	24.18	674	58.94

Source: Bureau of Labour Statistics. *Fatal and non-fatal Occupational Statistics from 1992–2004.* Online at http://www.bls.gov. Accessed on March 1, 2006.

Direct Costs

Workers' Compensation Insurance (WCI) premiums represent the greatest direct cost of accidents paid by any industry (Fullman 1984). The premiums can vary for each state and also among insurance companies. A contractor who is concerned about the high cost of such insurance premiums must recognize that these costs are directly related to each company's safety records. Insurance companies are usually willing to give substantial discounts to contractors who maintain good safety records. These discounts are often higher than the cost of a good safety program (Wilson and Koehn 2000).

Other direct costs associated with an occupational accident in a construction company include: an increment in liability insurance premiums, replacement or repair of lost or damaged equipment, and loss of key personnel time (laborer, foreman, superintendent, project manager, and others). Worker's compensation premiums come from the experience rate plans, based on the following formula:

$$\text{Standard premium} = \text{base} \times \text{payroll} \times \text{EMR}. \tag{12.2}$$

The base or manual rates are the average cost of accidents plus administrative costs and profit for insurers per $100 of straight time wages paid for each of 600 work classifications. The payroll units can be calculated by dividing an employer's straight time direct labor costs by $100. The experience modification ratio (EMR) of a company is based on its own accident records, and is calculated by rating bureaus or advisory organizations. Depending on the frequency and severity of the injuries suffered by the contractor's employees during a given year, the EMR can vary dramatically. Most states accept the EMR ratings provided by the National Council on Compensation Insurance (NCCI) but a few states have their own rating bureaus. The NCCI is a private corporation, created and funded by member insurance companies. California, Delaware, Hawaii, Indiana, Massachusetts, Michigan, Minnesota, New Jersey, New York, North Carolina, Pennsylvania, Texas, and Wisconsin have their own government-run rating bureaus that are separate from the NCCI.

Direct costs are defined as the benefits paid to and on behalf of injured employees by WCI (Everett and Frank 1996). According to worker's compensation data tabulated for Engineering News Record (ENR) by broker Marsh USA Inc., New York City, the national average premium for the three key crafts—carpenters, masons, and structural ironworkers per $100 payroll in the year 2004 was $23.74 (Bureau of Labor Statistics 2006).

In a report obtained by Everett and Frank in 1996 from the NCCI, the mean EMR for the 37 states was 96.3%. On an average, 25% of the total project costs were attributable to labor.

Therefore:

$$\text{WCI}_{\text{(construction industry 2004)}} = \$463.6 \times 23.74\% \times 25\% \times 96.3\% = \$26.5 \text{ billion}. \tag{12.3}$$

Where $463.6 billion is the construction volume of work in the year 2004 (Bureau of Labor Statistics 2006), 25% is the percentage of construction volume that is direct labor, 23.74% is the average manual rate for the three key trades, and 96.3% is the mean EMR. However, the insurance benefits received by injured workers represent only about

65% of the insurance costs (Fullman 1984). Everett and Frank (1996) evaluated the BR report that established this percentage and they found that this value still remains the same (65/35).

$$\text{direct cost}_{(WCI\ insurable)} = \$26.5 \times 65\% = \$17.22\ \text{billion.} \tag{12.4}$$

Indirect Costs

Indirect costs of accidents in the construction industry are normally expressed as a function of the direct costs (Hinze, Coble, and Haupt 2000). Everett and Frank (1996) estimated that the indirect costs are roughly double the direct costs; a research study at the Stanford University indicated that this ratio is 1.6 (Fullman 1984), and the Hinze and Applegate (1991) found this ratio to be approximately 4.2.

The indirect costs are all costs resulting from the injuries that are not covered through insurance coverage. Some indirect costs normally borne by a construction company are:

1. Lost productivity:
    a. Job shutdown at the time of injury
    b. The injured worker's reduced capacity upon return to work
    c. Coworkers at the time of the injury: watching and helping the injured
    d. Coworkers who are shorthanded following the injury
    e. Coworkers who must train a replacement worker
    f. Supervisor/management time hiring or retraining a temporary or permanent replacement worker
    g. Management time lost in investigating and reporting the incident (to government, insurance, and news media representatives)
2. Fines
3. Extra wage costs
4. Damaged equipment and the costs of repairing or replacement
5. Clean up
6. Lawsuits
7. Damage to the company's image and reduced competitiveness
8. Reduced workers' morale
9. Transportation costs
10. Cost to reschedule the work
11. Liquidated damages for rescheduling the work

Safety and Lean Construction

Safety has a significant impact on construction-related waste. Labor hours lost to illness or job-related injuries do not add value to the construction process. In fact, there is the added penalty of increased insurance costs to builders that have an imperfect safety record, even if they have a record of otherwise successful performance. Although significant improvements have been made in the United States since 1996, construction work still results in the injury or death of over 8% of its workers. The improvement of construction safety is therefore an important component of lean construction practices.

Conventional construction approaches result in self-interest by the parties that leads to optimization at the task level as individual subcontractors seek the highest level of productivity. This action tends to reduce the predictability with which work is released downstream. In turn, this reduces trust, complicates coordination, and increases project durations. It also reduces the safety of workers on site. Lean construction, on the other hand, improves the predictability of work flows and the effectiveness of work planning, and reduces workers' exposure to hazards on site

Research has shown that 45% of all accidents are associated with planning-related failures. In Danish research, projects that are managed with The Last Planner® System have been shown to experience approximately 50% of the accident rate and up to 70% fewer lost work days due to illness, when compared with traditional projects.

Best practices have improved national accident statistics (Howell et al. 2002), in 2000 OSHA reported an injury rate of 4.3 per 1000 employees versus an industry-wide rate of 8.2. Lean construction advocates recommend that further improvements need to be made, beyond the prevailing best practices.

The first run studies approach in lean construction is based on simulation and experimentation. It can measure the impact of better planning of work assignments to ensure they are capable of meeting safety, quality, cost, and time standards before they are due to be implemented. There are several strategies for improving safety in construction projects. They are discussed next.

Protection through Design

Construction site safety has been an area of concern for employers of construction workers. This concern is extended to the employees of subcontractors as well, and it has intensified in the past two decades due to several factors:

- The escalating costs of workers' insurance
- The rising number of liability suits
- Intensification of safety regulations
- The decision by owners to address construction workers' injuries

This has led to a steady decline in the incidence rate of fatalities and disabling injuries. The attitude that construction work is inherently unsafe and that a project's design has no effect on safety, is of particular concern. This attitude must be dispelled.

Physical work conditions do not have to harbor numerous safety hazards in order for an unsafe condition to exist. Additional hazards may exist due to no fault of the contractor; in some cases the contractor is bound by a design that places specific hazards on the construction site. If designers fully understood their effect on construction safety, that would help in making job sites safer for construction work. They are often unaware of the extent to which their design decisions may impact construction safety. Construction worker safety is often overlooked until the start of the construction phase. As owner concerns about construction worker safety increase, it is anticipated that designers and contractors will be more heavily involved in ensuring safety in the future. While the involvement of contractors in safety is well established, very little information exists on how designers can be involved to a greater extent.

Role of Design Professionals in Construction Safety

The most significant role of a design professional has traditionally been to design a building or structure such that it conforms to accepted engineering practices, local building codes, and is safe for the public. The safety of construction workers is generally considered to be in the hands of contractors. However, design professionals can improve construction safety by making safer choices in the design and planning stages of a project. It would result in fewer on-site decisions by the contractor that may create hazards for the work force.

Research presented by Behm (2005) suggests that designers can have a strong influence on construction safety. In 1985, the International Labor Office recommended that designers give consideration to the safety of workers that were involved in erecting buildings. In 1991, the European Foundation for the Improvement of Living and Working Conditions concluded that about 60% of fatal accidents in construction were the result of decisions made before the beginning of site work. In 1994, a study of the United Kingdom's construction industry found a causal link between design decisions and safe construction.

Figure 12.1 depicts the ability to influence construction safety at various stages of a project. The ideal time to influence construction safety is during the concept and design stages. The ability to influence safety diminishes as the schedule moves from concept toward start-up. Unfortunately, safety usually is not addressed until construction begins.

The clearest example of design professionals' influence on safety is in the design of a parapet wall. The International Building Code paragraph 704.11.1 requires that a parapet wall be at least 30 inches high. OSHA 1926 Subpart M requires a 42-inch guardrail or other fall protection when working at elevated heights. This means that if the parapet wall were

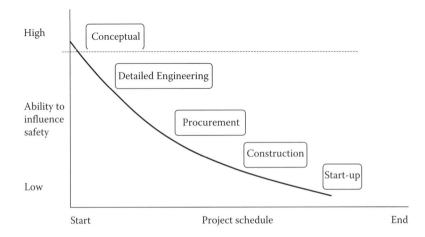

FIGURE 12.1
Time/safety influence curve. (Adapted from Behm, M., *Safety Science*, 43(8), 589–611, 2005. With permission from Elsevier.)

designed to be between 30 inches and 42-inches high, a temporary guardrail at a height of 42 inches or other fall protection would have to be used during construction and future roof maintenance. A decision would have to be made at the site concerning fall protection. This leaves open the possibility of an injury if fall protection is inadequate, workers are not trained, or if fall protection is not used at all. However, if the designer specifies a 42-inch high parapet wall, not only does the design comply with the building code (safe for the public), but the risk of a fall injury during the lifetime of the structure is eliminated because fall protection would not be required.

Designing for Safety can be termed as a formal process that adopts hazard analysis at the beginning of the design stage. Identifying hazards is the first step of the procedure. Engineering measures are then applied to eliminate these hazards or reduce their risk. The design measures start with trying to eliminate the hazards by engineering design. If the hazards still prevail, then safety devices are incorporated. If these methods do not eliminate the risk then warnings, instruction, and training are used as a last resort. This process has been applied to the design of products, equipment, machines, facilities, buildings, and job tasks.

Designing for Construction Worker Safety (DfCS) is the extension of the DFS process to construction projects. This process applies to the design of a permanent building, facility, or structure. It does not address methods to make construction safer, but helps in making a project safer to build. The use of a fall protection system, for example, is not part of the DfCS process. It would come into play in influencing design decisions that could eliminate or significantly reduce the need for fall protection systems during construction and maintenance. An ability to identify potential hazards associated with construction and maintenance workers in the design stage of a project is required. It is the skill of the design professional to eliminate the hazard by incorporating the appropriate design features.

The involvement of design professionals, specifically engineers, is not totally new to construction safety. Many of OSHA's construction regulations require an "engineer" or "engineering controls." The design expertise would be extended to include the safety aspects of permanent structures, including maintenance, rather than designing temporary structures and systems for construction. The most important feature of this process is the input of site safety knowledge into design decisions. Progress reviews would ensure that safety is considered throughout the design process. The key feature of this process is the input of site safety knowledge into design decisions (Figure 12.2).

Enhancing Design To Improve Site Safety

In order to improve the design process, suggestions can be taken from targeted resources such as safety design manuals and checklists, ideas generated by the researchers, and interviews with industry personnel. Manuals and checklists typically address hazards during the startup, operation, and maintenance stages of a project. The on-site workforce and others who visit the job site generally provide the most design suggestions. They are involved in the day-to-day construction activities and are frequently exposed to job site hazards. Any effort that designers put to addressing construction worker safety is voluntary, since OSHA does not place any great responsibility for safety on them.

The great diversity of the construction industry has led to the accumulation of design suggestions that reflect all types of design disciplines. The suggestions that deal with both the architectural and construction management disciplines constitute most of the design suggestions. Hence, architects may have the most to gain from the database of design suggestions. The suggestions can be sorted according to the project components, construction

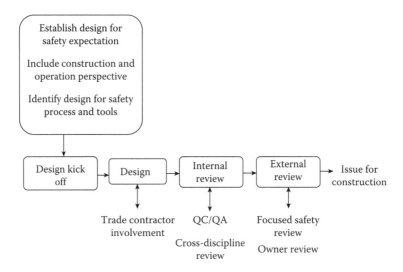

FIGURE 12.2
A typical design for safety process flowchart.

site hazards, and project systems. The following list documents possible design suggestions their purposes.

1. Suggestion: The design of prefabrication systems should be based on units that can be built on the ground and erected in the required place.

   Purpose: It would reduce the worker's exposure to falls and being struck by any falling objects.

2. Suggestion: Design underground utilities such that they can be placed using trenchless technologies.

   Purpose: This would ensure the elimination of the safety hazards associated with trenching.

3. Suggestion: There should be adequate clearance allowed between structures and power lines.

   Purpose: Overhead power lines are hazardous when operating cranes.

4. Suggestion: Permanent anchorage points should be designed in the residential roofs.

   Purpose: Provide fall protection anchorage for roofing contractors during future maintenance.

5. Suggestion: Cable type lifeline system are required to be designed for tower structures.

   Purpose: These will allow workers to hook onto the structure and move up and down during any future maintenance.

6. Suggestion: The window sills should be designed 42 inches above the floor.

   Purpose: This will eliminate need for fall protection during construction and future maintenance.

7. Suggestion: Design permanent guardrails around skylights.

   Purpose: Prevent workers from falling through skylights.

8. Suggestion: Design components to facilitate prefabrication in the shop or on the ground so that they may be erected in place as complete assemblies.

   Purpose: Reduce worker exposure to falls from elevation and the risk of workers being struck by falling objects.

9. Suggestion: Design steel columns with holes in the web at 0.53 and 1.07 m above the floor level to provide support locations for guardrails and lifelines.

   Purpose: By eliminating the need to connect special guardrail or lifeline connections, such fabrication details will facilitate worker safety immediately upon erection of the columns.

10. Suggestion: Design beam-to-column double connections to have continual support for the beams during the connection process by adding a beam seat, extra bolt hole, or other redundant connection point.

    Purpose: Continual support for beams during erection will eliminate falls due to unexpected vibrations, misalignment, and unexpected construction loads.

11. Suggestion: Minimize the number of offsets in a building plan, and make the offsets a consistent size and as large as possible.

    Purpose: Prevent fall hazards by simplifying the work area for construction workers.

12. Suggestion: Design underground utilities to be placed using trenchless technologies.

    Purpose: Eliminate the safety hazards associated with trenching, especially around roads and pedestrian traffic surfaces.

13. Suggestion: Design roadway edges and shoulders to support the weight of construction equipment.

    Purpose: Prevent heavy construction equipment from crushing the edge of the roadway and overturning.

14. Suggestion: Position mechanical, piping, and electrical controls away from passageways and work areas, but still within reach for easy operation.

    Purpose: Controls that protrude into passageways and work areas, or are hard to operate, hidden, or inaccessible, create safety hazards for construction and maintenance workers.

15. Suggestion: Allow adequate clearance between the structure and overhead power lines. Bury, disconnect, or reroute existing power lines around the project before construction begins.

    Purpose: Overhead power lines that are in service during construction are hazardous when operating cranes and other tall equipment.

16. Suggestion: Route piping lines that carry liquids below electrical cable trays.

    Purpose: Prevent the chance of electrical shock due to leaking pipes.

17. Suggestion: Do not allow schedules with sustained overtime.

    Purpose: Workers will not be alert if overtime is maintained over a sustained period.

Behavior-Based Safety

This term is used for programs focused on changing the behavior of workers in order to prevent occupational injuries and illnesses. Such programs make workers accountable for their injuries and illnesses, and rely on the belief that most workplace safety and health problems are the result of voluntary acts. Behavior-based programs target specific worker behaviors, enlist hourly employees and management in monitoring these behaviors on the shop floor, and use a checklist to document workers actions. Related to behavior-based safety are safety incentive programs that reward individual employees, entire departments, and/or workplaces for lower injury and illness rates.

There has been a rise in adapting behavior-based safety as a way of side-stepping the safety and health risks associated with increased line speeds, work duties, mandatory overtime, and other forms of work restructuring. In the year 1931, H. W. Heinrich, an insurance company executive, reviewed accident reports completed by plant supervisors. Because supervisors tended to blame workers for injuries and illnesses, their conclusions backed Heinrich's belief that most industrial accidents are caused by unsafe acts. He later developed a model for explaining the factors contributing to workplace accidents (1959). Often, union members and representatives look to energize existing safety programs and are eager for any attention from management about safety. Their desire to improve cannot be held at fault. They may be unaware of the negative consequences associated with behavior-based programs. Behavior-based safety systems may vary from one organization to the next, depending on its form and complexity. At the basic level they have the following elements in common.

- Identifying behaviors that impact safety
- Defining these behaviors precisely
- Development and implementing mechanisms for measuring those behaviors in order to determine their current status and set reasonable goals
- Providing appropriate feedback
- Reinforcing progress

Due to the failure of behavior-based safety to address the root causes of injuries and illnesses, it is important for union members and representatives to demand resources for safety programs targeting workplace hazards. Behavior-based safety is a bargaining issue just like other management proposals. To substantially improve safety, one must also improve the larger organization.

Worker Attitudes toward Safety

It is commonly thought that the best way to control and prevent accidents is to have close control over workers. This approach might work with children but it will not work with any workforce. Behavioral scientists have suggested that employees are more committed if they have a voice in setting work goals and controlling safety. Working people always want to contribute to the accomplishment of a worthwhile effort. They seek purposes and principles that lift, inspire, empower, and encourage them to do their best. In order to meet these needs, more companies are moving toward employee empowerment and self-directed work units in the decision-making process. Along with improving safety performance it will also provide positive momentum to the overall work and quality.

Companies must acknowledge that employees can perform the job provided they receive appropriate training.

There is a need for a new approach that blends safety attitudes and behavior patterns with production and profit realities. With a change in priorities this approach must stress safety as a human value rather than as a priority. Even if this approach is easy to understand it takes time to apply. The paradigm needs to be shifted along with a change in culture. Culture dictates how employees act and how they are treated. It also demonstrates the values of the company and its employees, and determines whether a job will be performed safely. Culture influences business functions like safety, construction, quality, and scheduling.

There are many possible ways to improve this culture. Top management's philosophy flows down through an organization and impacts all functions. Therefore an organization's philosophy is a projection of the CEO's personal values, beliefs, and ideals as understood and applied by middle management. The deployment of these attributes to middle management and throughout the organization requires careful strategic planning; it is often not done with the rigor it deserves.

Worker attitudes can best be shaped through a culture that involves everyone, at all levels, in developing safety practices and procedures for the organization. Each person within the organization must be visible in the safety process and empowered to do what is needed, and that promotes the sense of self-worth and value. In such an environment safety achieves a value status, hence these values must be based on a mutually accepted set if principles that effectively govern all the actions.

The strengthening of employee safety values requires a supportive approach to promote positive attitudes and behavior modification. It has gained much more importance in today's workplace, as more people are working without direct supervision. The following methods should help to develop a safety-oriented environment.

- Safety awareness: Promote the importance of critically observing safety practices, and understanding how to identify and heed early warnings of safety.
- Changing risk behavior: Increase awareness of attitudes, beliefs, and values to promote behavioral change.
- Thinking process: Learn to manage automatic responses, and how to manage attitudes for the appropriate behavior.
- Individual responsibilities: Increase the level of personal responsibility for safety. This process is bound to intensify the concept of being accountable for one's actions.
- Commitment and values: The employees must be reminded consistently what is at stake for one's self, family, and colleagues and how to make the workplace safe for themselves and others.
- Leadership: The most important aspect of leadership is the acceptance of responsibility, which is absolutely true in the field of safety management.

Contractor's Role and Contribution to Safety

Contractors have been the focus of many studies involving the construction industry, construction management firms, and design/build firms. But the reality is that specialty contractors, often working as subcontractors, perform most construction. The chances of a construction project involving an injury to a construction worker or to a member of the public are continually increasing, Construction industry professionals have to assume that any resulting litigation is likely to involve a jury trial, leading to an uncertain outcome. Moreover, in cases of this type there is always an expert that is willing to advance the opinion that the defendant was negligent. The key to improving this state of affairs is to draft each construction contract in a manner that shifts the responsibility for site safety and the safety of the public to the organization or individual that is best able to control the associated risk. In typical projects, the contractor organization is the most appropriate entity.

Several contractor associations provide a variety of safety-related services and information; these resources have been effective in reducing the incidence of worker injuries. In projects where trade associations have been involved, safety performance has been better than in cases where this support was unavailable. Specialty contractors and subcontractors such as mechanical contractors may benefit especially from this approach, given their comparatively smaller size and structure.

The AGC and ABC are the largest of these associations, with nationwide offices; depending on the respective geographic location, their capabilities include a broad variety of services including non-safety issues such as skills testing and technical training. They are also equipped to screen workers and conduct drug testing, as a reliable third party. The use of these services increases the likelihood that costly staff turnover can be minimized, as the screening and testing process identifies those people that are best suited for working with the contractor organization

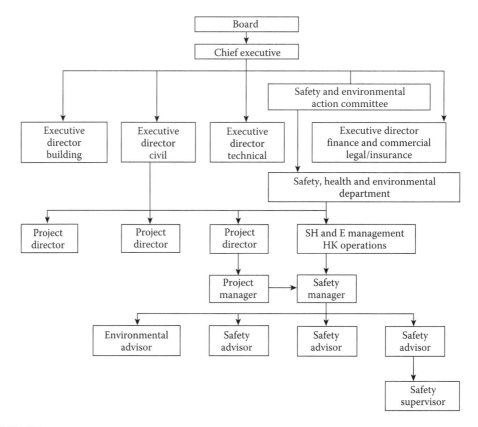

FIGURE 12.3
Typical safety organization structure of a contractor.

Research findings suggest that safety incentives are not a guarantee of good safety performance, and should be employed with caution. As construction activity increases, the injury rate tends to increase as well. Formal safety programs should be introduced in keeping with construction activity, so that site safety can be managed before it becomes a problem.

Safety through Regulatory Practices

A failure to manage site safety effectively results in worker injuries and leads to financial losses, human conflicts, and civil penalties. The government sets regulations to provide a safe working environment for workers. There are two kinds of safety regulations: prescribed safety regulations, where the government states the minimal working conditions that must be provided in a construction site; and self-regulatory, where the government states general duties for employers, thus allowing them to determine the best way of achieving the objectives of the legislation in their organization. The knowledge of a safety engineer is of great importance in both situations. In a prescribed system, government regulation cannot address all of the possible safety hazards that might occur at construction sites. In a self-regulatory system, on the other hand, an organization must develop its own approach to providing adequate working conditions and therefore the experience of safety engineers is of equally great importance.

To capture knowledge, it is important to know the typologies of knowledge. Two such typologies have been defined. Explicit knowledge is defined as knowledge that is precisely and formally articulated and is often codified in databases of corporate procedures and best practices, whereas tacit knowledge is understood and applied unconsciously. Tacit knowledge is based on practical experience, while explicit knowledge is theoretically based.

Explicit Knowledge

Explicit knowledge of construction site safety exists in accident records, safety regulations, and safety guidelines. Accident records contain information on actual accidents reported at construction sites. These are useful for risk assessment by categorizing the safety hazards in terms of frequency-consequence. An organization can use this knowledge to prioritize its responses to these respective hazards, and better manage its resources to address them.

In Figure 12.4, hazards are categorized as Type I must be prioritized while Type IV hazards can be assigned a lower priority. Type I hazards are those that may have major

		Frequency		
		Often	Moderate	Seldom
Consequences	Major	I	I	II
	Moderate	I	II	III
	Minor	II	III	IV

FIGURE 12.4
Frequency-consequence level for categorizing safety hazards.

consequences and occur frequently in building construction projects, such as objects falling through an unprotected opening in a floor slab. Type II hazards may have major consequences but do not occur as frequently as Type I hazards. A worker's fall from the upper floor of a building would be such an example. Type III hazards may have moderate consequences and seldom occur, such as a worker being struck by a tower crane. The categorization of hazards may vary in different organizations since the frequency and consequences of accidents depend on several factors.

Safety regulations can also represent a type of explicit knowledge in site safety. Examples of these are the Occupational Safety and Health Acts from the United States and Construction Site Safety Regulations (CSSR) of Hong Kong. These regulations affirm the minimum required conditions that must be followed in a construction project; however these conditions might not be enough to provide a safe working environment. This is especially true for contractors working in countries encouraging self-regulation through the implementation of a safety management system. The scheme provides general duties for employers, allowing them to achieve the objectives of the legislation in an approach best suited to their organizational culture.

Tacit Knowledge

Tacit knowledge of construction site safety is of paramount importance for organizations in the self-regulatory system, since the knowledge of safety engineers and managers influences the level of on-site safety. Hence the knowledge of these professionals must be harnessed; it relates to the senses, instinctive bodily responses, individual perception, experience, rules of thumb, and intuition. Safety hazard recognition is an important actualization of tacit knowledge in construction site safety. Safety hazard recognition is considered a tacit knowledge because it relies on the safety engineer's experience. If management does not recognize the hazards that may occur on a site, then management cannot provide relevant training or procedures to handle uncertain conditions.

Knowledge management is usually manifested as a business system that is enabled by an array of technologies. It requires both the explicit and tacit knowledge of construction site safety personnel to derive the following benefits:

1. Developing effective safety programs that recognize actual safety hazards. After capturing the available knowledge, an organization can ensure that safety engineers and operational units develop a common definition of the respective hazards. This common definition enables practical safety programs that directly address the hazards that are likely to be met.

2. Establishing an effective training program that improves workers' skill related to the actual safety hazards identified. Once the knowledge is captured by the safety management team, they can provide an effective training program for workers that is tailored to the hazards that are likely to be encountered.

Problems with Capturing Knowledge

Explicit knowledge is easier to capture than tacit knowledge. Explicit knowledge can be captured from existing theories and axioms written in books, regulations, company records, and guidelines. However, tacit knowledge is obtained from the experience of an individual (i.e., safety engineer or manager), therefore it is difficult to capture since the knowledge is stored in an unique way in each individual's mind. Two conventional mechanisms are available to capture the knowledge of safety engineers or managers. First, it can be explored by conducting discussions. Second, it can also be observed at an actual site and provide better safety information related to the actual construction components and the processes that may have safety hazards.

This approach has the advantage of reflection, reminding safety engineers or managers to of an earlier experience they have had. But this mechanism has the following limitations:

1. It may be unsafe for workers since they are exposed to construction operations.
2. It may obstruct workers in carrying out their tasks.
3. It can only be undertaken during the construction stage.

Management Responsibility

The Occupational Safety and Health Administration (OSHA) specifically requires that employers be responsible for providing workers a place of employment free from recognized hazards. This may be interpreted as a safe place in which to work. The type of work in which the company is engaged influences a construction firm's safety practices. A contracting firm may employ hundreds of workers as employees and subcontractors. Consequently, the general contractor may have the responsibility of managing the safety of its own employees as well as the employees of the various subcontractors utilized for the project. This often places the general contractor in an awkward position, since the firm may not be competent in managing the safety practices of all subcontractors. In most cases the general contractor defers the responsibility of safety to the individual subcontractors and may never take part in ensuring that they are actually utilizing all the measures necessary for a safe working environment.

Subcontractors face similar problems to those of general contractors as they may employ large groups of workers. They are challenged to implement a safety program that will satisfy both the requirements of the general contractor and the standards outlined by OSHA. However, a subcontractor's emphasis on safety is often influenced by the size of its organization. Although some smaller firms may have excellent safety programs and records, many find it a difficult process because of the economies of scale. Safety training is often left to an on-the-job learning exercise or taught by the employees' union or trade organization.

According to J. Hinze at the University of Florida, the owner of the project and the project designers can also have an impact on the safety practices encountered on the job site. Moreover, engineers have a professional and moral obligation to take safety, health, and welfare under consideration. As described above, today's trend in the construction industry is to enhance safety by design.

However, it is important to note that the owner is best positioned to make safety a priority in a project by including safety management in the design and performance criteria. In fact, there may be a decrease in cost due to reduced insurance rates, which in turn will lower the contractors' expenses. As this method of design gains acceptance, it is almost certain to improve overall construction safety along with the safety records of construction companies.

Safety Practices

The safety practices that are encountered on construction sites are as diverse as the sites themselves. All construction sites have their own distinctive aspects of safety. The larger the construction projects the better they tend to be organized from a safety standpoint. These types of projects are often high profile. Companies involved in such visible construction projects have reputations to sustain as well as safety records to maintain and hence are generally better prepared to manage the safety aspects of a project. Having full-time safety personnel relieves the pressure on the on-site construction project team. They can keep team members informed of possible safety problems and ensure that subcontractors are implementing their safety programs adequately, freeing the project personnel to focus their efforts on the project itself.

Small to medium projects tend to hold the greatest potential for improvement in safety management. Projects of this type usually involve smaller firms that may not have an adequate safety program or the personnel to oversee safety criteria. Implementation of their safety management programs is usually left up to the foreman or the project superintendent. It is on these types of projects that good safety management practices, implemented properly, can have a very positive impact on overall project safety and improve the company's profits by keeping injuries and claims to a minimum.

Case Study: Safety Implementation at the Workspace

An effective method of safety management was used on a small-to medium-sized project that experienced many of the problems characteristic of smaller projects. While the project was not difficult to construct, it had a tight schedule, which resulted in the project team directing the majority of their efforts in attempting to meet construction deadlines. As a result, the safety management on the project did not obtain the attention that it should have received. At peak times, this firm had approximately 20 employees working on the project. The company required daily "tool box" safety talks each morning for their employees. These talks lasted about 10 minutes and included a description of the work to be accomplished that day. In addition, the foreman or job superintendent discussed appropriate safety information for the task(s) with the team.

This approach worked well at the beginning of the project, since the small number of subcontractor employees on the job generally attended the contractors' morning safety sessions. However, as construction progressed it became increasingly difficult to maintain control of the on-site safety practices. The company had a full-time safety department, but, as it was a branch office of a Canadian construction firm, the site was often not visited by the safety officer. Therefore, little safety support was given to the superintendent. The difficulty encountered was that subcontractors on the project did not, at times, place the proper emphasis on safety. In fact, they sometimes overlooked safety violations in order to maintain the project schedule.

As the total number of workers on the site increased to approximately 50, and subcontractors began holding their own safety talks, it became clear that changes needed to be made. In addition to the tool box meetings, a weekly safety conference of foremen was also held by the general contractor. Appropriate safety information and project scheduling were reviewed in those meetings. In addition, possible conflicts between various trades were discussed in order to decrease potential problems and increase the safety level at the job site. It was first suggested at this weekly safety/project meeting to assign different trades the task of conducting safety inspections. Therefore, instead of the general contractor being solely responsible, the enforcement of safety was to be shared. The weekly safety inspections would be conducted on a rotating basis by the various on-site subcontractors.

This method also educated employees on the safety concerns of the various trades and how they could affect the project. The overall results achieved by this method of safety management were excellent. In particular, the safety inspections were generally conducted the day before or the morning of the weekly meeting with the foremen. With this approach, any needed corrective action could be quickly taken. In addition, the foremen and superintendent were exposed to the concerns of various trades that increased their knowledge of a variety of different safety practices. It should be noted that the subcontractors did not request payment for the time spent making the inspections, and they were not concerned with possible liability problems.

As with any new concept, there was some difficulty in implementing the process. Older workers, in particular, perceived that they were being criticized for their actions and initially rejected the plan. As the project progressed, the various trades actually looked forward to their assigned week to conduct the safety inspections and, surprisingly, most of the subcontractors welcomed the feedback on their safety practices. Another positive benefit of this approach was that it required the involvement of the workers, not just the foremen and superintendents. This meant that the apprentices' comments and observations were just as important as those of the journeyman or the foreman on the project.

Ways to Improve Workplace Safety

There are several factors that can help to improve safety in the workplace. These factors, on their own or combined, provide a comprehensive method for developing an improved safety program.

Stretching Exercises

Resse and Edison (2006) explained that scientific research has not yet determined whether exercise programs are an effective prevention technique. However, many companies that have implemented exercise programs report that warm-up programs have helped them reduce back injuries and other cumulative trauma disorders. Stretching exercises can help to keep involved muscle groups from tightening up; workers may be unaware of gradual changes that eventually lead to chronic medical problems. Stretching exercises are most effective if they are done on a daily basis, as such routine activity help workers to strengthen major muscle groups, especially the ones that are important for maintaining a healthy back.

Daily Safety Huddles

Safety huddles are short safety meetings covering a specific safety topic or type of hazard. They have been used in the industry for many years. A strong safety culture is created by keeping employees focused on safe work practices and behaviors. A daily safety huddle program is the tool that helps to keep the workforce talking and thinking safety. A huddle can be held with a group of workers to review:

1. Hazards of a job that is about to be started
2. Causes and corrective action for a recent accident or "near miss"
3. Job hazards that should be emphasized along with safety rules that will prevent accidents
4. General problems concerned with both on-the-job and off-the-job safety.

The ideal size for a safety huddle is 6–10 workers. Larger groups can be used, however, the workers will be less likely to enter into any discussions. A huddle normally lasts approximately 10 minutes if workers are not seated, if the topic is well chosen, and if the discussion does not get off on a tangent. In preparing for a huddle, if it appears that the interest will last more than 10 minutes, consider a series of three or four huddles or hold the meeting in a room where the workers can sit comfortably and where noise and other distractions are absent. Normally a supervisor leads a huddle, simply because it is his or her responsibility to know the employees, the nature of the job to be undertaken, and its associated hazards. There is no reason that a worker cannot conduct a huddle provided he or she (1) takes the assignment seriously, (2) is given sufficient time to prepare, (3) is not too nervous about the idea, and (4) is given encouragement and help, if it is needed. A small huddle is generally better for a person who is not used to speaking to a group. There is no one best way to conduct a huddle. However, the following suggestions should give a good chance of success.

1. Make a clear announcement of the time, place, and reason for the huddle. Start on time.
2. Explain why the huddle is being held.
3. Keep the huddle from going off on a tangent. If a worker hits on a thought that has merit, promise to have a huddle on that topic at a later date. Keep the promise!
4. Allow time for discussion and questions. If you don't know the correct answer, don't guess at it. Call your safety representative who probably has the answer.
5. Appeal to the workers' desire for approval. Point out things that are being done right as well as problem areas. Also, you can appeal to the workers' competitive instinct. Build a good "sales pitch" and show some enthusiasm!
6. Use a real accident case to emphasize a point. The more recent and the more close (geographically) to your location, the more effective the example will be.

Safety Officer/Liaison

Duties vary, but generally an OSHA safety officer oversees the planning, creation, and implementation of the occupational health and safety program of a company or governmental organization. Duties include administrative, supervisory, and hands-on tasks.

Program Design

The officer designs a safety program consistent with OSHA requirements that sets up processes to reduce accidents and work-related injuries. The program should assign safety responsibilities to various employees. The officer designs procedures and forms for internal reporting of compliance with safety procedures as well as program violations, curative steps, accidents, and injuries. She/he also maintains Materials Safety Data Sheets (MSDS) and other OSHA forms.

The officer analyzes workforce accidents and injuries to determine where process changed, communication or training needs improvement, and develops and manages reports for company management. He/she also recommends and implements program changes where necessary. He/she ensures OSHA requirements are followed and prepares the company for audits. The safety officer acts as a liaison between the company and OSHA.

The OSHA safety officer supervises staff safety officers and ensures coverage when the company operates. An OSHA safety officer may be responsible for developing safety procedures for new company products, services, and manufacturers.

Incentive Programs

Peer and 360 reviews measure employee strengths and development areas (Campbell 2008). Peer reviews offer candid feedback to employees from those that work with them the most. A 360 review is the most comprehensive type of review and includes peer reviews, self-ratings, and higher management reviews (skipping the employees' direct manager). Many companies conduct peer and 360 reviews on employees at the senior management levels.

Reward programs such as annual or off-cycle promotions, salary increases, and bonuses that are tied to employee performance are used by most firms. In addition, there are certificate, anniversary, and achievement award programs. Depending on the structure of the company's human resources corporate initiatives, reward programs may be used weekly, monthly, quarter, biannually, or annually to recognize outstanding employee contributions.

Skip level lunches are a tool employers use to encourage, recognize, and motivate employee performance. During skip level lunches, senior managers meet with top employees within their department or division. Employees are encouraged to talk freely with senior managers, ask questions, network and, of course, enjoy a free meal.

Another form of an employee motivation incentive program is the employee networking program. Employee networking programs bring in prominent guest speakers to kick off events such as a women's conference or a department manager's off-site meeting. Some companies create employee networks that focus on culture, history, and current achievements, physical, and other employee differences (Black employee network, Hispanic employee network, employees with disabilities network). The networks are generally open to all employees at the firm.

Effects

Employee motivation incentive programs increase employee awareness about how performance criteria are measured at the company. The programs also motivate an employee to take responsibility for the success of his/her career.

Summary

There has been a growth of interest in construction worker safety. Rather than being just a concern of constructors that is enforced by OSHA regulations, job site safety has also attracted the attention of owners. These owners have recognized that the cost of construction accidents and injuries eventually increases construction costs, and they have begun to place safety on a higher level of priority. However, this attitude has not yet been adopted by the design community, and designers typically avoid any responsibility for the safety of construction workers to minimize their liability exposure. Existing codes and design standards reflect this attitude, and worker safety rests on the constructors' shoulders. Moreover, designers are not traditionally educated about DfCS. No reference standards exist to bridge the gap between existing design standards and construction worker safety.

Design professionals can enhance project safety, mitigate common safety hazards, and reduce the number of worker injuries by using the DfCS (Design for Construction Safety) approach. It offers the benefit of reducing the need for redesign work, and can help to avoid the types of accidents that could lead to the need for liability claims on all parties, including the designers themselves. Design suggestions have been accumulated that provide designers practical examples of how to design for construction worker safety. And since designers traditionally have not been educated on site hazards and do not continually observe them, they have not been able to provide as much input into the list of design suggestions.

A gap exists between constructors and designers in the knowledge of and commitment to job site safety. As a result, information on designing for construction site safety must ultimately be drawn from construction personnel and transferred to designers. Construction workers are unique facility users and their safety warrants the attention of designers. Addressing safety in the construction phase not only requires soliciting and publishing design suggestions, but also requires designers to change their traditional mindset. Thus, owners must provide the initial impetus—by requesting or requiring by contract terms—that designers consider construction worker safety in their designs. Lean design and construction facilitates this paradigm shift through a closer collaboration between designers and contractors, and through a relational form of contract that apportions risk fairly.

Questions for Discussion

1. Why is it important to consider the safety aspect in the construction industry?
2. What role do design professionals play in the workers safety? Elaborate.
3. What connection is there between safety and lean construction?
4. What part do contractors and owners play in the safety of the construction project?
5. How do workers' attitudes to safety affect the overall safety of the construction?
6. Describe behavior-based safety and its role in worker safety.
7. Discuss the pros and cons of attempting to build a project using lean methods without a specific safety component?

Appendix: OSHA Checklists (Selected Items)

Note: Due to space limitations, an abbreviated checklist is provided below purely for illustrative purposes. It was abstracted from a more comprehensive checklist developed by the Occupational Safety and Health Administration (OSHA 1992). Readers are advised to consult with safety professionals to gain access to checklists that are appropriate for field application.

Employer Posting

- Is the required OSHA workplace poster displayed in a prominent location where all employees are likely to see it?
- Are emergency telephone numbers posted where they can be readily found in case of emergency?
- Where employees may be exposed to any toxic substances or harmful physical agents, has appropriate information concerning employee access to medical and exposure records and "Material Safety Data Sheets" been posted or otherwise made readily available to affected employees?
- Are signs concerning "Exiting from buildings," room capacities, floor loading, biohazards exposures to X ray, microwave, or other harmful radiation or substances posted where appropriate?
- Is the Summary of Occupational Illnesses and Injuries posted in the month of February?

Recordkeeping

- Are all occupational injuries or illnesses, except minor injuries requiring only first aid, being recorded as required on the OSHA 200 log?
- Are employee medical records and records of employee exposure to hazardous substance or harmful physical agents up-to-date and in compliance with current OSHA standards?
- Are employee training records kept and accessible for review by employees, when required by OSHA standards?
- Are operating permits and records up-to-date for such items as elevators, air pressure tanks, liquefied petroleum gas tanks, etc.?

Safety and Health Program

- Do you have an active safety and health program in operation that deals with general safety and health program elements as well as the management of hazards specific to your worksite?
- Is one person clearly responsible for the overall activities of the safety and health program?
- Do you have a safety committee or group made up of management and labor representatives that meet regularly and report in writing on its activities?

- Do you have a working procedure for handling in-house employee complaints regarding safety and health?
- Are you keeping your employees advised of the successful effort and accomplishments you and/or your safety committee have made in assuring they will have a workplace that is safe and healthful?

Medical Services and First Aid

- Is there a hospital, clinic, or infirmary for medical care in proximity of your workplace?
- If medical and first-aid facilities are not in proximity of your workplace, is at least one employee on each shift currently qualified to render first aid?
- Have all employees who are expected to respond to medical emergencies as part of their work: (1) received first-aid training, (2) had hepatitis B vaccination made available to them, (3) had appropriate training on procedures to protect them from bloodborne pathogens, including universal precautions, and (4) have available and understand how to use appropriate personal protective equipment to protect against exposure to bloodborne diseases?
- Where employees have had an exposure incident involving bloodborne pathogens, did you provide an immediate postexposure medical evaluation and follow-up?
- Are medical personnel readily available for advice and consultation on matters of employees' health?
- Are emergency phone numbers posted?
- Are first-aid kits easily accessible to each work area, with necessary supplies available, periodically inspected, and replenished as needed?
- Are means provided for quick drenching or flushing of the eyes and body in areas where corrosive liquids or materials are handled?

Fire Protection

- Is your local fire department well acquainted with your facilities, its location, and specific hazards?
- If you have a fire alarm system, is it certified as required?
- If you have a fire alarm system, is it tested at least annually?
- If you have interior stand pipes and valves, are they inspected regularly?
- If you have outside private fire hydrants, are they flushed at least once a year and on a routine preventive maintenance schedule?
- Are fire doors and shutters in good operating condition?
- Are automatic sprinkler system water control valves, air, and water pressure checked weekly/periodically as required?
- Is the maintenance of automatic sprinkler systems assigned to responsible persons or to a sprinkler contractor?
- Are sprinkler heads protected by metal guards, when exposed to physical damage?

- Is the proper clearance maintained below sprinkler heads?
- Are portable fire extinguishers provided in adequate number and type?
- Are fire extinguishers mounted in readily accessible locations?
- Are fire extinguishers recharged regularly and noted on the inspection tag?
- Are employees periodically instructed in the use of extinguishers and fire protection procedures?

Personal Protective Equipment and Clothing

- Are protective goggles or face shields provided and worn where there is any danger of flying particles or corrosive materials?
- Are approved safety glasses required to be worn at all times in areas where there is a risk or eye injuries such as punctures, abrasions, contusions, or burns?
- Are employees who need corrective lenses (glasses or contacts) in working environments having harmful exposures, required to wear only approved safety glasses, protective goggles, or use other medically approved precautionary procedures?
- Are protective gloves, aprons, shields, or other means provided and required where employees could be cut or where there is reasonably anticipated exposure to corrosive liquids, chemicals, blood, or other potentially infectious materials. See 29 CFR 1910.1030(b) for the definition of "other potentially infectious materials."
- Are hard hats provided and worn where danger of falling objects exist?
- Is appropriate foot protection required where there is the risk of foot injuries from hot, corrosive, poisonous substances, falling objects, crushing, or penetrating actions?
- Are approved respirators provided for regular or emergency use where needed?
- Is all protective equipment maintained in a sanitary condition and ready for use?
- Do you have eye wash facilities and a quick drench shower within the work area where employees are exposed to injurious corrosive materials?
- Where food or beverages are consumed on the premises, are they consumed in areas where there is no exposure to toxic material, blood, or other potentially infectious materials?
- Is protection against the effects of occupational noise exposure provided when sound levels exceed those of the OSHA noise standard?

General Work Environment

- Are all worksites clean, sanitary, and orderly?
- Are work surfaces kept dry or appropriate means taken to assure the surfaces are slip-resistant?
- Are all spilled hazardous materials or liquids, including blood and other potentially infectious materials, cleaned up immediately and according to proper procedures?
- Is combustible scrap, debris, and waste stored safely and removed from the worksite promptly?

- Is all regulated waste, as defined in the OSHA bloodborne pathogens standard (29 CFR 1910.1030), discarded according to federal, state, and local regulations?
- Are accumulations of combustible dust routinely removed from elevated surfaces including the overhead structure of buildings, etc.?
- Is combustible dust cleaned up with a vacuum system to prevent the dust going into suspension?
- Is metallic or conductive dust prevented from entering or accumulating on or around electrical enclosures or equipment?
- Are covered metal waste cans used for oily and paint-soaked waste?
- Are all oil and gas fired devices equipped with flame failure controls that will prevent flow of fuel if pilots or main burners are not working?
- Are the minimum number of toilets and washing facilities provided?
- Are all toilets and washing facilities clean and sanitary?
- Are all work areas adequately illuminated?
- Are pits and floor openings covered or otherwise guarded?

Walkways

- Are aisles and passageways kept clear?
- Are aisles and walkways marked as appropriate?
- Are wet surfaces covered with non-slip materials?
- Are holes in the floor, sidewalk, or other walking surface repaired properly, covered, or otherwise made safe?
- Is there safe clearance for walking in aisles where motorized or mechanical handling equipment is operating?

Floor and Wall Openings

- Are floor openings guarded by a cover, a guardrail, or equivalent on all sides (except a entrance to stairways or ladders)?
- Are toeboards installed around the edges of permanent floor opening (where persons may pass below the opening)?
- Are skylight screens of such construction and mounting that they will withstand a load of at least 200 pounds?
- Is the glass in the windows, doors, glass walls, etc., which are subject to human impact, of sufficient thickness and type for the condition of use?
- Are grates or similar type covers over floor openings such as floor drains of such design that foot traffic or rolling equipment will not be affected by the grate spacing?
- Are unused portions of service pits and pits not actually in use either covered or protected by guardrails or equivalent?
- Are manhole covers, trench and similar covers, plus their supports designed to carry a truck rear axle load of at least 20,000 pounds when located in roadways and subject to vehicle traffic?

Stairs and Stairways

- Are standard stair rails or handrails on all stairways having four or more risers?
- Are all stairways at least 22 inches wide?
- Do stairs have landing platforms not less than 30 inches in the direction of travel and extend 22 inches in width at every 12 feet or less of vertical rise?
- Do stairs angle no more than 50° and no less than 30°?
- Are stairs or hollow-pan type treads and landings filled to the top edge of the pan with solid material?
- Are step risers on stairs and stairways designed or provided with a surface that renders them slip resistant?
- Are stairway handrails located between 30 and 34 inches above the leading edge of stair treads?
- Do stairway handrails have at least 3 inches of clearance between the handrails and the wall or surface they are mounted on?
- Where doors or gates open directly on a stairway, is there a platform provided so the swing of the door does not reduce the width of the platform to less than 21 inches?
- Are stairway handrails capable of withstanding a load of 200 pounds, applied within 2 inches of the top edge, in any downward or outward direction?
- Are signs posted, when appropriate, showing the elevated surface load capacity?
- Are surfaces elevated more than 30 inches above the floor or ground provided with standard guardrails?
- Are all elevated surfaces (beneath which people or machinery could be exposed to falling objects) provided with standard 4-inch toeboards?
- Is a permanent means of access and egress provided to elevated storage and work surfaces?
- Is required headroom provided where necessary?
- Is material on elevated surfaces piled, stacked, or racked in a manner to prevent it from tipping, falling, collapsing, rolling, or spreading?

Exiting or Egress

- Are all exits marked with an exit sign and illuminated by a reliable light source?
- Are the directions to exits, when not immediately apparent, marked with visible signs?
- Are doors, passageways or stairways, that are neither exits nor access to exits and that could be mistaken for exits, appropriately marked "NOT AN EXIT," "TO BASEMENT," "STOREROOM," etc.?
- Are exit signs provided with the word "EXIT" in lettering at least 5-inches high and the stroke of the lettering at least 1/2-inch wide?
- Are exit doors side-hinged?
- Are all exits kept free of obstructions?

- Are at least two means of egress provided from elevated platforms, pits, or rooms where the absence of a second exit would increase the risk of injury from hot, poisonous, corrosive, suffocating, flammable, or explosive substances?
- Are there sufficient exits to permit prompt escape in case of emergency?
- Are special precautions taken to protect employees during construction and repair operations?
- Is the number of exits from each floor of a building and the number of exits from the building itself, appropriate for the building occupancy load?
- Are exit stairways that are required to be separated from other parts of a building, enclosed by at least 2-hour fire-resistive construction in buildings more than four stories in height, and not less than 1-hour fire-resistive constructive elsewhere?
- Where ramps are used as part of required exiting from a building, is the ramp slope limited to 1 foot vertical and 12 feet horizontal?
- Where exiting will be through frameless glass doors, glass exit doors, storm doors, etc., are the doors fully tempered and meet the safety requirements for human impact?

Exit Doors

- Are doors that are required to serve as exits designed and constructed so that the way of exit travel is obvious and direct?
- Are windows that could be mistaken for exit doors, made inaccessible by means of barriers or railings?
- Are exit doors openable from the direction of exit travel without the use of a key or any special knowledge or effort when the building is occupied?
- Is a revolving, sliding, or overhead door prohibited from serving as a required exit door?
- Where panic hardware is installed on a required exit door, will it allow the door to open by applying a force of 15 pounds or less in the direction of the exit traffic?
- Are doors on cold storage rooms provided with an inside release mechanism that will release the latch and open the door even if it's padlocked or otherwise locked on the outside?
- Where exit doors open directly onto any street, alley, or other area where vehicles may be operated, are adequate barriers and warnings provided to prevent employees stepping into the path of traffic?
- Are doors that swing in both directions and are located between rooms where there is frequent traffic, provided with viewing panels in each door?

Portable Ladders

- Are all ladders maintained in good condition, joints between steps and side rails tight, all hardware and fittings securely attached and moveable parts operating freely without binding or undue play?
- Are non-slip safety feet provided on each ladder?

- Are non-slip safety feet provided on each metal or rung ladder?
- Are ladder rungs and steps free of grease and oil?
- Is it prohibited to place a ladder in front of doors opening toward the ladder except when the door is blocked open, locked, or guarded?

References

Behm, M. 2005. Linking construction fatalities to the design for construction safety concept. *Safety Science* 43(8): 589–611.

Bureau of Labor Statistics. 2006. *Fatal and non-fatal occupational statistics from 1992–2004*. Online at http://www.bls.gov. Accessed on March 1, 2006.

Dorman, P. 2000. The economics of safety, health, and well-being at work: An overview. In *Focus program safe work, international labor organization (ILO)*. Geneva/Olympia, WA: The Evergreen State College.

Everett J. G., and P. Frank, Jr. 1996. Cost of accidents and injuries to the construction industry. *Journal of Construction Engineering and Management, ASCE* 122(2): 158–64.

Fullman, J. B. 1984. *Construction safety, security, and loss prevention*. New York: John Wiley and Sons, Inc.

Heinrich, H. W. 1959. *Industrial accident prevention*, 4th ed. New York: McGraw-Hill.

Hinze, J., and L. Applegate. 1991. Costs of construction injuries. *Journal of Construction Engineering and Management, ASCE* 117(3): 537–50.

Hinze, J., R. Coble, and T. Haupt. 2000. *Construction safety and health management*. Upper Saddle River, NJ: Prentice-Hall Inc.

Howell, G. A., G. Ballard, T. S. Abdelhamid, and P. Mitropoulos. 2002. Working near the edge: A new approach to construction safety. Proceedings of the 10th annual conference of the international group for lean construction (IGLC-10). Gramado, Brazil.

OSHA Checklists. 1992. Adopted from OSHA Publication #2209—*OSHA Handbook for Small Businesses* (Revised). Washington, DC: OSHA.

Weinstein, M. B. 1997. *Total quality safety management and auditing*. Boca Raton, FL: Lewis Publishers, New York.

Wilson, J., and E. Koehn. 2000. Safety management: Problems encountered and recommended solutions. *Journal of Construction Engineering and Management, ASCE* 126(1): 77–79.

Bibliography

Gambatese, J., J. Hinze, and C. Haas. 1997. Tool to design for construction worker safety. *Journal of Architectural Engineering, ASCE* 3(1).

Kelley, R. 1996. Worker psychology and safety attitudes. *Professional Safety, Journal of the American Society of Safety Engineers* July: 14–17.

Krizan, W. 2000. Insurance: Party is over for cheap workers' compensation coverage. *Engineer News Record ENR* 245(13): 44–45.

OSHA Work zone safety Quick Card. 2007. 3284-05R-07. Working Safely in Trenches Safety Tips QuickCard™ U.S. Department of Labor. Available at http://www.osha.gov.

Reese, C. D., and J. V. Eidson. 2006. *Handbook of OSHA construction safety and health*, 515, 2nd ed. Oxford: Taylor & Francis Group.

13

Management and Worker Factors

This chapter addresses management-worker factors in two distinct dimensions that leaders of construction organizations must consider if they wish to implement lean construction and Integrated Project Delivery (IPD). It describes:

1. Managing and motivating the work force: These issues influence work attitudes and behaviors in construction organizations that directly affect performance, both at the organization level and at the project level.
2. Improving worker performance with Ergonomics-based Strategies.
   a) The study and redesign of the construction workspace based on the principles of ergonomics can increase the efficiency of human labor while minimizing on-the-job injuries and worker health impacts
   b) Managing personal environmental impacts: It explains the effect of the environment on workers—temperature, wind, humidity, and noise.
   c) Managing the impacts of scheduled overtime on productivity. It discusses the relationship between overtime hours scheduled for construction work and the productivity of workers.

Managing and Motivating the Work Force

Introducing Behavioral Change in Construction

Although technology is being gradually incorporated in construction processes, the industry is still driven largely by the human factor. All levels of the supply chain are people-dependent: designers, contractors, subcontractors, and suppliers all interact to take a construction project from the owner's project requirements (OPR) to the finished product. Improving performance, therefore, secures the greatest return on investment by addressing the human element. Performance improvement initiatives in construction encompass a wide range of alternatives. These include total quality management, Kaizen, Six Sigma, ISO 9000, and more recently, lean construction.

Although these initiatives vary widely in their approaches, they share one important characteristic; they require a new paradigm in which people who work in the construction industry adopt new behaviors. This need exists whether the people involved perform design work, manage construction projects, or apply physical labor in construction activity.

Improving Management–Worker Relations through the Malcolm Baldrige National Quality Award Criteria

Performance in the construction industry is linked to motivational issues in the workforce. Project success is largely attributable to the human factor. When a worker is satisfied with the job he or she is assigned to, then that individual tends to be more productive and pleasant to work with as motivation levels are high. Job satisfaction among trades and forepersons is improved in projects that are well planned and where all the facilities are available to allow them to be productive. Such conditions provide an incentive for workers to increase their productivity. The success of lean construction depends heavily on having workers engaged and actively involved in improving work processes.

Benchmarking against the Criteria

Construction leaders can secure significant gains in performance by benchmarking their organizations against the Baldrige Criteria, even if they do not wish to apply for an award. This applies equally to all construction-related organizations—architects, engineers, general contractors, construction managers, subcontractors, and suppliers.

The Malcolm Baldrige National Quality Award is based on seven criteria that serve as a model for organizations that aspire to be world-class (see also Chapter 9). Candidate organizations that apply for a quality award are evaluated against the seven criteria, and may succeed only if they have embodied them in their business model and strategic plan for a number of years.

Construction projects are accomplished by people, hence the greatest benefit to projects would be derived from the fifth criterion, termed the Workforce Focus category. It is concerned with how well an organization engages its workforce, manages and develops it to utilize its full potential. It also examines how an organization creates an environment that enables workers to maximize value to the organization.

Nonprescriptive Nature of the MBNQA

It is important to note that the Baldrige standards are not prescriptive. They do not dictate specific means and methods for improving performance. Rather, they provide a framework for performance improvement by asking questions that point to behaviors necessary for superior performance. Furthermore, awards are based on sustained activity, not ad hoc applications of a particular technique.

The MBNQA Criteria refer to "workforce" as the people actively involved in accomplishing the work of one's organization. In keeping with the construction environment it includes permanent, temporary, and part-time personnel as well as any contract employees supervised by the organization. Section 5.1 of the criteria addresses workforce engagement, and Section 5.2 is concerned with workforce environment.

The MBNQA Criteria are designed primarily for permanent, ongoing organizations. They have to be modified for construction activity as construction projects are carried out by teams that are assembled on a project-by-project basis. A typical construction organization would be reviewed under the following human resource areas

1. Motivation
2. Training
3. Communication
4. Work systems
5. Work environment
6. Employee well-being

The evaluation factors of work systems (Category 5.1) pose questions that address the following:

1. Workforce enrichment
   - Key factors that affect workforce engagement
   - Key factors that affect workforce satisfaction
   - Promoting open communications
   - How the organization harnesses diverse ideas, cultures, and thinking from the workforce
   - How the workforce performance management system supports high performance work and workforce engagement
2. Workforce and leader development
   - How a learning and development system is accomplished for the workforce and leaders
   - Core competencies, strategic challenges, and accomplishment of action plans
   - Performance improvement and innovation
   - Ethics and ethical business practices
   - Education, training, coaching, and mentoring
   - Knowledge from departing or retiring workers, and reinforcement of new knowledge and skills on the job
   - How effectiveness and efficiency of learning and development systems are evaluated
   - How leaders manage the career progression of the workforce and succession planning at the management level
3. Assessment of workforce engagement
   - How workforce engagement and workforce satisfaction are assessed
   - Use of indicators including absenteeism, retention, grievances, and productivity
   - How workforce engagement links with business results

Category 5.2 addresses workforce capability and capacity to carry out work and maintain a safe, secure, and supportive work environment.

1. Workforce capability and capacity
   - How workforce needs are assessed, skills, competencies, and staffing
   - How workforce members are recruited, hired, placed, and retained. How diversity is observed
   - How the workforce is managed and organized to conduct work
   - How it is acclimated to changing staffing levels
2. Workforce climate: employee well-being and satisfaction
   - How environmental factors are addressed to ensure and improve health safety and security
   - How policies, services, and benefits are tailored to a diverse workforce

The leaders of construction organizations can develop operational procedures from the MBNQA criteria to improve organizational performance. For example, they could review their processes in relation to their workforce focus by evaluating their responses to a number of questions including the following:

1. How does the organization encourage the workforce to achieve organizational success?
2. What types of motivation techniques does the organization have?
3. Are rewards varied to match them with contribution levels?
4. How does the organization improve workplace health, safety, and security?
5. Are employees enabled to learn from errors, or is there a punitive environment?
6. How do leaders create an environment that is conducive to high performance?
7. What type of training do managers need to support/improve employee performance?

By acting on the answers to the questions posed by the work systems criteria, a construction-related organization can develop a high performance environment. Readers should refer to the Malcolm Baldrige National Quality Award Criteria for more information.

Motivating Workers

Many techniques can be used to motivate workers in the construction environment. They include: goal setting, incentives, positive reinforcement, and worker participation. Worker motivation and job satisfaction are highest when a project is well planned and all the resources are available to enable everyone to be productive. Goal setting has succeeded in improving workforce productivity in the manufacturing and service industries, productivity has increased by up to 20% with a programmatic use of goal setting and feedback. Challenging and specific goals have produced better performance than medium, easy, or do-your-best goals or no goals. Despite the difficulty of applying goal setting in construction, productivity improvements of 10–20% are possible with that technique. Goal setting should be based on schedules that are realistic and are built on work standards that are historically based. Attainable targets should be established, based on the best historical performance.

Incentives

Construction organizations generally use incentive plans based on profit sharing. One significant limitation in such plans is that market conditions and factors beyond the organization's control may have a major impact on the profits generated in a particular firm.

Positive Reinforcement

Recognition programs have been used extensively in both the manufacturing and service industries; these typically include the "employee of the month" or "employee of the year."

More sophisticated suggestion award programs involve detailed review of employees' suggestions, and implementation if benefits can be substantiated. When suggestions are implemented the employees are recognized very openly; in some cases they receive monetary rewards as well. Such programs have proven to be very effective in improving morale. Despite this fact the programs are rarely implemented in the construction industry.

Changing Management Attitudes

The success of construction-related organizations is closely linked to the desire of the workforce to excel. Lean construction and integrated project delivery are highly dependent on top management's commitment to lean thinking; these initiatives can succeed only if leaders "buy in" to the concept and have their enthusiasm permeate through the organization and motivate the workforce at all levels. The strategic plan and each company's business model should reflect a commitment to adopt and implement lean principles to satisfy both internal and external customers. There is a wide spectrum of skills and abilities in the construction industry. Designers, on one hand, are highly trained architects or engineers with professional registration. Contracting organizations have formally educated managers as well as a workforce that is generally vocationally trained.

Regardless of their background, all providers are human beings and their performance is influenced by the extent of their motivation. In turn, the leadership of their respective organizations sets the tone for the level of motivation experienced by workers. In particular, workers' willingness to innovate and adopt new ideas and behaviors is positively correlated with their motivation. Similarly, workers' effectiveness in providing high quality products or services to an organization's external customers is closely linked with their level of motivation. Without motivation, workers (and many managers) resist change as they fear a loss of the ability to control their lives. Dr. Edwards Deming places the responsibility for the system at the feet of management. Deming takes the view that workers at all levels wish to experience a feeling of achievement and to take pride in their jobs. This desire is often thwarted by faulty systems that do not allow people to experience pride in their jobs and their workmanship. As management controls the system, it is up to them to remove the barriers to good performance.

Deming was distinguished for exhorting western managers to transform themselves to "improve and innovate the system of independent stakeholders of an organization over the long term to allow all people to experience joy in their work and pride in the outcome," and to improve and innovate the condition of society. While this advocacy may sound to some like altruistic rambling, it gets to the heart of the source of quality—the worker. The power of human resource is contained in the intrinsic motivation from the sheer joy of an endeavor, not just the extrinsic reward of a paycheck, no matter how large that may be. In his 14-point system of profound knowledge, Deming prescribes continuous improvement of product and service as the best route to competitiveness, to stay in business, and to provide jobs. He prescribes leadership as a substitute for management by numbers. Deming also points out that quality can only be built in, not inspected in, hence the worker is a critical component in the creation of quality construction. Traditionally, owners have deferred to the designer's best judgment and have not sought out ideas from field trades that actually execute construction work.

Top management must recognize that performance improvement is a journey, not a destination; it involves a delicate, harmonious interplay between such disparate factors as high technology, state of the art erection systems, and employee training and empowerment programs. For the transformation to occur, top management must establish an organizational culture that promotes quality principles, customer satisfaction, and continuous improvement as the business paradigm. Through strategic planning and implementation, this paradigm must be ingrained in the organization's vision statement, mission statement, guiding principles, broad strategic objectives, and specific tactics. Managers must exert leadership by example, and ensure that the paradigm is known and understood at all organizational levels.

Communication is an essential component of motivation. It involves not just imparting information, but verifying that the communication is understood by the receiver. Effective leaders are able to communicate in language that is familiar to the recipient, but also understand how the message is likely to be interpreted. As the old adage goes, "meanings are in people, not just in words." Good leaders also have the ability to inspire others, motivating them to perform at a higher level. Leaders can also bring about transformation by coaching, providing feedback, and rewards.

Moving beyond Traditional Thinking to Lean Thinking

Today's managers are generally aware of Theory X as described by Douglas McGregor (*Management Review*, November 1957) with respect to workforce attitudes. Theory X was based on three propositions:

1. Management is responsible for organizing the elements of productive enterprise in the interest of economic ends.

2. Management has to direct the efforts of the workforce, motivate them, control their actions, and modify their behavior to fit the organization's needs.

3. People are either passive or resistant to organizational needs and have to be persuaded, rewarded, punished, or controlled.

McGregor advanced the philosophy of Theory Y that shifted from a reliance on external control of human behavior to self-control and self-direction, essentially moving from treating people as children to treating them as mature adults. Theory Y held to a more positive view of human nature—that people are not passive by nature or resistant to organizational needs, but rather had been conditioned through negative experiences with organizations. It assumed that people have the capacity to grow and assume responsibility; management's essential task is to create the organizational conditions that enable people to self-actualize and direct their own efforts toward organizational objectives.

In the developed countries, Theory Y thinking has influenced the management of manufacturing and service organizations as these sectors have responded to the challenge of global competition. Quality management initiatives have created a cadre of workers in those sectors that have adopted a culture of "building in" quality as opposed to "inspecting in" quality. In the United States, The Malcolm Baldrige National Quality Award and state-level awards that are based on it, charge management with the responsibility of creating the conditions in the workplace that encourage workers to take the initiative in adopting quality-oriented behaviors. And yet, the construction industry as a whole has not yet broadly practiced Theory Y thinking. The attitude prevalent on many construction sites is still rooted in Theory X—workers are conditioned to do only what they are directed to do, with little room for initiative. It is not surprising, therefore that many work sampling studies have identified as much as 45% waste in direct labor activity.

Lean Thinking: Another Level beyond Theory Y

Theory Y did not require that production workers be trained in management techniques, but lean construction does. In the Last Planner® approach, the people closest to the work (i.e., the crews) are empowered to make decisions on detailed work assignments, as they know their own capabilities as well as the need for materials and equipment resources. Lean construction requires the crews to commit to and promise to carry out specific work assignments. To do this the crews ensure that the required resources are available and that the sequence is appropriate. Weekly meetings are held to review the work accomplished for the past week and plan the work for the following week.

Very importantly, these weekly meetings emphasize learning from experience with the most recent assignments—what went well and what did not. The purpose of the exercise is not to apportion blame for mistakes, but rather to determine how best to plan assignments to interact effectively with other trades and disciplines for optimal project results.

Lean construction requires a different culture in the workforce. Traditionally, less than 50% of the tasks planned for a given week may be accomplished in that week. Stops and starts typically occur or "making do" (i.e., keeping occupied with noncritical tasks), and that approaches significantly lowers work productivity. It also lowers the expectations of workers.

The concept of having work "pulled" from each crew by the downstream crew is an important lean principle, but it is counterintuitive. Tradition has conditioned workers to complete tasks as quickly as possible, to advance the accomplishment of their specific crews. Such actions may exceed the capacity of upstream trades and simply create work in process that cannot be completed without creating delays. By the same token, they may "push" their production to downstream crews faster than their capacity allows.

Lean coaching is an important activity in the implementation of lean construction. The *Toyota Way Fieldbook* describes the functions of a lean coach that include both process-related and training activity. They include:

- Promoting the lean transformation within the organization
- Coaching leaders at all levels
- Teaching lean tools and philosophy
- Leading value stream mapping
- Developing the lean operating system for the organization as well as metrics
- Leading Kaizen events

Kaizen events, in particular, are very successful in engaging the workforce. As described in Chapter 4, these events are usually of a short duration—approximately 1 week—but they provide an intense focus for everyone in the organization to investigate opportunities for improvement and implement them. One contractor that has successfully applied Kaizen is Tweet/Garot Mechanical Inc. of Wisconsin. They held a Kaizen event to clean up their main facility in order to smooth workflow and material storage activities.

Rank and file workers voiced a number of comments that indicated an interest far beyond wanting to carry out minimum job requirements. Examples of this include:

- Everyone brought a lot of quality ideas to this event
- Teams do have power (they have powerful ideas)
- It was hard to get started because of the disorganized yard
- There are no dumb ideas
- We save too much stuff in our everyday activities
- Work together as a team … not against each other
- We all have a better understanding what is a 5S

Managing Diversity

This section addresses two important contexts of diversity management; first with U.S.-based construction, and second with international projects involving U.S. managers in other countries.

Diversity may be defined as the variability in the workforce. Rasmussen (1996) states that the workforce is comprised of a variety of backgrounds, styles, perspectives, values, and beliefs as assets to the groups and organizations with which they interact. Some of its characteristics are inherited, while others are acquired. The management of diversity is essential for creating an environment in which people from different walks of life can come together as a unified team to produce high quality construction. This is sometimes very difficult to accomplish with workers of different cultural, racial, and ethnic backgrounds. An organization cannot have satisfied external customers until it has satisfied internal customers. It cannot meet, let alone exceed external customers' needs when internal customers are at war with each other, as is often the case with construction projects.

In his book *The New Economics*, Dr. Edwards Deming saw the organization as a system, with people representing components that work cooperatively with each other. Dr. Deming cited the importance of informal communication between people in various components of the organization, regardless of level of position. He recommended teamwork as a way of promoting joy in work, whereas he saw the competition that often occurs between people in the workplace as demoralizing to the individual and counterproductive to the organization as a whole.

Diversity Management in the United States

Several events over nearly four decades have influenced the status of the diversity question today: The Civil Rights Act, Affirmative Action, Equal Employment Opportunity, Sexual Harassment, and the Americans with Disabilities Act. These events, combined with increasing immigration have drastically changed the profile of the workforce; resulting in a very heterogeneous mix of workers, far different from the way it was in 1965. The workforce of the future will become increasingly multiracial, multiethnic, and multicultural, furthering the need for diversity management. The management of diversity is about good business, not just meeting state or federal mandates; it involves optimizing the interaction between people from all walks of life.

Model of Diversity

A typical model of diversity comprises two concentric circles. The inner circle represents attributes that can be seen readily, such as age, race, ethnicity, physical qualities including disabilities, and gender. Although not necessarily visible, sexual orientation is also included. These factors cannot be changed and form the basis of stereotypes. The outer circle has other factors that individuals have the ability to change, such as marital status, religious beliefs, education, income, geographic location, and work experience (Figure 13.1).

Discriminatory behavior is a fact of life, despite any hope for its eradication. Many large corporations have been the subject of lawsuits in recent years, charged by employees (or customers) with gender, racial, age, and other types of discrimination; many have settled out of court to avoid publicity. Historically, women and minorities have been underrepresented in the industry.

Construction companies do not usually interact with the general public and are driven by specific projects that may last from a few months to 1, 2, or 3 years. Discriminatory practices in their ranks scarcely reach the public view, yet their negative behaviors affect construction quality performance in subtle but tangible ways.

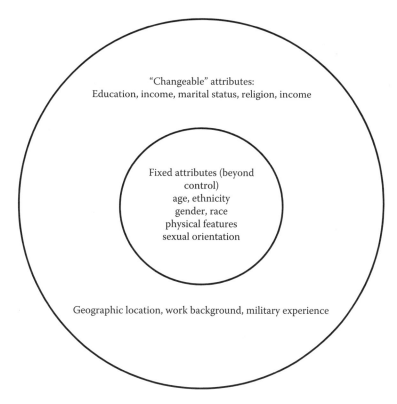

"Changeable" attributes:
Education, income, marital status, religion, income

Fixed attributes (beyond control)
age, ethnicity
gender, race
physical features
sexual orientation

Geographic location, work background, military experience

FIGURE 13.1
Representation of diversity factors.

Lean construction offers an excellent framework for overcoming this performance gap. It depends on a cohesive team, regardless of discipline or background, focused on delivering a project with the least amount of waste (i.e., without losses due to delays, quality lapses, cost overruns, or safety incidents). This requires working seamlessly, using the "pull" discipline for total project optimization, not local optimization. Every actor in the chain must be "in the loop" for this to occur.

Diversity in International/Overseas Projects

These projects are a very important market sector; at one point, 50% of the value of projects carried out by U.S. companies has been due to overseas projects, but there is growing competition from other countries. A typical project could be the construction of a hotel in the Cayman Islands with an American construction manager, a British architectural firm, a Trinidadian engineering firm and a Venezuelan contractor, using Latin American and Caribbean workers. Choudhury and Haque (2001) of Texas A&M University investigated 35 U.S. contractors working on international projects in Africa, Asia, Europe, Latin America, and the Middle East. The study findings pointed to the importance of cross-cultural training to project success; critical factors were the physical environment, language, economic environment, labor conditions, and social environment. The nature of sociocultural interaction appears to have a major impact on the productivity of project personnel; mistakes in this area can be costly, indeed.

The management of staff with multiple languages makes integration especially challenging because of the difficulty in translating heavily nuanced and complex technical concepts. The use of bilingual project managers appears to be the best strategy. A thorough understanding of the abilities and availability of foreign labor forces is critical in order to have realistic expectations of their production. Also importantly, the cross-cultural training of project staff should include cognitive studies of the host country population and their social and cultural norms. Swedish contractors have excelled in Saudi Arabia because of extensive preproject training on social values.

Decision making in some countries is far different from the way it is done in the United States. Whereas we have become accustomed to shallow or flat organizations and staff can collaborate openly and informally in making decisions, other cultures may approach the process in a far more complex manner, based on their power-distance index. Countries with a high index have a small middle class, high social stratification, and high levels of privilege for the powerful, as exemplified by such countries as India, Kuwait, Saudi Arabia, Nigeria, The Philippines, Mexico, Venezuela, and Panama, to name a few. Low index countries include the United States, Canada, Great Britain, Australia, Israel, Germany, and Costa Rica. In high index countries, the decisions tend to be more centralized and less democratic. These patterns are crucial in the conduct of negotiations. U.S. managers need to be patient and respect that approach.

The host country's customs may be very unfamiliar, yet U.S. managers must observe them. For example, Americans tend to use first names freely, even in formal communication. Some cultures take umbrage at this perceived overfamiliarity.

Attitudes to time may diverge widely throughout the world. North Americans and Northern Europeans treat time as a very precious commodity; they observe monochronic time, which involves a linear, orderly time progression. According to Seelye et al. (1995), polychronic time is the norm especially in many agrarian countries. Time is treated far more flexibly as an objective, not a precise commitment. Several tasks are pursued simultaneously, adapting to changing circumstances. An American project manager needs to understand this in order to manage successfully.

Taking Action on Diversity Management

Leaders should take the trouble and time to understand the potential of diversity; it goes far beyond hiring a portion of the workforce from "other groups" simply because it is said to be a good practice. To be meaningful, diversity needs to be applied across the organizational spectrum. Unfortunately, many organizations that desire to have a diverse workforce actually have a number of women and minorities in clerical or low-level professional ranks. That situation is a beginning, and leaders should build on it to tap people's potential by assembling the best team based on talent regardless of the individual's background.

Effective Diversity Management

A design or construction organization should set up a formal structure for managing diversity, so that it does not get taken for granted; the degree of complexity should be based on the organization's size. Several major U.S. corporations have established formal programs with diversity criteria integrated into their strategic plan, with associated measurement systems. Texas Instruments and Kraft Foods/Philip Morris Companies are well-established examples that have the important feature of measurement systems. As the old quality adage goes, "what gets measured gets done." Without measurement, the success (or failure) of a diversity management program may be impossible to determine.

- Communicate a commitment to staff about providing an environment that is conducive to diversity. This fosters cooperation, inclusiveness of all staff, and recognizes value of knowledge, talent, and skills from all sources.
- Diversity/sensitivity training for all staff is an absolute necessity. Role-playing is an indispensable component of that training;
- Managers should remember the fact that no ethnic group has a monopoly on natural mental ability or leadership. They should communicate belief in dignity and respect for all, identify and remove barriers to minority groups in the organization, eliminate judgmental attitudes, and learn to listen.
- Managers should practice employee development in order to make diversity successful: Establish and communicate performance expectations, use appropriate motivational approaches, provide working conditions they appreciate, provide meaningful feedback, both positive and negative, but respect the individual's dignity. Use appropriate rewards and recognition, preferably things that the employee values, not just a routine award.
- For international projects, train management staff on the social and cultural values of the respective host country. Use bilingual, bicultural staff to interface where possible.
- Develop conflict resolution procedures that allow a win–win outcome. Differences in viewpoints and cultural factors must be respected and addressed.

- Recognize that the management of diversity is an ongoing long-term process. Like all quality endeavors, a culture of continuous improvement is the best prescription for a design or construction organization to ascend to world-class performance.

Everyday Practices to Promote Harmony

Managers and staff employees can use the following ideas alike to improve harmony in the workplace, especially in design-related organizations.

Take a "vacation" without leaving home. We vacation in distant lands in order to see other cultures and broaden our knowledge of the world. Why not take advantage of a talk with a coworker who may be from a particular geographic region; it costs far less than an airline ticket and hotel stay, and can even equip you to better enjoy your visit if you decide to go.

Try to learn at least a few phrases of another language. View it as preparation for future opportunities. Students of international business pay universities handsomely for the privilege of learning about other cultures while they learn about marketing, finance and project management, and often go on to lucrative assignments with firms that have foreign affiliates. Interactions with coworkers of other backgrounds cost nothing, and at the very least increase the quality of teamwork in your organization. Better teamwork improves organizational performance and leads to greater competitiveness.

Nurture employees' potential. It is a good business practice to hire the best people. It is also good business to recognize the potential of individuals and provide them with an opportunity to grow. In turn, people who have grown with an organization often develop a high level of loyalty to it, but even more importantly, are willing to take ownership to provide customers with service beyond their expectations.

Actively create a mentoring culture. Mentoring is a critical factor in retaining minority professionals. While they may be hired on the basis of their training and skills, their advancement and longevity in an organization depends heavily on their ability to gain mission-critical experience and credibility. It is important to note that minority employees do not necessarily need to be mentored by other minorities; they simply need someone with more experience to show an interest in their welfare, provide some career guidance, and make them feel like a valued part of an organization.

Practice sensitivity. Managers in the construction arena should avoid appearing to "talk down" to people from less developed countries. People in host countries do not want to be made to feel their culture is inferior; they often regard external influences as irrelevant. Company representatives in a host country should be careful not to flaunt the fact that their compensation is higher than that of locals. This information may simply alienate others and destroy any hope of having a cohesive team.

Construction professionals should recognize that unwitting actions may offend others. Some managers from developed countries occasionally put up their feet to rest their shoes on desks or other furniture, sometimes facing their audience. Some cultures are deeply troubled by this action.

Ultimately, a project team can only excel if its people are able to take pride and joy in their work, regardless of their background. Those people must work together harmoniously, as a system, and treat each other as important internal suppliers and customers without reservation. Effective diversity management helps to accomplish that end.

Improving Worker Performance: Ergonomics-Based Strategies

Ergonomics is a practical discipline that deals with the design and improvement of productivity and safety in the workplace. Ergonomics involves studies of human perceptions, motions, workstations, machines, products, and work environments. The main methodology of ergonomics involves the mutual adaptation of the components of human-machine-environment systems by means of human-centered design of machines in production systems.

The application of the principles of ergonomics in the construction environment has many benefits. Physical work input is a major component of the construction process. Building components are typically large, heavy, and cumbersome. Building blocks (CMUs) and bricks are used in large quantities and are installed by hand. In the course of a day a bricklayer may handle large quantities of bricks or building blocks; this handling imposes a physical workload on the worker. Reinforcing steel is also used in the construction process and its weight is physically stressful as well.

Construction involves the use of tools and equipment in tight and awkward spaces, requiring workers to assume uncomfortable postures. Especially in cases where such aids are used for prolonged periods of time, workers' effectiveness and capacity to work with high levels of concentration is reduced. These situations often lead to on-the-job injuries and worker health impacts. Workers cannot be expected to "build in" quality in constructed facilities if they are subjected to excessive physical stress caused by tools and equipment that are difficult to use.

Reducing Musculoskeletal Stresses

Musculoskeletal disorders (MSDs) are different from other occupational injuries and illnesses because they develop over a long period of time. Musculoskeletal diseases (mostly low back problems) represent 27% of nonfatal injuries and illnesses in construction. In 2001 disorders resulting in lost work days were:

Overexertion while lifting	45%
Overexertion without lifting	32%
Bending/twisting	17%
Repetitive motion	7%

Overexertion injuries are mostly observed in the following trades: roofing/siding/sheet metal, masonry, carpentry, and mechanical trades (U.S. Department of Labor 2001).

Work-related musculoskeletal disorders (WMSDs) account for 34% of all lost workday injuries and illnesses according to the Bureau of Labor Statistics (BLS) and account for between $15 and $20 billion in worker's compensation costs in the United States each year. These disorders have a critical impact on the construction industry as workers with severe injury may experience permanent disability that renders them unable to perform their job tasks in keeping with standard output levels.

The WMSDs include: carpal tunnel syndrome, tenosynovitis, tension neck syndrome, and lower back pain. Carpal tunnel syndrome accounts for more days away from work than any other workplace injury. All body parts involving joints, muscles, tendons, nerves, and blood vessels are susceptible to WMSD. Overuse disorders include: strains, which are injuries to muscles or tendons; inflammation, as a protective response to limit bacterial invasion; and sprains involve the stretching, tearing, or pulling of a ligament.

Carpal Tunnel Syndrome (CTS)

CTS is one of the most common and well-known WMSDs. It receives its name from the eight bones in the wrist, called carpals. The disease occurs with the gradual swelling of the tendon sheaths that surround each of the tendons. CTS symptoms are:

- Painful tingling in one or both hands
- Finger numbness
- Fingers are described as being swollen
- A weakened grip

There are nonoccupational factors that contribute to CTS. They include: systematic diseases such as rheumatoid arthritis, acromegaly, gout, diabetes, myxoedema, ganglion formation, and certain forms of cancer. Congenital defects may also lead to CTS. It can be caused by acute trauma to the median nerve inside the carpal tunnel that can may result from a blow to the wrist, a laceration, or a burn. Women tend to have more susceptibility to CTS due to their naturally smaller body frame.

Risk Factors for WMSD

The National Institute for Occupational Safety and Health (NIOSH) attributes WMSDs to a number of causes:

- Repetitive, forceful, or prolonged exertions of the hands
- Frequent or heavy lifting
- Pushing, pulling, or carrying of heavy objects
- Prolonged awkward postures
- Vibration

There is a relationship between workplace factors and MSD development. Some joints display significant susceptibility to work-related injury. They are: the neck and neck/shoulder area, the back, and the hand/wrist area. It is not clear that elbow injuries are generally work-related.

Back disorders may be associated with both occupational and nonoccupational factors. Back injuries account for almost 20% of all injuries and illnesses in the workplace, and the prevalence in the general population has been estimated at 70%. The costs of these injuries in the United States range from $20 to $50 billion annually, hence its impacts on the economy are significant. At the organization level such injuries exact high costs by both reducing workforce availability and inflate insurance costs.

Evaluating and Addressing MSDs

There are seven steps for evaluating and addressing MSDs in the workplace. Strategies are described in this section below for addressing their effects. The steps are:

- Be able to identify signs of WMSDs
- Set the stage for action
- Train personnel in order to build in house expertise
- Gather and examine evidence concerning WMSDs
- Identify effective controls for WMSD-risky tasks, and evaluate these controls to determine effectiveness
- Implement health care management
- Maintain a proactive ergonomics program

Work System Design

According to Albers et al. (2005), tools and their selection influence work postures and work stress. In order to reduce worker stress and improve productivity, work system design should anticipate the postures required to accomplish the respective tasks. Using electrical work as an example, the work system comprises:

- The work task itself (e.g., high level cable tray installation)
- The available equipment (e.g., scaffolding)
- Tools, power or hand such as electric drills
- Design processes that affect these components

Modular and Architectural Innovations

Modular and architectural innovations can have an impact on the construction process and worker activities. Modular innovations can be implemented by one trade with minimal effect on others. They reduce exposure to one or more risk factors (Slaughter 1998). Architectural innovations change the nature of the activity for a number of trades. For example, a hammer drill is often used to drill holes and embed concrete inserts for the hangers for piping, cable trays, or other devices. Instead, inserts may be cast into concrete beams by inserting them in the form work used for pouring concrete to form the beams. One trade would be relieved of the drudgery of drilling holes in beams to set the concrete inserts after the building structure has been erected. On the other hand, another trade would have the responsibility to place the inserts in the concrete form work.

NIOSH Recommended Best Practices for Work Tasks

In the absence of ergonomic standards for U.S. construction, NIOSH assembled a number of best practices from a 2003 stakeholder meeting (Albers, Estill, MacDonald 2005). These stakeholders included mechanical/electrical contractors/installers and trades; 39 industry representatives, 17 construction ergonomics researchers from government and academia, and four ergonomics consultants with construct experience. Common tasks were (a) drill holes and shoot fasteners, (b) place and install systems, and (c) lift and carry materials and equipment (Table 13.1). Table 13.2 suggested ergonomics intervention to reduce musculoskeletal loading during placing and installing mechanical/electrical systems.

Voluntary Ergonomics Guidelines Developed by OSHA

Design for Construction

Design processes should consider the construction process in terms of safety, health, and ergonomics as design decisions are being made (BNA 2002). The proximity of pipes, ventilation ducts, and cable trays can be a problem for construction workers. Workers are forced to work in very confined spaces, and to assume postures that result in very high musculoskeletal stresses. The inclusion of fastening anchors and hanging devices in the building structure can have a significant impact on the complexity and risk factor exposure of construction workers. The configuration of building elements that place more emphasis on mechanized transportation and handling rather than on manual methods can also have a significant impact on construction processes. It can also impact the speed of construction and the rate of productivity.

TABLE 13.1

Suggested Ergonomics Intervention to Reduce Musculoskeletal Loading During Drilling Holes and Shooting Fasteners

Problem/ Musculoskeletal Load	Electrical	Piping	Sheet Metal HVAC
Install anchors for hanging systems in concrete		Specify embedded concrete inserts (i.e., Unistrut, Anvil, etc., or beam clamps, caddy clips)	Specify embedded concrete inserts (i.e., Unistrut, Anvil, etc., or beam clamps, caddy clips) Place minimum number of hangers required
Operate rotary-impact hammer drill	Use clutch-driven drill to control torque Use sharp bits Use low vibration tools Use vibration damping gloves	Use clutch-driven drill to control torque Use sharp drill bits Use vibration damping gloves Implement a tool and bit maintenance program	Use drill with side arm to control torque Use low vibration tools Identify, purchase tools on performance criteria
Operate powered actuated tool overhead		Use extension pole and remote triggering	Use extension pole and remote triggering Use drill bit extender
Drill overhead	Support head with neck pillow Work on platform or scaffold	Use drill bit extender Use suspension and balance system for tool	Work on powered lift or scaffold
Work overhead	Work on powered lift or scaffold	Work on powered lift or scaffold	Work on powered lift or scaffold
Standing on concrete	Stand on antifatigue mat	Stand on antifatigue mat or use shoe inserts	
Working at floor level			Use knee pads

Source: Adapted from Albers, J., Estill, C., and MacDonald, L., *Applied Ergonomics*, 36, 428–39, 2005. With permission from Elsevier.

TABLE 13.2

Suggested Ergonomics Intervention to Reduce Musculoskeletal Loading During Placing and Installing Mechanical/Electrical Systems

Problem/ Musculoskeletal Load	Electrical	Piping	Sheet Metal HVAC
Intensive manual hand tool use	Use cordless power tools	Select tool with best design for task	Use power tool when possible
Work overhead		Work on powered work platform or scaffold	Work on powered work platform or ladder platform
Install anchors for hanging system in concrete	Work on powered work platform or scaffold	Specify embedded concrete inserts (i.e., Unistrut, Anvil, etc.) or beam clamps, caddy clips	Specify embedded concrete inserts (i.e., Unistrut, Anvil, etc.) or beam clamps, caddy clips
		Facilitate worker physical conditioning (i.e., stretching programs)	Place minimum number of hangers
		Take short ("micro") breaks	
Set threaded rod anchor with manual tools		Use ratchet with open socket to tighten nuts	Use power drill to tighten anchors and set nut
		Use split or button nut	Use ratchet with open socket to tighten nuts
			Use split or button nut
Manually lift, position, and hold system components	Use powered and mechanical devices to raise materials and equipment	Use powered and mechanical devices to raise materials and equipment	Use powered and mechanical devices to raise materials and equipment
	Use fixtures to hold, lift, and position materials	Use fixtures to hold and position work materials (i.e., "v" fixture for pipe)	Use fixtures to hold and position work materials (i.e., half-moon fixture for circular duct)
			Use magnets or suction cups with handles to position duct
Using floor as work surface	Use portable work table		Keep work off the floor

Source: Adapted from Albers, J., Estill, C., and MacDonald, L., *Applied Ergonomics*, 36, 428–39, 2005. With permission from Elsevier.

Interventions

Some interventions are within the reach of contractors and some are not. The use of embedded concrete inserts or beam clamps are determined during the design phase of the project, and are considered to be engineering interventions. Administrative controls are within the reach of contractors for reducing musculoskeletal loading on workers:

- Worker training programs in body mechanics
- Job rotation and microbreaks
- Physical conditioning (i.e., stretching programs)
- Prejob planning and hazard analysis
- Optimizing communication and coordination between contractors and trades

Mechanized Material Handling

Rough terrain forklifts facilitate the stressfree movement of materials from trucks or storage areas to the point of use. Some equipment can raise materials to several stories, speeding up material handling time, and eliminating or minimizing worker injuries.

Strategies for Manual Material Handling

- Observe weight limits—require weight labeling
- Optimize building site organization and housekeeping (this is complicated by having multiple contractors)
- Utilize lifting programs—training programs can significantly lower the occurrence of low back pain

Six recommended lifting methods are:

- Straight back, bent knees lift (most effective lifting method)
- The hip-flex lift
- The kinetic lift
- The dynamic lift
- The stooped lift
- The pelvic-tilt lift

Back injury prevention programs have three general parts. Preemployment and job-placement screening, training and workplace design. Back belts have been proven to be effective in preventing back injuries related to manual holding jobs. However, a disadvantage of back belts is a risk of homeostasis, in which a person believes that he/she can lift more than they are capable of when wearing the belt.

Tool and Equipment Design

In today's construction related tasks, it is almost impossible to accomplish any construction projects without the use of hand tools. Unfortunately the U.S. construction industry has the highest rate of nontraumatic soft tissue injuries to the neck, back, and upper extremities with incidence rates of 622.2 per 10,000 full-time employees. The use of hand tools alone, according to a report in 2003 by the National Safety council, accounts for 4.39% of all the compensation injuries related to work. For powered hand tools, drilling accounts for 17.6% of cases of work related injuries.

Much research has yet to be done in the design of construction-oriented tools and equipment. The factors that may cause fatigue include: weight, size, vibration, and operating temperature. WRMSDs generally include strains, sprains, soft tissue and nerve injuries; they are cumulative trauma disorders and repetitive motion injuries. The construction workers who are at the highest risk for these disorders are: carpenters, plumbers, drywall installers, roofers, electricians, structural metal workers, carpet layers, tile setters, plasterers, and machine operators. The top five contributory risk factors are: Working in a specific posture for prolonged periods, bending or rotating the trunk awkwardly, working in cramped or awkward positions, working after sustaining an injury, and handling heavy materials or equipment.

The use of a shovel is a very typical example of the labor-intensive material handling activities that are routinely carried out on construction projects. This activity requires workers to bend over, apply force to a shovel in different planes and rotate the trunk in a flexed position. Such movements impose biomechanical stress that may impose cumulative trauma risk. Freivalds (1986) studied the work physiology of shoveling tasks and identified the shovel design parameters that would increase task efficiency. Friedvald's two-phase experimental study addressed the following parameters:

- The size and shape of the shovel blade
- The lift angle
- Shovel contours—hollow and closed-back design
- Handle length
- Energy expenditure
- Perceived exertion
- Low back compressive forces

The recommended shovel design is as follows:

- A lift angle of approximately 32°
- A hollow-back construction to reduce weight
- A long tapered handle
- A solid socket for strength in heavy duty uses
- A large, square point blade for shoveling
- A round point blade for digging, with a built-in step for digging in hard soil

Ergonomics Applications in Structural Ironwork

The BLS reports that construction trades workers experience higher rates of musculoskeletal injuries and disorders than workers in other industries: 7.9 cases per 100 equivalent workers as compared with the industry average of 5.7 per 100 (BLS 2001). In overall injuries, construction workers registered 7.8 versus the industry average of 5.4. As pointed out by Forde and Buchholz (2004), each construction trade and task represents a unique situation; the identification and application of prevention measures, tools and work conditions is best derived from trade and task specific studies. This approach is the most likely to minimize the incidence of construction trades WRMSDs.

By way of illustration, Forde et al. studied construction ironworkers to identify mitigating measures in that group. Construction ironwork refers to outdoor work (not shop fabrication) as four specialties; the erection of structural steel (structural ironwork, SIW), placement of reinforcing bars (rebars; reinforcing ironwork RIW), ornamental ironwork (OIW), and machinery moving and rigging (MMRIW). Previous studies determined that construction ironwork involves the lifting, carrying, and manipulating of heavy loads; maintaining awkward postures in cramped quarters; working with arms overhead for extended periods; using heavy, vibrating pneumatic tools; and extensive outdoor exposure in temperature and weather extremes. Forde et al. (2004) made the following observations and recommendations on the various categories of ironwork.

Machinery Moving/Rigging

The erection of equipment such as a crane involves the pushing and pulling of large and heavy segments, and lining them up for bolting together. During an 8-hour shift, this activity was observed to require 1.3 hours of significant whole-body exertion. Workers in this scenario are most susceptible to overexertion of the back, legs, and shoulders.

Ornamental Ironwork

This work was observed to require arms to be above shoulder level 21% of the time. Trunk flexion and/or twisting and side bending were observed 23% of the time. These percentages indicate a high risk of overexertion of the involved muscle groups. Industrial engineers should review the work methods to increase the amount of preassembly at workbench height.

Reinforcing Ironwork

The preparation of reinforcement cages and tying of rebars was seen to cause nonneutral trunk postures up to 50% of the time. The handling of heavy loads (50 pounds or greater) was observed to occur for 1.9 hours of an eight-hour shift, representing significant long-term risk. Forde et al. identified a need to improve the design of hand tools used for securing rebars. Such redesign would reduce nonneutral hand/wrist postures such as flexion, extension, and radial and ulnar deviation. These postures put construction workers at risk of repetitive motion injuries.

Auxiliary Handling Devices

A number of research studies have shown that construction workers have suffered back, leg and shoulder injuries because of overexertion resulting from stooped postures, performing manual tasks above shoulder level, and the lifting of heavy objects. Such overexertion and injuries reduce worker productivity and may negatively affect the timeliness and profitability of construction projects. The use of auxiliary handling devices may reduce the degree of overexertion experienced by construction workers and enhance productivity. Sillanpaa et al. studied five auxiliary devices:

- Carpet wheels
- A lifting strap for drain pipes
- A portable cutting bench for molding
- A portable storage rack
- A portable cutting bench for rebars

The survey subjects utilized these devices to carry out typical construction tasks such as carrying rolls of carpet, mounting drain pipes, cut pieces of molding, and fashioned rebars. The results of the study were mostly positive but mixed, pointing to the need for further research. The auxiliary devices were found to reduce the muscular load of some subjects, but others experienced an increased load because of differences in anthropometric dimensions, work modes, and level of work experience.

Drywall Hanging Methods

Drywall lifting and hanging are extensively conducted in both residential and commercial building construction; drywall board has become the standard for interior wall panels. It is the standard for surfacing residential ceilings. Workers are required to handle heavy and bulky drywall sheets and assume and maintain awkward postures in the course of performing installation work. These activities often cause muscle fatigue and lead to a loss of balance; studies have identified drywall lifting and hanging tasks as causing more fall-related injuries than any other tasks. Pan et al. (2000) studied 60 construction workers to identify the methods resulting in the least postural stability during drywall lifting and hanging tasks.

The subjects' instability was measured using a piezoelectric-type force platform. Subjects' propensity for loss of balance was described by two postural sway variables (sway length and sway area) and three instability indices (PSB, SAR, and WRTI). The study was a randomized repeated design with lifting and hanging methods for lifting and hanging randomly assigned to the subjects. ANOVA indicated that the respective lifting and hanging methods had significant effects on two postural-sway variables and the three postural instability indices. The recommended methods were:

- Lifting drywall sheets horizontally with both hands positioned on the top of the drywall causes the least postural sway and instability.

- Hanging drywall horizontally on ceilings produces less postural sway and instability than vertically.

Case Study in Overhead Drilling

This case is based on a study carried out by Nabatilan et al. at Louisiana State University (2007). Drilling is one of the most physically demanding construction activities. At different positions, overhead drilling accounts for 68.8% of nontraumatic MSDs (Bjelle et al. 1979). Construction workers may be required to use high forces during drilling tasks such as, when sheet metal workers drill into concrete ceilings. There is strong evidence that combination of two or more risk factors, such as force and awkward posture, increase the risk of WRMSDs (NIOSH 1997).

In construction work, it is often necessary to work with arms in awkward positions such as overhead positions. Rosecrance et al. (1996) reported that 41% of a sample of construction workers in the pipe trades complained of work-related shoulder pain when tasks were performed in different postures including overhead postures. The shoulder is one of the most complex biomechanical structures of the human body. Awkward postures (shoulder elevation greater than 60°; NIOSH 1997) impose increased physical demand to the shoulders and can cause damage to the shoulder girdle.

Nabatilan et al. (2007) conducted a study at Louisiana State University to evaluate the effect of providing support on the muscular activity of the shoulder while using a drill with and without support. Five young males participated in the study, and the myoelectric activity was recorded from the bicep, anterior deltoid, and trapezius muscles for 180 seconds at the rate of 200 readings per second. The analysis of raw EMG that was collected during the experimentation process was done by Root Mean Square technique (RMS). The mean RMS values for all the subjects with and without support for different muscles were used to perform statistical analysis and to compare the mean difference of the RMS values. The following sections discuss the effectiveness of the overhead support stand on the shoulder muscular activity.

Evaluation of an Overhead Support Stand

The average RMS values of EMG of the five participants for each of the muscles were calculated from the raw EMG data. The RMS values were divided into intervals of 1 second, and the average RMS values for each second were calculated. Table 13.3 shows the average RMS values and the difference in RMS values for all the muscles for drilling tasks with and without support. It can be observed from RMS values in Table 13.3 that the overhead support stand was effective in reducing the load on the muscles for the drilling tasks used. The RMS values without support are greater than those with support. There is a reduction in muscle activity as measured through the EMG while holding the drill with support.

The mean RMS value for bicep muscles with support was less when compared without support (0.155 mv). This value is 47% lower in contrast to the without support. The mean RMS value for anterior deltoid was 0.264 and for trapezius muscle was 0.0529. Figure 13.2 shows a graph that compares the RMS values with support and without support for three muscles namely bicep, anterior deltoid, and the trapezius muscle.

The mean RMS values of anterior deltoid and trapezius muscles with support decreased by 49% and 42%, respectively, as compared without support. To get feedback from participants and to analyze the effect of the overhead drilling support system from a user's point of view, a subjective evaluation was conducted. After each trial of the experiment, the participants were asked to fill a subjective rating form to indicate in what segment of the body they felt stiffness, ache, pain, or discomfort. Also at the end of the trials, the participants were instructed to evaluate the overhead support stand. The subjective rating form consisted of a body map showing the pinpoint location of muscles where discomfort was felt. The participants were asked to rate the extent of pain or discomfort in the body segments according to the scale provided that is rated from 0 to 4 with 0 being the least and 4 being the highest. The results are shown in Table 13.4 and illustrated in Figure 13.2. Based on the subjective evaluation, the participants indicated neither pain nor discomfort on the waist. Figure 13.2 illustrates the results of the ratings using a bar graph.

Subjective Ratings

The participants evaluated the drilling position with and without support. A ratings scale was designed in order to measure the subjective ratings of body parts stress/fatigue that the participants experienced. The results of the subjective rating showed that all participants experienced the least pain when using the drill with the overhead support.

Based on the results of the participant's feedback, there was 53% less pain in the chest, 40 and 38% reduction in shoulder and upper arm pains, respectively. Based on the ratings scale, an average of 43% less stress was felt by the participants when holding the drill with

TABLE 13.3

Mean Difference and Percentage Decrease in RMS Values for Overhead Drilling With and Without Support

Muscles	Without Support RMS (mv)	With Support RMS (mv)	Difference in RMS (mv)	Percent Decrease in RMS
Biceps	0.332	0.176	0.155	47
Anterior deltoid	0.545	0.280	0.264	49
Trapezius	0.125	0.073	0.0529	42

Source: From Nabatilan, L., and Venkata, B., *Proceedings, Annual Conference, Institute of Industrial Engineers.* 2007.

TABLE 13.4

Subjective Rating Results With and Without Support

	Upper Chest	Shoulder	Upper Arm	Standard Deviation Upper Chest, Shoulder and Upper Arm
Without support	3	3	3.2	0.115
With support	1.4	1.8	2	0.3055

Source: From Nabatilan, L., and Venkata, B., *Proceedings, Annual Conference, Institute of Industrial Engineers*. 2007.

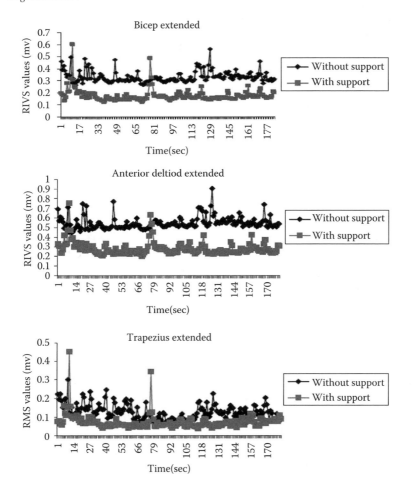

FIGURE 13.2

RMS values for the three muscles. (From Nabatilan, L., and Venkata, B., *Proceedings, Annual Conference, Institute of Industrial Engineers*. 2007.)

the designed support. The subjects rated the amount of discomfort in the waist region zero, which suggests that they did not experience any load in that part of the body while wearing the overhead support stand.

All participants indicated that the use of an overhead support mechanism made the overhead drilling tasks easier. This further supports the conclusion that an overhead support stand is effective in reducing the load on the specified muscle groups. The effectiveness of ergonomic interventions such as the one used in this experiment, should be studied as health benefit factors will play a major role in decisions about the use of interventions in the workplace. This will also guide ergonomists when implementing occupational health programs.

Managing Environmental Impacts

Effects of Vibration and Noise

The most common injury due to vibration is sound-induced hearing loss. Some of the effects that listening to music may have are:

- Loss of hearing sensitivity
- Immediate physical damage
- Interference
- Annoyance
- Distraction
- Contribution to other disorders

Noise is a major concern in the construction environment. Construction equipment and construction activity in general produce high noise levels. Unlike most harmful contaminants, the effects of noise may be unnoticed instantaneously and its accumulation can lead to an obvious physical, psychic, and social deterioration. Exposed people are usually unaware of the progressive sensory damage caused by noise exposure. Unless specific steps are taken to counter the effects of noise it may cause permanent deafness among construction workers. Noise exposure, therefore, may have a negative impact of the health and well-being of the construction workforce.

Reduced hearing sensitivity may render workers unable to hear speech and alarm sounds, representing a potentially dangerous situation. Voice problems, like nodules, loss of voice, and abnormalities in the vocal chords can be suffered by the workers that have to communicate within noisy environments with levels higher than 85 dBA if there is no other way to communicate but by speaking.

Properties of Sound

Sound may be described as a wave caused by a change in atmosphere pressure resulting from a vibration or turbulence. Hearing loss is an impairment that interferes with the receipt of sound, and with it the understanding of speech in sentence form. The most important frequencies for speech understanding fall between 200 and 5000 hertz. A young person with normal hearing can detect sounds with a frequency range that extends from 20 to 20,000 hertz. Tests have shown that hearing loss occurs as persons age due to continuous exposure to environmental noise of modern society rather than to simply aging.

Loud impulsive noises can have worse effects than the hearing loss due to aging. An impulsive noise is one that occurs suddenly, such as that due to the impact of a heavy steam hammer, an explosion, or a rifle shot. When an impulsive noise occurs, most of the mechanisms in the ear are incapable of providing self-protection.

Annoyance

- Noise annoys people.
- Noises that annoy need not be loud noises.
- A slow water drip in the silence of the night can be as annoying as a loud noise.
- An unexpected impulsive noise can be the most annoying.

Distraction

Sounds can distract workers from doing what they are supposed to.

Other Vibration Effects

- Noise is the most common vibration problem.
- High-intensity, low-frequency sounds can cause the skull, other bones, and internal organs to vibrate with injurious or annoying amplitudes.
- Resonance will occur at certain frequencies so that these effects become much more noticeable.
- Vibrations can be produced within the equipment itself or induced. It is transmitted more easily through solid materials than through air.

Effects of Tools

- The use of vibrating tools can lead to arthritis, bursitis, injury to the soft tissues of the hands, and blockage of the blood vessels.
- Raynaud's phenomenon involves paleness of the skin from oxygen deficiency due to reduction of blood flow caused by blood-vessel and nerve spasms.

Interference and Masking

- Any operation that requires oral communication will suffer from a noisy environment.
- Interference with communications can create misunderstandings about information transmitted from one person to another.
- Masking is the level by which a sound must be increased to be understood in the presence of another, interfering sound.

Elements of a Hearing Conservation Program (HCP)

United States OSHA regulations require that a hearing conservation program be instituted if the noise exceeds 85 dBA TWA in a workplace. This program should include record keeping of these activities: audiometric testing, monitoring of noise exposure, using hearing protection devices appropriately, employee training, and noise control engineering.

Audiometric Test Programs

A worker's hearing may have been defective before employed. It is important that accurate records of the tests be maintained. Each record should contain the audiogram itself, the employee's name, social security number, date and time, and so on.

Exposure Monitoring: Measuring Sound Levels

Sound surveys to assess the degree of exposure to hazardous sound levels should be conducted regularly. Noise measurements should include measurements taken close to where a worker's ear might be when working. Sound-pressure level meters are used for this purpose.

Weighted Sound Level Meters

Three weighting circuits are incorporated into the standard sound-pressure level meter. Because the ear is less sensitive to low frequencies, the A network attenuates very low frequencies to approximate the response of the human ear. Octave-band analyzers are used to determine at which frequency bands sounds are being generated. If a piece of equipment has been found, an octave-band analyzer can be used to determine the noise source.

Hearing Protection Minimizing Adverse Effects

The ear itself has natural protective mechanisms that help reduce possible effects of loud noises and their attendant pressures and vibration. The ear canal is curved so that sound waves cannot impinge directly on the eardrum. Eardrum muscles contract in response to a loud noise.

Hearing Protection Devices

- Plugs: Rubber or plastic devices fit snuggly against the ear canals, blocking the passage against transmission of sound.
- Foam plugs: Small foam rubber cylinders are compressed and twisted between the thumb and forefinger, before inserting into ear.
- Wools: The earliest ear protection devices were probably masses of cotton or wool forced into the ear canal as plugs.
- Muffs: These cover the entire ear and some of the bony areas around it through which sound might be conducted.

Employee Training

- Understand the danger to hearing that comes from noise exposure.
- Recognize noise exposures that are harmful.
- Evaluate noise levels of exposure in a practical way.
- Take action to protect themselves from harm from noise.

Engineering Control: Eliminating Vibration Causes

- Select equipment for installation that has low vibration and noise characteristics.
- Select and operate rotating and reciprocating equipment.
- Mount equipment that might vibrate on firm surfaces.

Isolating Sources

- Locate noisy activities such as individual machines in sound-absorbing enclosures, or provide barriers between such activities and other locations where personnel are present.
- Ensure that floors, walls, and other structural features do not vibrate and transmit vibrations and noise to other locations.
- Where vibrations of fixed equipment cannot be eliminated, mount the equipment on vibration isolators to prevent transmission of motion.

Isolating Personnel

- If only one or two workers must remain in a large, noisy area, determine whether they can be isolated in an acoustically quiet booth or other enclosure.
- If the noise level cannot be reduced to a maximum of 85 dB, provide workers with hearing protection devices.
- Attempt to schedule personnel so they remain in high-noise-level areas for periods that are as short as possible.
- Ensure that workers who must wear hearing protection devices such as ear plugs are initially fitted with them by approved audiologists and that each worker is aware of the proper type and size to use.

Temperature Effects on Personnel

Arousal theory explains the impact on a worker's productivity caused by the demands of a task and the physical environment. The physical environment creates physical stimuli that, in turn, create arousal in the human body, particularly heat and noise. The body regulates its equilibrium through thermal exchanges with the surrounding environment (i.e., evaporation, conduction, radiation, and convection).

If the approximate level of a worker's productivity is estimated, then that information can improve the accuracy of resource management activities. Variability in workforce performance can be compensated for by basing resource allocation on a forecast of seasonal weather conditions. Four major parameters can affect human performance: ambient temperature, relative humidity, wind speed, and radiant temperature.

Workers are most effective when working in an environment where temperatures vary between 50 and 70° F. In this temperature range the human body gives up heat by evaporation from the skin and by respiration, without the need for much sweating. As the temperature and humidity rise, the body is less able to lose heat, and its stress level rises as well. This results in lower levels of productivity. There are a number of thermal comfort indices:

1. Effective temperature
2. Wet bulb globe temperature
3. Operative temperature
4. Heat stress index
5. Standard effective temperature

Normal body temperature is approximately 98.6° F (Table 13.5). There are a number of thermal and humidity effects that impact workers.

High Temperatures

- Continued exposure to heat and humidity is a common cause of heat cramps, heat exhaustion, or heat stroke.
- Heat evaporation is caused by excessive perspiring, because of a hot environment or strenuous physical exertion.
- Heat stroke is much more serious than heat cramps or exhaustion.

High Temperature and Performance

- Stresses generated by high temperature degrade performance.
- The degree to which temperature affects performance is a topic of constant debate.
- Five categories that determine the effects of heat:
  1. Intensity of heat
  2. Duration of exposure
  3. Tasks involved
  4. Persons performing the tasks
  5. Presence of other stresses
- High humidity can cause severe stresses, especially at high temperatures.

Effects of Cold

- Chilblaiwn: a relatively mild form of tissue damage.
- Local itching and swelling
- Wet cold syndrome: trench foot and immersion foot. These result from exposures below 53°F for several days.
- Frostbite: results from prolonged and severe vasoconstriction at temperatures below 32°F.

TABLE 13.5

ASHRAE Seven-Point Scale

ASHRAE Scale	Sensation	Physiological
3	Hot	Profuse sweating
2	Warm	Sweating
1	Slightly warm	Slight sweating, vasodilation
0	Neutral	Neutrality
−1	Slightly cool	Vasoconstriction
−2	Cool	Slow body cooling
−3	Cool	Shivering

Source: ASHRAE Fundamentals Handbook (SI), 8.1–8.28, 1997.

Additional Effects: Wind

- High and low temperature variations can all cause damage directly or generate other conditions that can result in damage.
- Wind impacts body temperature by removing heat from the skin surface. In effect for a given ambient temperature an increase in wind speed creates a lower "effective" temperature.

The National Electrical Contractors Association (NECA) sponsored a study in 2004 to examine the effects of extreme combinations of temperature on electrical labor productivity. The principal researcher was Awad S. Hanna, PhD, PE. A number of tables and charts were compiled to document the interaction of temperature, humidity, and wind, as may be encountered in field situations. NECA recommends that the measurement data compiled in these tables and charts be used in factoring of estimates, job planning and scheduling, and in estimating change orders related to extensions of time. Tables 13.6 and 13.7 were adapted from the findings of the NECA study. The general findings from the research were:

1. Worker efficiency of 100% can be achieved only when the temperature is between 40 and 70°F, and the relative humidity is below 80%.

2. While humidity of any percentage is not a significant factor in the 30–80°F range, it plays a very important role at elevated temperatures.

3. In extremely cold conditions, temperature is far more significant than humidity. Regardless of humidity, an effective temperature of –20°F or lower may justify a journeyman to stop working.

4. High humidity at colder temperatures lowers productivity but the effects may be reduced significantly by wearing appropriate clothing and by providing protection from the wind.

TABLE 13.6
Effective Temperature Vs. Wind Speed

Wind Speed MPH	Actual Temperature in –Degrees Fahrenheit														
	40	35	30	25	20	15	10	5	0	–5	–10	–15	–20	–25	–30
Calm	40	35	30	25	20	15	10	5	0	–5	–10	–15	–10	–25	–30
5	34	31	25	19	13	7	1	–5	–11	–16	–22	–28	–15	–40	–46
10	32	27	21	15	9	3	–4	–10	–16	–22	–28	–35	–33	–47	–53
15	32	25	19	13	6	0	–7	–13	–19	–26	–32	–39	–45	–51	–58
20	30	24	17	11	4	–2	–9	–15	–22	–29	–35	–42	–53	–55	–61
25	29	23	16	9	3	–4	–11	–17	–24	–31	–37	–44	–59	–58	–64
30	28	22	15	8	1	–5	–12	–19	–26	–33	–39	–46	–63	–60	–67
35	28	21	14	7	0	–7	–14	–21	–27	–34	–41	–48	–67	–62	–69
40	27	20	13	6	–1	–8	–15	–22	–29	–36	–43	–50	–69	–64	–71
45	26	19	12	5	–2	–9	–16	–23	–30	–37	–44	–51	–59	–65	–72
50	26	19	12	4	–3	–10	–17	–24	–31	–38	–45	–52	–63	–67	–74
55	25	18	11	4	–3	–11	–18	–25	–32	–39	–46	–54	–67	–68	–75
60	25	17	10	3	–4	–11	–19	–26	–33	–40	–48	–55	–69	–69	–76

Source: Adapted from Table 2.2, NECA study: The Effect of Temperature on Productivity, 2004.

TABLE 13.7

Effect of Temperature and Humidity on Productivity

RH.%	Journeyman Productivity Percentages at Various Environmental Conditions												
90	56	71	82	89	93	96	98	98	96	93	84	57	0
80	57	73	84	91	95	98	100	100	98	95	87	68	15
70	59	75	86	93	97	99	100	100	99	97	90	76	50
60	60	76	87	94	98	100	100	100	100	98	93	80	57
50	61	77	88	94	98	100	100	100	100	99	94	82	60
40	62	78	88	94	98	100	100	100	100	99	94	84	63
30	62	78	88	94	98	100	100	100	100	99	93	83	62
20	62	78	88	94	98	100	100	100	100	99	93	82	61
	−10	0	10	20	30	40	50	60	70	80	90	100	110
	Effective Temperature (F)												

Source:　Adapted from Table 2.1, NECA study: The Effect of Temperature on Productivity, 2004.

They cite the benefits of weather prediction to project planning and execution as:

    a. Time savings

    b. Improved job safety

    c. More efficient handling of materials

    d. Improved labor planning

    e. Other economies

The NECA points out that the shifting of job tasks from one season to another may have significant impacts on environmental conditions, and on the productivity of workers as a result. A quantification of these changes in productivity can serve as justification for adjustments in contract budgets and schedules.

Impact on Lean Construction

Weekly work plans are based on pulling activities or work assignments from a look-ahead schedule. As described in Chapter 4, eligible assignments are the ones that have no current constraints, and that have resources available and assigned. Workers are a major component of those resources, hence their capabilities must be known with some certainty in order for a foreman and a last planner to accept a crew's commitment to having a promised activity or task completed. Adverse weather conditions can significantly impact workers' productivity. Tables 13.16 and 13.7 can provide fair approximations of productivity for job planning purposes. Example:

- Air temperature:　　30°F
- Wind speed:　　　　10 mph
- Relative humidity:　70%

A simple 2-stage estimation is required. Table 13.6 reflects the relationship between relative humidity and perceived temperature. From Table 13.6, the effective temperature equals 21°F. Determining productivity from Table 13.7 at an effective temperature of 20° and 70% R.H, productivity equals 93%. At an effective temperature of 30° and 70% R.H, productivity equals 97%. Interpolating between these values yields a value of 93.4% approximately. A worker productivity level of 93.4% would have to be considered when sizing the work crews for the project.

The NECA points out that the reference standards from the study provide basic data for planning and cost estimation, but that individual contractors will need to apply additional job factors for the specific workers and projects involved.

Test Your Understanding

Assume that a carpenter can erect 160 square feet concrete wall forms per eight-hour day under ideal environmental conditions. The budgeted labor rate for a carpenter is $15 per hour. Unexpectedly, weather conditions change such that wind velocity is 25 mph, the temperature drops to 30°F, and the humidity is 65%. Calculate the number of square feet of forming the worker will place under these conditions. Use the tables supplied.

Managing the Impacts of Scheduled Overtime on Productivity

The Business Roundtable (BRT) conducted a study in 1974 on the impact of scheduled overtime operation on construction projects with respect to worker productivity and labor costs. The BRT repeated the study in 1980 and obtained similar results. The BRT study identified disproportionate increases in project costs to accomplish construction work when it was done with scheduled overtime, as compared with a nominal 40-hour week. This premium became larger as overtime was continued for several weeks; the BRT found that a schedule of 60 or more hours per week would potentially delay a project if it were continued for more than two months with the same crew size. Table 13.8 shows raising scheduled hours from 40 to 50 results in lower productivity for the entire 50-hour period.

More recent studies by other researchers obtained similar results to the BRT's findings. Thomas et al. (1997) observed losses of efficiency of 10–15% for 50- and 60-hour work weeks. For this analysis, disruptions in three categories—resource deficiencies, rework, and management deficiencies—were analyzed. The analyses showed that the disruption frequency, which is the number of disruptions per 100 work hours, worsened as more days per week were worked. This led to the conclusion that losses of efficiency are caused by the inability to provide materials, tools, equipment, and information. Hanna et al. (2005) observed behaviors similar to those identified in the BRT study. Mayo et al. (2001) included

TABLE 13.8

Relationship of Hours Worked to Productivity

1	2	3	4	5	6	7	8
50-Hour Overtime Work Weeks	Productivity Rate		Actual Hour Output for 50-Hour Week	Hour Gain Over 40-Hour Week	Hour Loss Due to Productivity Drop	Premium Hours	Hour Cost of Overtime Operation (at 2×)
	40-Hour Week	50-Hour Week					
0–1–2	1.0	.926	46.3	6.3	3.7	10.0	13.7
2–3–4		.90	45.0	5.0	5.0	10.0	15.0
4–5–6		.87	43.5	3.5	6.5	10.0	16.5
6–7–8		.8	40.0	0.0	10.0	10.0	20.0
8–9–10		.752	37.6	−2.4	12.4	10.0	22.4
>10		.75	37.5	−2.5	12.5	10.0	22.5

travel time in their analysis and observed that longer work days could be more productive. That special circumstance, however, does not apply to most construction projects.

Implications for Lean Construction

As indicated in Table 13.8, hours worked beyond the nominal 40-hour shift may result in lower productivity. More total labor hours are utilized to carry out scheduled tasks than would be needed with a 5-day, 40-hour shift. The lost hours as shown in the table are in effect waste, representing the very antithesis of lean construction. These losses may even be compounded by difficulties in keeping crews supplied with needed materials and instructions. Careful attention should be directed to planning work activities and monitoring accomplishment, at least on a weekly basis, to avoid bottlenecks and maximize crew performance. The methodology of The Last Planner® System combined with daily huddles can be effective in optimizing work output.

References

Albers, J., C. Estill, and L. MacDonald. 2005. Identification of ergonomics interventions used to reduce musculo-skeletal loading for building installation tasks. *Applied Ergonomics* 36: 428–39.
ASHRAE. 1997. *ASHRAE Fundamentals Handbook (SI).* 8.1–8.28. In American Society of Heating, Refrigerating and Air-Conditioning Engineers, Inc, 1791 Tullie Circle, Atlanta, GA 30392.
Bjelle, A., M. Hagberg, and G. Michaelsson. 1979. Clinical and ergonomic factors in prolonged shoulder pain among industrial. *Work, Environment and Health* 5:205–10.
BNA, 2002. Ergonomics: OSHA's plan for reducing ergonomic injuries. Occupational safety and health reporter. *Bureau of National Affairs* 32(15): 358.
Choudbury, I., M. Haque. 2001. A study of cross-cultural training in international construction using GLM procedure and neural network approach. *ICCM Conference Proceedings,* Singapore.
Forde, M., and B. Buchholz. 2004. Task content and physical ergonomic risk construction ironwork. *International Journal of Industrial Ergonomics* 34(4):319–33.
Freivalds, A. 1986. Ergonomics of shovelling and shovel design: An experimental study. *Ergonomics* 29: 19–30.
Hanna, A. S., C. S. Taylor, and K. T. Sullivan. 2005. Impact of extended overtime on construction labor productivity. *Journal of Construction Engineering and Management* 131(6): 734–39.
Malcolm Baldrige National Quality Award. 2010. U.S. Department of Commerce, Technology Administration, National Institute of Standards and Technology, Gaithersburg, MD.
Mayo, R., K. Knutson, G. Barras, and J. Pineda. 2001. Improving construction productivity with scheduled overtime. *Royal Institution of Chartered Surveyors Conference,* 53(2): 897–900. COBRA 2001 Conference organised by RICS Research Foundation, Glasgow, UK.
Nabatilan, L., and B. Venkata. 2007. Effectiveness of an ergonomic intervention in overhead work. *Proceedings, Annual Conference, Institute of Industrial Engineers.*
National Institute for Occupational Safety and Health. January 15, 2004.
NIOSH, 1997. Elements of ergonomic programs. US Department of Health and Human Services, Public Health Service, Centers for Disease Control and Prevention. National Institute for Occupational Safety and Health, Cincinnati, OH, DHHS (NIOSH) Publication No. 97–117.
Sillanpaa, J., A. Lappalainen, M. Kaukiainen, and P. Laippala. 1999. Decreasing the physical workload of construction work with the use of four auxiliary handling devices. *International Journal of Industrial Ergonomics,* 24(2): 211–22.

Thomas, H. R., and K. A. Raynar. 1997. The effects of scheduled overtime on labor productivity: A quantitative analysis. *Journal of Construction Engineering and Management*, 123(2): 181–188.

U.S. Department of Labor. 2001. Accident statistics for the construction industry. www.bls.gov/data/.

Pan, C. S., S. Chou, D. Long, J. Zwiener, and P. Skidmore. 2000. Postural stability during simulated drywall lifting and hanging tasks. *Proceedings of the XIV Triennal Congress of International Ergonomics Association, and the 44th Annual meeting of the Human Factors and Ergonomics Association, Ergonomics for the New Millenium*, pp. 679–82.

Rasmussen, T. 1996. *The ASTD trainer's sourcebook: Diversity.* New York: McGraw Hill.

Rosecrance, J., T. Cook, and C. Zimmermann. 1996. Work-related musculoskeletal disorders among construction workers in the pipe trades. *Work* 7:13–20.

Seelye, H., and A. Seelye-James. 1995. Culture clash: Managing in a multi-cultural world. Wausau, WI: NTC Books.

Bibliography

Aghazadeh, F., and A. Mital. 1987. Injuries due to hand tools. *Applied Ergonomics* 18(4):273–78.

Asfahal, R. 2004. *Industrial safety and health management.* Upper Saddle River, NJ: Prentice Hall.

Auliciems, A., and S. V. Szokolay. 1997. *Thermal comfort.* Brisbane: Dept. of Architecture, University of Queensland.

deCatanzaro, D. A. 1999. Motivation and emotion: Evolutionary, physiological, developmental, and social perspectives., Upper Saddle River, NJ: Prentice Hall.

Fanger, P. O. 1970. *Thermal comfort.* Copenhagen: Danish Technical Press.

Harris, C. 1991. *Handbook of acoustical measurements and noise control.* New York: McGraw-Hill.

Hinzie, J. 2005. Noise generated by construction power tools. *Proceedings of the Third International Conference on Construction in the 21st Century (CITC III), September, 15–17, Athens, Greece.*

Mohamed, S., and K. Srinavin. 2001. Impact of climatic environment on construction productivity. *Proceeding of the third International Conference on Construction Project Management*, pp. 120–29. Singapore

Oglesby, C., H. W. Parker, and G. A. Howell. 1989. *Productivity improvement in construction.* New York: McGraw Hill Inc.

Schaffer, R., and H. Thomson.1998. Successful change programs begin with results. *Harvard Business Review on Change*, HBS Press, pp. 189–213.

14

Systems Integration Approaches

Several chapters have described lean construction approaches for improving construction performance. These approaches work best when a number of prerequisites have been met for applying lean construction. As described in Chapter 6, the prerequisites include:

1. A willingness to change
2. A commitment to training and learning
3. A quality-oriented culture
4. A "Shared Vision"
5. A commitment to reducing or eliminating waste
6. A commitment to cost and performance measures
7. A willingness to implement lean during the design stages
8. Collaborative relationships
9. The effective use of information technology

Lean construction cannot be effectively applied to entire projects without the existence of these factors. The essence of lean construction is a willingness of the participants to collaborate for the optimization of the entire project, instead of obtaining local optimization. In the absence of project-wide commitment, improvement efforts tend to facilitate local optimization instead of benefiting the entire project.

Nevertheless, in the absence of a lean culture, it is possible to improve project performance measurably. Industrial engineers (IEs) can draw on a variety of tools to accomplish such improvements. They can also serve as a catalyst for the adoption of lean methods by working with other stakeholders to identify opportunities for improvement. Among the IEs that have made the transition to the lean construction environment are two U.S.-based members of the Lean Construction Institute: Dennis Sowards, principal of Quality Support Services (QSS), Inc., and Matt Horvat, a lean coach with the firm of Lean Project Consulting. They have adapted their IE skills to the construction environment very successfully.

The following cases show examples of systems improvement that benefited the stakeholders in the absence of a so-called lean environment. They represent the application of industrial engineering techniques by three IEs who were able to implement systems integration in projects that were not based on a lean methodology: Jeff Mason, of Integrated Facility Services, LLC; Al Attah, of MACVAL Associates, Inc.; and Jorge Cossio, of ITN de Mexico.

Industrial Engineering Solutions for the Construction Industry

A number of cases are presented in this chapter that represent work done by industrial engineers who are involved in the construction industry. In situations where a lean culture was lacking, they were able to use systems integration approaches based on the industrial engineering discipline.

Industrial engineers are trained as systems integrators, so they can provide strategies for having the involved parties work together for maximum effectiveness. IEs are also very concerned with improving quality and performance overall—they have the tools to reduce errors and the resulting financial waste that end up as a cost to society.

Professional Overview of Industrial Engineering

Not everyone is clear on what IEs do—they are mistakenly identified with manufacturing activity, which is only one facet of their professional reach. Industrial engineering is concerned with the design, improvement, and installation of integrated systems of people, materials, information, equipment, and energy. It draws upon specialized knowledge and skill in the mathematical, physical, and social sciences, together with the principles and methods of engineering analysis and design, to specify, predict, and evaluate the results obtained from such systems. Anecdotally, other engineering disciplines make systems, while industrial engineers "make systems better."

It has been said that construction may be the "last frontier of IE opportunity." IEs have been involved in virtually every other type of enterprise, with outstanding results—their earliest contributions were in the manufacturing arena and later they moved into the service sector. IEs have enabled the fast food industry to produce consistent quality at high volumes. The health care industry has also used IEs to enhance quality and reduce errors in service delivery. Package delivery organizations owe much of their competitive edge to the efforts of IEs. The same is true of virtually every type of enterprise, but the construction industry has yet to draw on the skills of IEs on a large scale. There is no doubt that the construction industry has many opportunities for improvement that could benefit significantly from IE involvement: research studies have shown that as much as 30% of the cost of construction may be wasted in typical projects. In a trillion dollar industry, IEs could translate their successes in other industries to the construction environment with significant dollar savings and performance improvement.

What are the traits of industrial engineering that it can contribute to the construction industry? IEs are multiskilled and multifaceted as a result of their training. They have a solid foundation in engineering principles, and very importantly, they are generally more oriented to the human element than are other branches of engineering. As we all know, design and construction projects can only be successful when all the parties work together harmoniously, and it is quite a challenge to orchestrate these complex activities. IEs are trained as systems integrators and lean facilitators, so they can help the involved parties work together for maximum effectiveness. IEs are also very concerned with improving quality and performance overall, and have the tools to reduce errors and the resulting financial waste that end up as a cost to society.

There is great untapped potential for improvement. Industrial engineering studies in the manufactured homebuilding industry improved labor productivity 59%, through a combination of improvements in the factory and better construction site supervision. Construction cycle time was reduced by 22% through tighter scheduling from 90 to 72 days. Such improvements represent major cost savings. Examples of areas for IE contribution are:

Facilities layout planning: Whereas construction professionals are often limited by laying out sites intuitively, IE techniques can be used to optimize the use of each construction site, thereby improving productivity and profitability. This helps to utilize manpower and space effectively, minimizing delays, backtracking, and multiple handling of material. IEs are equipped to design the layout of a facility so that it helps users to operate in a lean manner—such as lean manufacturing. A smaller, well laid-out building can be more usable than a larger, more expensive building.

Sustainable construction: The importance of sustainable construction has emerged as a major issue in the twenty-first century, requiring a paradigm shift in the skill sets and attitudes of today's design and construction professionals. For example, 36.4% of total U.S. primary energy consumption comes from construction. It also accounts for 36% of total U.S. CO_2 emissions, 30% of total U.S. global warming gases such as methane, nitric oxide, and hydrofluorocarbons, and 60% of total U.S. ozone depleting substances. Construction waste is a major problem; each residential building accounts for three to seven tons of waste. Nationally, the United States produces 136 million tons of construction waste, but only 20 to 30% of this is recycled or reused. This trend points to the need for change in the training provided by construction educators. Some IEs specialize in environmental issues and can support the implementation of sustainable construction.

Work measurement: These techniques can improve worker productivity. Many workers lack training in "best methods" and are less effective. Methods time measurement (MTM) can develop fair standards based on best methods for a number of tasks. These standards facilitate tracking and improving labor productivity. IEs have traditionally performed work measurement in manufacturing and can readily adapt it to construction.

Cycle time analysis: IEs can use operations research techniques to plan construction support activities like excavation work and dump truck operations to balance the use of expensive resources. Jobs can be planned so that such work can be done without wasted expenditure.

Simulation techniques: IEs can use simulation to plan large, expensive projects, and forecast the demand for work crews, materials, and equipment. Construction managers often use an experience-based approach that is suboptimal. Discrete simulation techniques can be used to compare alternative building methods and save labor, materials, and time.

Ergonomics: Construction workers use a wide assortment of tools and equipment to perform construction tasks. Workers cannot "build in" quality to constructed facilities if they are subjected to awkward positions and excessive physical stress caused by tools and equipment that are difficult to use. IEs have helped to develop both tools and work methods that enable workers to be more productive and injury-free.

Safety management: Accidents take a heavy toll on construction workers—21% of deaths and 11% of disabling injuries in private industry in 2000. IEs are adept at developing and implementing worker safety programs.

Quality management systems (QMS): IEs have been in the forefront of the quality movement in a variety of industries. They have helped many organizations to be more efficient, effective, and more quality-oriented. This system has led to greater customer satisfaction resulting from products and services of higher quality. Quality management systems run the gamut from TQM, ISO9000, Six Sigma to Lean Production and Kaizen. Some of these systems have been applied in construction, but on a limited basis.

Cutting-edge approaches such as computer technology and radio frequency identification are already being used by IEs. Automation and robotics are still in the experimental stage in construction, but hold much promise for the future.

Overall, the IEs' greatest strength for improving construction may be the fact that they are trained as systems integrators. With a solid foundation in engineering concepts while understanding the human element more than other engineering disciplines, IEs are highly equipped to help the wide variety of individuals involved in a construction project to achieve the integration of skills, material, and other resources needed for superlative performance.

This next section follows three cases and addresses strategies for IE participation in the construction industry.

Case 1: Seattle Area Coffee Company

This case describes a holistic approach by a consulting organization to design a facility to meet the owner's project requirements (OPR) and to design for constructability and minimal disruption to the owner's operations. The assignment was carried out by Integrated Facility Services LLC (IFS), a consultant based in Portland, Oregon. Jeff Mason, CEO, is an industrial engineer and uses IE-based solutions for many design projects. A project was undertaken for an organization called Seattle Area Coffee Company (SACC) for purposes of confidentiality.

Coffee is a hot commodity in the Pacific Northwest. In the early 2000s it was difficult *not* to make money in the coffee business. Green coffee was at an all time low price and the demand for quality coffee was rocketing. The SACC was enjoying these prosperous times, but facility, equipment, and distribution center constraints were soon going to be limiting their growth potential.

The SACC asked Integrated Facility Services, LLC (IFS) to assist with analysis of bottlenecks and identification of labor and equipment requirements to keep up with forecasted sales growth. This led to a creation of a five-year business plan driven by a dynamic modeling tool that would chart the necessary facility and equipment requirements to grow their business commensurate with the explosive demand.

This plan showed that within two years, new equipment requirements would necessitate expansion of various buildings footprint. IFS used systematic layout planning techniques to develop and evaluate alternative expansion scenarios. Flow, adjacency requirements, and from-to charts were developed to illustrate the relative travel distances and times, congestion, and relative adjacency benefits of each scenario. Material handling automation and storage methodologies were also reviewed to determine if better space utilization would result in the reduction or elimination of expansion space.

Systematic Layout Planning

SLP is a process developed by Richard Muther (1974) to determine optimum layout alternatives. The technique combines relationships between functional areas (adjacency diagrams, from-to charts), space requirements (space relationship diagrams), and other considerations such as flow and frequency analysis to develop desired layouts.

Adjacency diagrams are used to explore the key relationship between functional areas. These can take several forms, often a bubble diagram with lines between areas (bubbles) showing where proximity is valuable. In manufacturing processes, it can sometimes be important to have areas not be adjacent (an example of this could be chemical storage and electrical rooms). The diagram below is an example of an adjacency chart that quantifies the importance of proximity by applying scores to the following factors:

- Absolutely necessary
- Extremely important
- Important
- Ordinary importance
- Unimportant
- Avoid

The value at the intersection of the two rows is the relative importance of those areas being situated adjacent (or close) to each other. Figure 14.1 shows FGI and Shipping intersect at the value of I (important), while Receiving and QA Hold get a value of E (extremely important). The ratings can be given for safety, environmental, frequency of interaction, and many other reasons.

The studies, coupled with the sales forecast indicated that the expansion space and the capital requirements would be extensive. This led to adding greater sophistication to the model to allow modification of key variables and the calculation of detailed operating costs such as labor, energy, and direct and facility costs. This in turn led to the realization that operating costs were dramatically increased because of the company's current locations.

SACC's production facilities were on an island accessible only by boat. SACC was dependent upon the not always reliable ferry system to deliver raw materials, packaging materials, equipment, supplies, and services to their production facility. Even most of their workers were dependent upon the ferry to get to and from work. The model highlighted that the impact of being on-island increased SACC's operating costs by at least 20%.

Once again, IFS tuned the model to not only analyzing expansion of existing facilities and conditions, but also yielded the ability to craft alternative operating scenarios for off-island scenarios. This was used to identify the impacts on total system costs, including materials, equipment, labor, facilities, and logistics. This allowed us to isolate the incremental costs of being located on the island and understand the extremely negative impact it had on total systems costs.

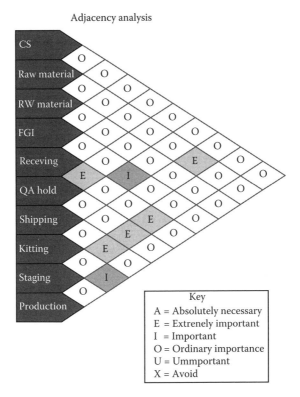

FIGURE 14.1
Adjacency analysis of functional areas.

Once it became clear that moving the production facility off the island would provide tremendous cost benefits, SACC asked IFS to perform an additional study to identify potential site alternatives. Based upon the model inputs, it was determined that east coast sales would increase and would dominate their sales within three years. This led to a plan to size the production and distribution capabilities to handle the entire growth requirements for the next three years. At that time logistics benefits would drive the creation of an additional facility to serve the east coast sales.

Evaluating Alternatives

Prompted by suggestions for layout changes in SACC's new facility, SACC manufacturing management requested a process for creating and evaluating potential layout alternatives. The results of this process are summarized in this document. The process steps were as follows:

1. Develop criteria for evaluating alternatives
2. Create a relative importance rating (weight) for each criteria
3. Develop layout alternatives based upon criteria strengths
4. Evaluate each layout alternative against the criteria matrix
5. Provide recommendations based on the analysis results

Evaluation Criteria

In any evaluation model, the selection of the criteria is a crucial step. The resultant list of criteria should be discrete, yet complete, to ensure the best results. The criteria used for this evaluation are listed below, with a brief description of each.

WB proximity to silo: The travel distance from roasters to whole bean silos, and from whole bean silos to whole bean packaging lines. It is extremely important to minimize this distance … quality deteriorates as transport distance increases (the roasted beans are fragile and suffer breakage).

Product flow: The ease with which finished goods flow from the palletizing area to finished goods warehouse.

Raw material flow: The ease with which green bean pallets can flow from green bean storage to the dumping station and packaging materials (film, boxes) can flow to the packaging lines.

Personnel flow: The accessibility of the equipment, the ability to move safely around the production area, and the impact to labor productivity.

Green bean storage: The capacity of the green bean storage area.

Production space: Space efficiency of the production equipment.

Flexibility: The ease with which future additions or modifications to the layout may be executed.

Control room location: The proximity of the control room to all roasters.

Tour location: How well the layout lends itself to viewing/touring production.

A total of nine different layouts were created, including the original concept (Figure 14.2). Each layout contains the same equipment, but the orientation, location, and/or adjacencies of the equipment varies. In addition to reviewing the production layout, the new alternatives show a designated palletizing area, with conveyors leading from the packaging lines to this area.

Revise Layout Alternatives

To ensure that all the concepts were fully explored, each layout was reviewed and revised to produce a more polished version of each, and thus a more realistic view of just how it would ultimately look if implemented. This exercise also led to the development of some

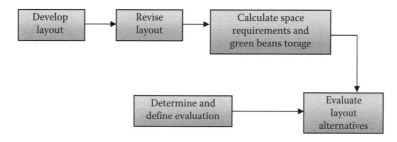

FIGURE 14.2
Development of layout alternatives.

of the alternatives. IFS collected data on local and eastward sites and used the model to forecast the relative benefits of geographic scenarios. Some of the parameters analyzed included:

1. Labor rates
2. Real estate costs
3. Transportation/logistics costs
4. Energy costs
5. Taxes

Finalists were further reviewed to determine proximity to competitive industry, and possible local incentives such as tax abatement and training assistance. The near term facility to serve the next three years growth was selected based upon a combination of these factors and also on site readiness, as the benefits for getting off island as quickly as possible were now understood to be substantial.

The model was once again used to determine whether the strategy should include combining the distribution facilities with the manufacturing facilities or to keep these as separate sites as they do today. Although the lease term would overlap that of the new facility commencement it was determined the most cost-effective solution was to combine the distribution and manufacturing facilities.

Analysis of the existing systematic layout data showed that the existing factory was very space inefficient due to suboptimum building geometry, including the height of the building. A short list of potential existing buildings were researched but not wanting to relive the inefficiencies created by the last facilities poor geometries, the focus quickly shifted to the design of a new factory.

Two of the alternatives follow; first the highest ranked alternative 2, then the lowest ranked alternative 4A. To allow for greater detail, only the primary processing areas are shown. Figures 14.3 and 14.4 illustrate these two layouts. The points of interest are colored lines that illustrate the travel routes and distances taken by

a. The product flow (green lines)
b. The people flow (blue lines)
c. The material flow (magenta lines)

An analysis was carried out to compare all layouts, including the ones in Figures 14.3 and 14.4; the results are listed in Table 14.1.

Evaluating Layout Alternatives

The evaluation model used in this study requires each alternative to be scored, on a scale from 1 to 10, for each of the aforementioned criteria. Two of the criteria (green bean storage and production space) were quantifiable, while the remainder were discussed and scored qualitatively. In addition to applying a score for each layout and criteria, the criteria were weighted so that certain criteria had a greater impact to the overall score than others.

Tables 14.1 through 14.6 show the final scoring. The top table shows unweighted scores. The values in the bottom table are the scores multiplied by the importance assigned to the

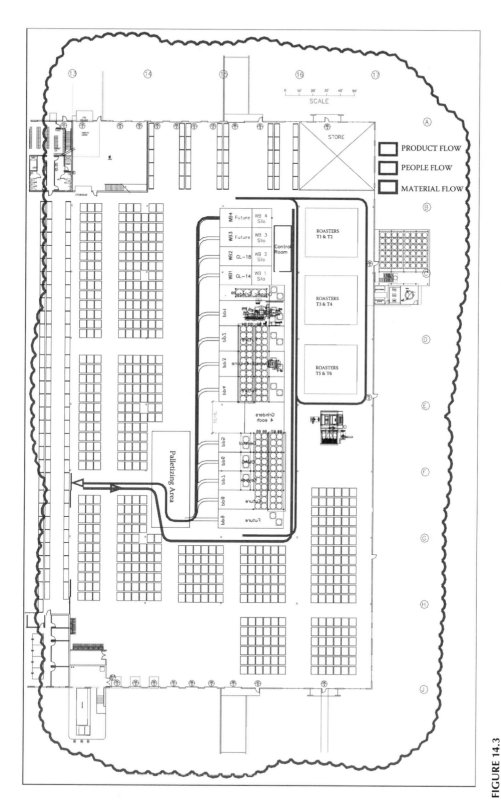

FIGURE 14.3
Alternative facility layout #2.

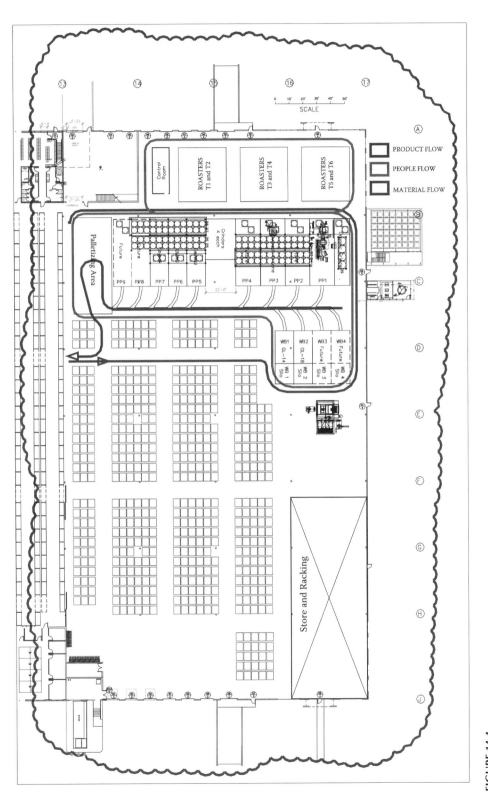

FIGURE 14.4
Alternative facility layout #4A.

TABLE 14.1

Evaluation of Layout Alternatives

	Importance	Original	Alt 6	Alt 5	Alt 4B	Alt 4A	Alt 3	Alt 2	Alt 1B	Alt 1A
WB Proximity to silo	10	10	5	9	3	4	8	7	9	9
Product Flow	8	9	7	4	7	7	4	9	4	4
Raw Material Flow	7	4	8	9	9	6	5	6	5	5
Personnel Flow	5	4	2	4	3	7	9	8	7	4
Green Bean Storage	3	9	6	5	7	7	6	5	8	8
Production space	4	5	5	5	7	3	9	7	7	9
Flexibility	6	10	8	6	6	8	8	10	6	6
Rel. Loc. Control 2 Rstrs	7	10	7	5	5	7	6	9	7	5
Tour Location	4	8	3	8	3	8	10	10	9	9
Total Score	**54**	**69**	**51**	**55**	**50**	**57**	**65**	**71**	**62**	**59**
WB Proximity to silo	18.5	185	93	167	56	74	148	130	167	167
Product Flow	14.8	133	104	59	104	104	59	133	59	59
Raw Material Flow	13.0	52	104	117	117	78	65	78	65	65
Personnel Flow	9.3	37	19	37	28	65	83	74	65	37
Green Bean Storage	5.6	50	33	28	39	39	33	28	44	44
Production space	7.4	37	37	37	52	22	67	52	52	67
Flexibility	11.1	111	89	67	67	89	89	111	67	67
Rel. Loc. Control 2 Rstrs	13.0	130	91	65	65	91	78	117	91	65
Tour Location	7.4	59	22	59	22	59	74	74	67	67
Weighted Score	**100**	**794**	**591**	**635**	**548**	**620**	**696**	**796**	**676**	**637**
Rank		**2**	**8**	**6**	**9**	**7**	**3**	**1**	**4**	**5**

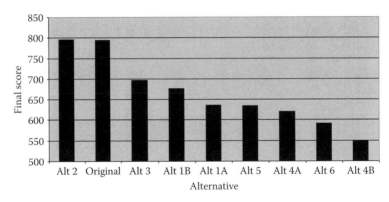

FIGURE 14.5
Pareto Chart of alternative layout scores.

criteria. The resultant values are then summed to calculate the final score. Alt 2 and the original layout ranked the highest. The two-point difference between them is statistically insignificant. Alternative layout #2 would be expected to minimize the long-term operating cost of the building users/owners.

The Pareto Chart (Figure 14.5) shows the alternative ranking from highest to lowest. It is important to note that there are other factors that need to be applied to a final decision, such as implementation cost, timeline impact, and difficulty of implementation. That is to say, the final scores do not necessarily signify absolute best/worst alternatives, but rather give an idea of which layout(s) represent the best production scenario.

Systematic layout planning (SLP) techniques were used for a Greenfield site designed to handle the manufacturing and distribution for a factory with a capacity four times the current sales. Since the equipment requirements varied substantially with the sales (mix for whole bean versus ground, bulk versus small package units, and varieties of bean types required), a sensitivity "what-if" analysis was performed to try and anticipate production and storage areas that might need to be expanded over time. This analysis was used to create the most flexible layout that minimized the need for expensive infrastructure change in case the model shifted from what was planned.

We were also able to provide for automated handling systems that took green coffee in sacks and conveyed them through all the process steps automatically until the product was ready to be placed in the finished goods section. This coupled with optimizing storage methodologies allowed us to accomplish this extra volume expansion in 330 k square feet or roughly twice the current space.

Starting from the bare dirt afforded us the opportunity to optimize the building utility infrastructure and reduce installation costs. Even though the chosen design was a 40 foot high concrete tilt, the building was arranged to allow the use of clerestory lighting to reduce the reliance on high bay lighting. Special dock doors were used to provide improved insulation from winter chill while allowing pest resistant screening for summer time circulation. Concrete floors were designed with high-strength, temperature-resistant coatings, and sanitary areas were designed so they could be maintained with the minimum use of water introduction to the sewer.

Much of the equipment being purchased for the new facility was being sourced from Europe and structural steel was being purchased from Canada. The detailed engineering and design teams were from various locations in the United States and Europe. Integrated Facility Services LLC (IFS) built a web-based collaboration site that the team could use to review specifications, drawings, submittals, and conversation threads to keep the project moving as quickly and efficiently as possible.

The SACC was on their way to becoming a very low-cost competitor in the national coffee market and our fast track project was slated to be ready for the equipment installation four months from the point of site preparation. This apparently attracted the attention of an even larger Seattle area coffee company as they purchased SACC before we even got the first walls up! Even though SACC was bought out, this was a great example of IE application to the big picture of construction—it was linked to new construction and would have used lean principles.

Case #2: Systems Integration and the Application of Lean Methods in Construction

This case illustrates a holistic approach to improving small construction projects by the use of IE-based systems integration.

MACVAL Associates, LLC is a small engineering and construction firm based in Dallas, Texas. About 60% of the firm's business is construction of municipal engineering projects such as highways/roadways; water, wastewater, and drainage systems. The firm obtains its contracts primarily through competitive bidding. Two techniques that have proven valuable during the construction phase are concurrent engineering and lean production.

One of the obstacles to improved quality and productivity in public works projects is government regulation with regard to bidding requirements. Because of this, public works projects are developed and brought forward in a number of sequential or serial steps. In this approach, projects are completely designed and then offered for bidding. This linear procedure is not always an efficient method with regard to the total time required for the design and construction of a project. Input from the prime contractor into the design process does not usually occur. Although the function of the prime contractor does not include project design, his/her knowledge of constructability can make him/her a valuable member of the planning and design team.

Once a project enters the construction phase, the contractor's work is generally the same regardless of the contract form or delivery method selected. The same activities are necessary to get the project built: the subcontractor's contracts must be executed, the project staffed, a detailed schedule of values and work schedules developed, and the job site assembled. In a typical construction organization, the activities listed above, among others, are accomplished through a serial iteration process as stated earlier. It involves methodically handing work over from one work group to another until the project is finally delivered. However, work-group barriers prevent customer-to-supplier relationships from developing. One disadvantage of this approach is that errors are not usually detected until after upstream activities are completed. In the end, this approach leads to costly rework. Although the project eventually gets delivered, the process usually takes too long and costs too much and the delivered project may be of questionable quality. This case study discusses how concurrent engineering and lean production techniques can minimize schedule delays and cost overruns during construction of public works projects, while at the same time achieving high quality and improved productivity.

Concurrent Engineering Approach

MACVAL uses a modified concurrent engineering approach during the construction engineering and operations phase to help minimize construction project costs. Here, an integrated team that consists of members from all the groups needed to build and deliver the project develops each construction process. Everyone involved has a basic understanding of the entire process and his or her role in it. All stakeholders are included early in the construction planning process. A typical team includes representatives from the city (project engineers/managers, inspectors, fire department, water utilities, storm water division, and traffic), from the utilities (gas, phone, and cable), from suppliers (equipment and materials), and others (surveyors, material testing firms, subcontractors, and property owners). Areas covered by the team include but are not limited to:

- Project planning
- Scheduling
- Value engineering
- Constructability issues
- Field drawings/design changes
- Procurement
- Inspection
- Surveying and layout
- Occupational safety and health management
- Materials testing coordination and/or supervision
- Coordination of construction activities

It is important to identify customer value by analyzing all activities required for project delivery. The entire process is optimized from the customer's viewpoint in order to increase value and reduce waste. Below is a typical sequence that guides the construction engineering and operations phase in MACVAL's lean production technique:

Concurrent construction process planning (see concurrent engineering above)

- Optimization of worker utilization and capabilities
- Continuous focus on the customer
- Development of mutual trust relationships
- Continuous challenge of existing methods
- Decision making at the construction site
- Flexible scheduling
- The "Lean-O-Mobile"

The company employs multitask workers and/or cross-trains workers to reduce the need for hiring specialists at higher rates. For example, on one project, a sprinkler line needed to be repaired. Instead of hiring a plumber at $120 per hour for this simple job, one worker repaired the line. The supervisor on the job site also performs jobs such as operating heavy equipment and directing traffic.

Overtime is used wisely to reduce expenditures while advancing the project schedule. Most of the expense in small municipal construction projects is in the material itself (concrete, rebar, etc.). Since 60% of the firm's contracts are of this type, children in the neighborhood (and sometimes adults) love to write something or make handprints/footprints on freshly poured concrete. When this happens, rework is quite expensive. To minimize possible rework, the contractor not only pours concrete during low-pedestrian times, but also has someone stay at the job site until the concrete cures. It is much less expensive to pay three hours of one person's overtime versus having the crew tear out the damaged concrete and re-pour the job. This simple action has resulted in a 90% reduction in rework costs.

Through experience, the company has learned the importance of stability in dealing with subcontractors. It developed a plan to form alliances with preferred subcontractors; those selected receive a percentage of the savings resulting from quality and improvement ideas and early completion of projects. This approach results in a stable workforce and also reduces variation in similar projects.

The company has turned a $7,000.00 moving truck into the so-called Lean-O-Mobile. Actually, this truck started out as a simple storage unit on the job site. After applying 5S principles, the truck was organized with shelves and proper storage locations. The truck enables frequent material delivery in small quantities, allows the team to gain better control of material and portable equipment, reduces the distance traveled to storage and staging areas, and supports minimum inventory required to keep the construction operation running. This is the company's version of JIT (Just-In-Time).

Described below is an example of a project that benefited from concurrent engineering and lean production techniques.

Project Name:
Alley between Country Club Circle and Gaston Avenue from Pearson Drive to Gaston Avenue—PB06U000
Location:
Northeast of City of Dallas, Texas

- MAPSCO 36 V
- Council District: 14

Description:
Construction of a reinforced concrete alley, drainage improvements and other miscellaneous items necessary to complete the project in accordance with the intent of plans and specifications.
Project Owner:
Department of Public Works and Transportation, City of Dallas, Texas

- Owner's Project Manager: Ashok Patel, P.E.
- Project Prime Contractor: MACVAL Associates, LLC, Dallas, Texas

Completed Project Results

- Quality: Zero items were noted for correction on the punch list; that is, no defects were noted.

- Cost: Utilized only about 67% of the budget; about 33% savings for the taxpayer.
- Schedule: Utilized only 50% of the contract schedule to complete the project. Project was scheduled to be completed in June 2009 but it was instead completed ahead of schedule in December 2008.
- Safety: There were no safety issues.

IE-Based Bid Strategy for New Projects

MACVAL uses an industrial engineering approach to bidding on new construction projects. By analyzing past projects they have compared the scope of work at completion with the scope originally defined in the plans and specifications. This difference was generally due to oversight in the design process, often because of designers' unfamiliarity with field conditions.

As each project progressed, it was often necessary for project owners to grant a change order to cover the unanticipated changes. An understanding in the categories of work that were most frequently affected allowed MACVAL to structure their bids most competitively, with a reasonable expectation of added revenue at a later point in certain projects. At the same time, their competitive pricing on the projects that they won helped to reduce the owners' budgetary needs for capital projects. Furthermore, an improved understanding of a project's cash flow requirements allowed the company to better manage their lines of credit and keep their overhead costs low.

Table 14.2 illustrates a sampling of frequently recurring work items. For example, Item 607A had a completion rate of 153%; that is, on the average, the owner increased the scope by approximately 50%. This item accounted for 5% of the cost of a typical project, as shown in the extreme right column titled Typical % Dollar Volume. By comparison, item 790A, adjust traffic pullbox, had a similar completion rate, but only represented 1.5% of project cost. The unit pricing for item 790A would be treated differently from that of item 607A.

A4 Chart

Problem solving involves the use of an A4 Chart (Figure 14.6) and was used to address a recurring problem on a project. The technique used is similar to that used in the A3 Chart, except that less space is available on the chart. The lower left quadrant lists the contractor's concerns. In this case gas lines were repeatedly hit during excavation, and construction was halted. Proper precautions had been taken such as: preconstruction meetings to

TABLE 14.2

Summary of Observed Trends from 33 Completed Projects =

Item No.	Description	Completion Rate	Typical % Dollar Volume
201	Remove concrete. Pave drive and apron	245%	3.50%
504	Asphalt concrete. Fine grade surface. Course	162%	5.00%
607A	Bermuda/St. Augustine grass sodding	153%	5.00%
623	Sidewalk lug	246%	3.20%
790A	Adjust traffic pullbox	156%	1.50%
1225A	Inlet protection for existing inlet	200%	3.80%
		Total ⟶	22.00%

Use of A4 Chart for Status Reporting

Example from Street Reconstruction Operation

Owner: Al Attah, P.E.

Problem Statement	*Problem Resolution Method*
• Hit gas line three times at different stations and elevations during excavation operation. Construction was suspended until the root cause of the problem was identified. __NOTE: Use of Traffic Light Coding Symbols for Problem Resolution__ – *see upper right corner of chart.* • Problem with unknown solution is coded with large **red dot**. • Problem with possible solution but not final solution is coded large yellow dot. • Problem with definite solution is coded large green dot.	• Formed problem solving team from people who were knowledgeable in the specific construction processes • Initiated brainstorming session • Used Cause and Effect (C&A) diagram to determine the root cause of the problem. • Identified the following major cause categories on the C&A diagram: people, equipment, method and engineering plans. • Using dispersion analysis technique, the team placed individual causes within each major cause category. Then asked of individual cause, why did the cause occur? The question was repeated for the next level of detail until the team ran out of causes.
Concern	*Findings/Actions Taken*
Could not determine immediate cause though the following actions were taken: • Held pre-construction meeting with key stakeholders to review the project and identify all potential issues before resuming construction operation • Located underground utilities per established procedure • Assigned competent professionals/operators to the excavation operation • Utilized appropriate excavation equipment • Followed all safety procedures and established methods • Excavated per cut sheet supplied by the surveyor	• After going through the major causes in detail, the root cause of the problem was identified as misinterpretation of an engineering note by the surveyor. The surveyor interpreted an 8-inch cement treated sub-grade as an 8-inch concrete pavement. With this in mind, the surveyor provided the equipment operator with a cut sheet showing that extra 8 inches of dirt would be removed instead of treating the dirt with cement in place prior to paving the street with 8-inch concrete. • Corrective measures were taken and lessons learned were communicated to the design engineer and other key stakeholders. Similar engineering plans were reviewed and corrected to prevent future problems.

FIGURE 14.6
The Use of the A4 chart for status reporting.

identify problems beforehand, locate underground lines with proper procedures, use of proper excavating equipment, and use of surveyor-provided details.

The upper right quadrant lists the problem resolution method. In this case a problem-solving team was formed to find the root cause of the unexpected encounters with underground gas lines. They jointly prepared a cause and effect diagram based on the categories of: people, equipment, methods, and engineering plans. Possible causes were assigned to these respective branches using dispersion analysis. The Why-Why method was used to identify possible root causes.

The lower right quadrant lists the problem resolution findings and actions taken. The root cause was traced to a surveyor who missed an engineering note, resulting in instructions for the equipment operator to excavate eight inches too deep resulting in the damage to the gas lines. The lessons learned were implemented through changes to other involved engineering drawings for the project.

End Note

The firm's founder, Aloysius A. Attah, P.E. is a trained civil engineer and industrial engineer. Prior to founding MACVAL, he was in the aerospace and manufacturing industries for 17 years. In the subsequent 12 years he has been involved in the construction of public works and transportation projects and has successfully applied industrial engineering techniques to these projects.

Case #3: Industrial Engineering Applications in the Mexican Construction Industry

The construction industry has a tremendous impact on the economy and on the rate of employment in Mexico. It is also a major component of the gross national product (GNP), hence construction productivity is very important. According to government figures, 8.3% of the jobs are in the construction industry in Mexico and the GNP of the construction industry accounts for $160,000 million annually. These figures do not include informal, private work that employs a significant number of people.

Trying to describe the main characteristics of the Mexican construction industry is not an easy task, mainly because projects are executed with different approaches; some of them are done by fairly large companies, others by medium size enterprises, and the rest by very small contractors.

ITN de Mexico, S.A. de C.V. is a consulting firm that has specialized for many years in the application of industrial engineering techniques in the Mexican construction industry. ITN CEO Jorge Cossio, an industrial engineer, has been involved in several such projects; he provides a background on important factors in the industry and describes techniques that were applied in three projects:

  a. Construction of a GM car dealership
  b. Remodeling of a luxury hotel in Mexico City
  c. Construction of a courthouse complex

Human Resources

The backbone of the industry is the versatile Mexican construction worker that is commonly referred to as *albanil*. Similarly, the supervisors or leaders of the workers are often small entrepreneurs who manage the labor force, and are generally known as *maestros*. In general, construction-related skills are often handed down from parents or relatives. Most of the people in the industry have a primary school education and in some cases a secondary education. On the other hand, there are university-educated engineers and architects who represent the remainder of the industry—they are capable of competing at an international level.

Project Organization

Project execution is accomplished through a chain of companies, contractors, subcontractors and maestros. In the case of big projects this chain has many links, ranging from the director of the project to the trade worker.

One factor that must be considered very carefully is the cost of labor in Mexico. On average, a Mexican construction worker's income is between 1/6 and 1/7 the income of a U.S. or

European worker. Nevertheless, there is a concern for the low yield of workers in Mexico. Low wages are often offset by a tendency to use a larger number of workers for a given project. This may lead to unfavorably high project labor costs. There is also a tradition of placing more emphasis on the execution phase of projects and less on the design and planning phases. This occurs even though it is known that for each dollar saved in the design stage, there is a cost of several dollars during the construction phase. This cost results from frequent interruptions due to unexpected events, material shortages, and rework to correct construction errors.

Project Planning and Scheduling

As a general rule, all construction projects have a plan that documents the main stages of the work to be carried out. When bids are requested for construction projects, the specifications call for the submission of a project program in order for bidders to qualify. This is especially true in the public works arena. The problem is that these programs are very general in scope and are often not updated. In many cases the initial program remains during the life of the project, and sometimes when it is updated the line supervisor and the contractors are not invited to participate. It is at the line supervisor (*residente*) level that there is a need for aggressive training programs to ensure that the people in the field, including the contractors, make their own work programs for their respective areas of responsibility. Such activity would also enable them to understand the project programs made by the project directors, and to participate in the follow up and update of the programs. ITN describes the benefits of applying industrial engineering to the construction processes as follows.

Work Scheduling

Innovative measures are applied in order to change the way in which construction work is scheduled. Instead of following traditional programming methods, ITN uses the options mentioned in Table 14.3.

Methods Analysis and Work Simplification

Work study and process improvement, usually used in industrial engineering, have been applied in several construction projects. ITN has developed a methodology termed the

TABLE 14.3

Work Process Comparisons

Traditional Programing Method	Method Used to Improve the Effectiveness of the Work Programs
Broad programs with lack of detail.	Much more detailed programs.
Very small participation of the line supervisors and subcontractors in the program preparation.	Very intense participation of the line supervisors and subcontractors in the program preparation.
The project program is fixed.	Frequent revisions and changes to the project program, keeping the delivery time.
Activities with long time duration periods.	All the activities are short in duration. If there are some with long duration, they will be decomposed into shorter activities.
Only one general program is used for the whole Project.	Each area has its own program, of three weeks duration.

video productivity system (VPS). The VPS is used to analyze the methods, tools, and other factors used by workers in a project; these factors can be adjusted to improve work performance. Analysis is carried out with the participation of workers to identify the best alternatives. In using the VPS, ITN makes a video recording of the construction processes, codes the operations in a computer, then discusses the involved work methods with the workers. Appropriate changes are discussed with workers and are implemented to improve the amount and quality of the work accomplished.

The flow chart in Figure 14.7 represents the use of the VPS. The results from its application indicate that Mexican workers are highly collaborative and willing to improve their project outcomes. These traits emerge after initial objections—with comments such as: "I have been doing this job for many years and nobody can teach me anything better." For example, the results obtained after the use of the VPS in some drywall jobs reduced the time needed to perform the same job by 22% and with excellent results in quality improvement. The analysis and discussions with the line supervisors, subcontractors, and workers led to the introduction of changes in work methods to avoid rework and reduce materials consumption. Table 14.4 is a brief example of the information captured by videotape and processed in the computer.

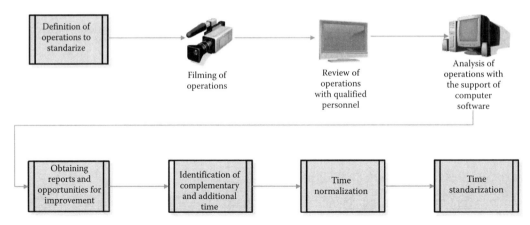

FIGURE 14.7
The ITN video productivity system (VPS).

TABLE 14.4

Process Improvement Information

#	Description	Time (min)	Comments
23	Install frame post 50 cm distance	1.52	Distance between frame posts should be 61 cm
47	Install the square (90° angle) to fix both walls in the bathroom	8.37	Use jigs to form the 90° angle
	This modification reduced the number of posts by 20%, and the manufacturer's specifications were met.		The use of jigs to perform this operation was instrumental to avoid the reworks and reduced the installation time by 60%.

Distributed Information

The distributed information model has been used successfully to handle the information requirements of several projects. It is comparable to the visualization technique that makes project information transparent. Frequently, there is a lack of necessary construction information in the field, especially where construction work is being carried out. Project information normally resides in the central construction offices. It is difficult for the line supervisors and the workers to gain quick access to that information to resolve uncertainties and problems at the work site itself. In order to solve this problem ITN developed a method to distribute the information, called the Information Quadrant. In each construction area an Information Quadrant is placed for use by the line supervisors and subcontractors.

The Information Quadrant has four information spaces:

Quadrant 1   Project scheduling: Three actual weeks and two in advance
Quadrant 2   Architect and engineering drawings
Quadrant 3   Specifications and special instructions
Quadrant 4   List of information and materials needed at each site

Quality Assurance

Quality assurance is probably the area where the best results can be achieved. Some of the quality systems and methodology in use within the manufacturing environment can be applied to the construction environment. The method of "Quality at the Source" was used to review and inspect fabricated materials prior to their shipment to the construction site for a GM dealership. Members of the construction team visited the fabrication plant and reviewed the shop drawings in detail to assure the quality of the structural steel frame. The manufacturing processes were examined closely and the fabricated steel frames were checked before shipment. This action was the key to avoiding rework and problems in the assembly stage.

Use of Checklists

ITN used checklists in a remodeling project for a luxury hotel in Mexico City. The method of Process Approval was used to validate each job to be done in electrical, plumbing, and drywall installation. Another approach called Process Liberation involved identifying a best practice and then having all workers adopt it. For example, if one were to observe 10 electricians at work they might be seen using three or more different approaches for the same job. The best method would be identified through analysis, and then the process would be liberated by having everyone adopt that method. Excellent results were obtained. The hotel supervisor approved the installation and work quality was improved, significantly enhancing the image of the construction company. This approach also reduced the rework that is normally associated with finishing a remodeling project of this type.

Table 14.5 is an example of a check list that is used to verify the quality of work done by electricians, in pipes, boxes and fittings. During the first inspections several problems were detected. Each one was discussed with the workers and corrective actions were implemented. At first the workers rejected this inspection procedure, but as soon as they saw the results obtained (less rework and faster production), they collaborated and helped to follow the recommended methods.

TABLE 14.5

Project Check List

| Room # | Wall -(Bedroom) | | | Remarks | Correction Date | Responsible |
	Fittings	Piping	Wiring			
1		x		Fix the cover plate	Feb-10	J. Garcia
2						
3	x			Correct attachments	Feb-11	G. Murillo
4						
5		x		Modify the path of the pipe	Feb-10	J. Garcia
6						
7						
8			x	Wrong wiring, review drawings for bath switch	Feb-12	G. Murillo
9						
10						
11						
12						

Project: Hotel ABC - Level 25

Quality Assurance Inspector: Juan Cortes

Production Control

On those jobs that were behind schedule, ITN implemented a well-established industrial engineering tool known as Short Interval Scheduling (SIS). Alexander Proudfoot pioneered this type of scheduling in the 1930s. The technique has been used since the 1950s to provide greater control of routine and semiroutine processes, through regular checks on individual performance over short spans of time. SIS has been extensively used in the manufacturing arena but it required adaptation to the construction environment.

SIS enables productivity to be improved on complex projects. It is based on the principle of assigning a planned quantity of work to a workstation, to be completed in a specific time. All delays can be identified and corrected at an early stage. To accomplish this, a foreperson or other crew leader discusses the conduct of the work with the crew on a scheduled basis. The involved workers know what performance is expected of them and obtain daily feedback for all scheduled tasks.

In ITN's projects the foreman and project manager were trained in the use of the SIS technique. In using SIS, time intervals are adjusted to match the involved construction tasks. For short-duration construction tasks, one-hour periods may be appropriate. For example, cover plates can be installed on electrical outlets at a rate of 10 minutes each. A one-hour control period would be based on a production output of six cover plates. Longer tasks would be appropriate for two-or four-hour control periods.

SIS can be very helpful in detecting interruptions in planned work. For example, if a worker is expected to install six cover plates but is only doing three, then that might indicate a problem with the supply of materials or the availability of the required tools. In such a situation, SIS facilitates a very quick response to problems with work assignments as they are discovered during a work shift instead of after it has ended.

SIS was used selectively in the hotel renovation project with short-duration tasks that were most appropriate for the technique. Its application yielded increases of as much as 40% in work production on specific days. First, ITN defined the work to be done by a crew

TABLE 14.6

Example of Short Interval Scheduling

Time Elapsed	Scheduled Production
10:00	Install the post in axis 5A
12:00	Install the drywall in axis 5A and 5B
14:00	Coating drywall in 6C walls
16:00	Install the wood stripes for reinforcing

within a fixed amount of time. A duration of two hours was selected as the most appropriate interval. In Mexico, the normal work shift starts at 8:00 a.m. and ends at 5:00 p.m. with a half-hour break for lunch at 1:00 p.m. The shift was divided into four intervals of two hours each: from 8:00 to 10:00, from 10:00 to 12:00, from 12:30 to 2:30, and finally from 2:30 to 4:30 p.m. Every morning, 15 minutes before the shift began, a meeting was held with the supervisor of the jobs to be controlled by SIS. In this meeting the participants clearly defined the amount of work and the tasks to be performed during each two-hour interval.

Table 14.6 is an example of the resulting daily schedule:

Trade: Tablaroca (drywall)   Team: Juan and Pedro.
Date: Monday, November 16

In each one of the SIS implementations the results were: more production, better coordination, and reduced schedules that were well within the projected dates.

Just-In-Time Technique

As described in earlier chapters, Just-In-Time (JIT) is a lean principle developed by Toyota for the supply of materials to the production lines, just when needed and in the amount needed. The benefits of using JIT include:

a. Less space is needed to store "idle" materials, or materials that may be waiting for assembly as "work in process." (In fact, with true JIT implementation, the need for a warehouse may be eliminated.)
b. Quality is improved and rework is significantly reduced.

ITN successfully adapted the JIT concept to the construction of a courthouse complex in Mexico. This application involved large, heavy, and expensive building elements, specifically:

- The steel structure
- Prefabricated concrete facades

In both cases, the project team coordinated very closely with the fabricators to produce the elements of the steel structure and the concrete facades in the precise sequence required by the on-site installation work (Figure 14.8). Significant improvements were obtained:

- Very little space was needed for in-process inventory (it was reduced to a 72 hour inventory on site)

FIGURE 14.8
A view of the courthouse complex with façade panels being installed.

- Improved efficiency and reduced erection time
- Higher quality and fewer "punch list" items

In the case of the prefabricated concrete facade, the assembly was completed one month earlier and with better quality than other comparable projects.

Defining the Industrial Engineer's Role in Construction

In order for IEs to contribute meaningfully to the construction environment they need to become familiar with construction processes. They need to be able to speak the language of construction in order to gain the credibility of project managers, designers, contractors, subcontractors, construction attorneys, and other stakeholders. This often requires that IEs obtain additional training and education on the subject.

Construction management education comprises courses that provide a basic understanding of building design and field-based construction methods as well as the planning and scheduling of the logistics of the construction process. As described by Ahmed, Zheng, and Ahmad (2005) the most popular core course topics in contemporary construction management programs include: Project/Construction Management, Finance/Economics/Accounting, Cost Estimating, Computer-related (Computer Applications, Web-based Systems), Scheduling/Planning, Law/Legal Practice. Other frequently offered courses include Construction Procurement Systems, Scheduling/Planning, Productivity (Construction Productivity Improvement), Equipment and/or Facility Management, Construction Delivery Systems (Design-Build, etc.), and Land Development.

While these programs are continually evolving, they do not seem to include the tools and techniques that have made manufacturing and service industries high-level performers. Unless significant changes occur in construction education, graduates will be constrained to continuing the prevailing business models, without major improvements in performance. Many recent studies have been directed at the construction industry to address the technological and philosophical lag experienced in the industry relative to others. The areas of interest include: Lean Construction, Construction Supply Chain Management, Automation, Information Technology, Simulation, and Construction for Sustainable Development. These subjects represent a knowledge gap that IEs can fill, if they have the necessary preparation in the basics of construction.

Construction industry performance can be improved by changes on two fronts:

a. Enhancing project managers' expertise with Lean methods and other modern methods for construction improvement.

b. Preparing a cadre of industrial engineers that are trained in the elements of construction management, and who will work as systems integrators and facilitators to implement Lean and modern methods.

1. Project Managers: Their syllabus could be enhanced with select industrial engineering courses, as indicated in Table 14.1.

2. Industrial Engineers: IEs would require a foundation in a number of construction management courses as indicated in Table 14.6. IEs would benefit greatly from construction safety training for three important reasons:

   a. They need to understand the site environment so they can avoid unsafe behavior themselves.

   b. IEs that have safety certification such as an OSHA (Occupational Safety and Health Administration) certification can help contractors to avoid being cited for unsafe practices of their staff, and even from having sites shut down by OSHA inspectors. They can conduct training for construction workers.

   c. Safety certification may provide an avenue for employment. Once hired, IEs may explore other opportunities, such as lean facilitation.

Implementation of Curriculum Enhancements

Several options worth exploring: Universities that have both construction management and industrial engineering programs, students in each respective department should be encouraged to enhance their knowledge by taking the courses indicated (see Table 14.7). Although in most such institutions there are clear boundaries between departments,

TABLE 14.7

Construction Curriculum Requirements

	Masters Degree Curriculum Enhancements for Project Managers to Implement Lean Methods	Masters' Degree Curriculum Enhancements for Industrial Engineers to Perform Construction Systems Integration
1	Work design/work measurement	Plans interpretation
2	Quality engineering/total quality management	Construction estimating
3	Simulation models of industrial systems	Construction scheduling
4	Just-in-time/lean methods	Construction equipment
5	Ergonomics/human factors	Construction materials and methods
6	Enterprise engineering	Site work
7	Management of innovation and technology	Building codes
8		Construction productivity
9	_____	Safety management (for construction)

the possibility of collaboration should be explored. This would benefit all concerned—graduates would be uniquely equipped to improve the competitiveness of construction-related enterprises. Both departments would have a stake in this success.

In recognition of the fact that the courses listed in Figure 14.1 could extend a student's program unduly, a certificate program could be developed that packages the courses as a separate qualification that may be taken after the completion of a degree program. On-line delivery of this course material would increase the access to working professionals. At the university level there should be an exploration of collaboration between both the industrial engineering and construction management departments in the interest of providing value to the construction industry.

Preparing IEs for Lean Project Delivery

In Chapter 6, Matt Horvat, an industrial engineer, describes his experiences working with project teams to implement lean principles. It is clear that lean coaching and facilitation calls for special skills—the coach has to be able to communicate comfortably with people at all levels. CEOs and project executives are increasingly college educated; architects and engineers are highly accomplished as registered professionals. Field crews of contractors and subcontractors often lack a formal education, but are extremely street smart and their acceptance has to be earned. The ability to persuade is a natural gift to some and a lifetime challenge to others; nevertheless, it can be learned and the aspiring IE must learn to use this valuable tool in the construction environment in order to be successful.

The IE must develop an ability to encourage others to brainstorm and think analytically. In deploying the last planner, performance has to be tracked (generally weekly) and the reasons for underperformance have to be analyzed using the 5-Why technique and other root cause analysis tools. The IE has to guide project team members toward identifying for themselves how improvements can be made. This is analogous to teaching people to fish instead of always handing out fish from someone else's catch.

The role of lean facilitator offers many possibilities.

a. A facilitator may work with a consulting firm and move from project to project.

b. Large and well-established contractors may wish to have in-house staff that can perform the facilitation role on an ongoing basis—with a specific project or with several.

c. Similarly, design organizations may wish to have those skills in-house.

d. Large owners may have ongoing projects that involve several phases, and they recognize the advantage of having a liaison that can interface with outside consultants. This model has been used by the Sutter Health organization in California, which established a number of lean projects in their expansion/modernization program.

e. A facilitator can benefit by having a multiplicity of skills, such as a facility with project planning and scheduling software. The ability to use information management tools and Building Information Modeling (BIM) software would be a distinct plus. The possession of a safety certification, such as OSHA certification, would enable the facilitator to have a highly desirable skill that can find daily application in the construction environment.

The professional who is so equipped would be able to serve in many different roles as workloads fluctuate.

Implementation Issues

For successful implementation in the construction arena, a support system will need to be provided. This support system must include professionals who have a thorough understanding of such techniques as supply chain management, Just-In-Time principles, value stream mapping, quality-based continuous improvement, root cause analysis, performance measurement, and post-occupancy evaluation. Most of these techniques lie within the skill set of industrial engineers.

A quality culture is needed for successful application of such techniques as Just-In-Time and Supply Chain Management to function. JIT, in particular, demands discipline, as there is no room for unreliable suppliers. JIT does not work in an atmosphere of suspicion, distrust, and internal competition. Consequently, all parties need to come together and share a unified mission for a project. Integration cannot be mandated.

There is generally a culture of dependence on inspection and on meeting codes. In other arenas, meeting specifications would be a minimum requirement. One cannot inspect in quality. It has to be built in by knowledgeable workers who are motivated to produce quality. IEs can change that culture; they are highly suited to become lean facilitators for construction projects provided that they undergo the preparation needed to understand the construction environment.

Questions for Discussion

1. Describe the main areas of the design and construction processes where industrial engineering techniques may be used. Name the techniques and explain briefly what they can accomplish.
2. How can an IE's skills support lean construction?
3. In the SACC case, how would IE applications have benefited the project owner?
4. Explain how systematic layout planning (SLP) was applied in case #1.
5. Explain the benefits of MACVAL's modified concurrent engineering approach to construction projects.
6. With ITN de Mexico, explain how short interval scheduling was applied to a project. What were the benefits?
7. Discuss posible roles that industrial engineers (IEs) can play in design and construction activity.
8. How can IEs best prepare for those roles?

Reference

Ahmed, S. M., J. Zheng, and I. Ahmad. 2005. An analysis of master's curricula in construction engineering and management programs in the US. *Proceedings of the eighth UICEE conference on engineering education*. Kingston, Jamaica.

Bibliography

Ahmad, I. and S. M. Ahmed. 2001. Integration in the construction industry: Information technology as the driving force. *Proceedings of the 3rd international conference on construction project management*, ed. L. K. Robert Tiong. Taipei, Taiwan: Nanyang Technical University Press.

Elshenawy, A., M. Mullens, and I. Nahmans. 2002. Quality improvement in the homebuilding industry. *Conference proceedings, industrial engineering research conference*, Orlando, FL.

Forbes, L. H. 2004. How industrial engineers can improve construction productivity and profitability. *Proceedings of the 2004 institute of industrial engineers' annual conference*, Houston, TX.

Forbes, L. H. 2005. Industrial engineering applications in construction. In *Handbook of industrial and systems engineering*, ed. A. Badiru. Boca Raton, FL: Taylor & Francis.

Forbes, L. H. 2007. A model for collaborative industrial engineering support and performance improvement of construction projects. *Fourth international conference on construction in the 21st century (CITC-IV) Accelerating innovation in engineering, management and technology, July 11–13, 2007*, Gold Coast, Australia.

Forbes, L. H., and S. M. Ahmed. 2003. Construction integration and innovation through lean methods and e-business applications. *Proceedings of the conference-winds of change*. ASCE, Hawaii.

Macomber, H., and G. Howell. 2005. Using study action teams to propel lean implementations. Louisville, CO: Lean Project Consulting Inc.

Muther, R. 1974. *Systematic layout planning*. 2nd Ed. Boston, MA: Cahners Books (1st Ed. 1961).

Pocock, J., C. Hyun, L. Liu, and M. Kim. 1996. Relationship between project interaction and performance indicators. *Journal of Construction Engineering and Management* June: 165–76.

Womack, J., D. Jones, and D. Roos. 1990. *The machine that changed the world*. New York: Harper Collins.

15

Learning from Projects and Enhancing Lean Project Delivery and IPD

Introduction

This final chapter draws on the knowledge provided in the preceding chapters in the context of learning from projects and using those lessons to improve future performance, especially with lean-based projects.

    a. For both lean and traditional projects, post occupancy evaluation (POE) is discussed as a means of learning those lessons that are only available after the completion of a project or facility, i.e. those that focus on how well it meets the everyday needs of its users. "Post mortem" surveys are presented as a subset of POE. An improvement is recommended to adapt POE to lean-based projects.

    b. An enhanced model of the Lean Project Delivery and IPD processes is presented, based on an aggregation of the knowledge gained by observers or participants in a cross section of lean-based projects.

Overview of Post-Occupancy Evaluation (POE) for Continuous Improvement in Construction

The process of post-occupancy evaluation (POE) provides a structured and systematic method for learning lessons from past construction projects. Preiser et al. (1988) define POE as the process of evaluating buildings in a systematic and rigorous manner after they have been built and occupied for some time. Although one may intuitively suspect that a facility succeeds (or fails) in serving its users well, one cannot fully appreciate and measure to what extent this occurs without conducting a POE. Construction professionals have come to realize that those who forget the "sins" of the past are doomed to repeat them again and again. POE can provide redemption from these mistakes if it is used correctly.

    The quality of constructed facilities is important to its users. People experience more satisfaction within their environment if it is kept in an esthetically pleasing manner. Preiser et al. (1988) noted that "spatial attributes, the sequence, location, relationships, size, and detail of a facility's spaces have been shown to affect occupant behavior." POE provides a means for improving the quality of future facilities by learning from past projects.

A POE involves the measurement of the functioning of a facility as compared with its purpose as defined in a formal program, and by the objectives of the architect/designer. As these objectives should mirror the agreed-upon wants and needs of the client/owner, POE requires that a systematic methodology be used to compare these expectations with the effects of the facility on its users. The results of the POE can identify the extent to which the design intent has been met; this feedback can also help to identify "best practices" that can be used to improve future designs. The quality advocate, Peter Senge, in his book *The Learning Organization*, advocates that organizations follow a cyclical process in order to learn from their actions and use that knowledge to improve their performance.

In order to apply POE effectively, it is good for design and construction professionals to start with introspection, to look inward and ask why POE is being considered, and what one hopes to derive from it. As with so many other endeavors, if one does not plan where one is going, one is likely to arrive at an unexpected or unwanted destination. The experiences of world-class organizations indicate that analysis can provide invaluable information, enabling one to manage analytically, not just intuitively. In the design and construction arena, this analytical approach leads stakeholders to examine and improve the processes involved in both construction design and delivery.

Scope of Evaluations

The scope of POE studies varies with the evaluation objectives. The evaluation may focus on:

- A specific part of a facility
- An entire facility
- A building program
- Programming/design
- Construction procurement and delivery

Categories of POE: Historical, Comparative, Longitudinal, Quasi-Experimental

Ex-ante evaluation involves an analysis of facility performance "before the fact." It may be informal, as conducted by some architects in visioning how a facility may be used. It is a standard part of the design cycle—imaging, drawing, and testing. It often does not include users' inputs but it should, as the users can project themselves into the use of a facility with more realism than others can. On the other hand, simulation methods evaluate the utility of a facility in terms of travel distances, and so on, as well as perceptual criteria.

There are several categories of POE. They include:

- Historical: Studying the facility in retrospect to determine if actions taken during the design/construction process have been effective.
- Comparative: Contrasting two situations such as two similar facilities after one has been specifically changed.

- Longitudinal: Taking baseline measurements before changes are made. Changes are then initiated and differences attributed to them.
- Quasi-experimental: Using statistical approaches to compare experimental and control situations.
- Post-mortems: A revisitation of the design and construction processes themselves provides critical process-related lessons.

Many different models have been developed for the process of POE. A frequently used one is the environmental design model that is based on the performance concept. It involves comparing specific performance criteria that have been established for a given type of facility, such as a hospital or a school, with the actual performance that is observed or measured by the building's occupants and evaluators. The performance measures generally include: the spatial aspect, the physical interaction between the individual and the facility; the social aspect, the social significance of the environment; and the service aspect, the environmental conditions necessary for safety and health.

Role of Programming in POE

Programming is a critical reference point for POE studies.

- Programming is a statement of a project's problems or constraints as visualized before design work is begun.
- Programming is analysis and design is synthesis.

A well-designed program treats users' requirements in terms of form, function, and economy. It also provides a methodology to evaluate the success of the design procedure. The subsequent POE compares project outcomes with the program and its influence on design.

Using an educational facility by way of an example, a POE of facilities may comprise four headings: user satisfaction, economy, function, and performance. User satisfaction will address the question: How satisfied are the users with the facility? Are there distinct likes and dislikes with regard to a particular building feature? Economy is concerned with whether or not the facility is economical to operate and maintain. It considers the efficient use of space, time, and resources, ease of cleaning, maintenance, and repair; the use of appropriate materials and technology for the educational environment; the relative quality of materials and construction. Function addresses whether or not the facility functions as intended; if it helps or hinders the essential activities such as learning and physical activity. The factors considered are: flow, orientation, accessibility, and supervision of circulation of students; comfort, safety, and convenience; the actual use of space for programmed and nonprogrammed instructional activities. Performance addresses the application of creativity and innovation in design and construction: the performance of electrical, mechanical, structural, and other building systems; the quality of temperature and humidity controls, light and sound controls; color, texture, space, and perceived aesthetics; the social image and character of the facility; user satisfaction, and the adaptability of the facility for changes in use.

Planning for the POE

A facilitator must be designated to coordinate the POE; this should be a seasoned professional with a background in engineering, architecture, or construction. A background in quantitative analysis would also be helpful. It is highly recommended that this facilitator should not be a member or affiliate of the design team or the construction team, in order for the POE process to be unbiased and objective.

As the facilities to be studied are likely to belong to institutions or large owner organizations such as retail chains, it should be feasible for them to use a member of their technical staff to serve as a POE facilitator. In projects based on lean-based delivery methods, an independent facilitator such as a lean coach could manage the POE process.

A POE survey instrument is developed to address the factors of user satisfaction, economy, function, and performance in the project/facility being reviewed. It uses a five-point Likert scale ranging from "very dissatisfied" to "very satisfied," with a midpoint indicating "neither satisfied nor dissatisfied." One version of the survey is developed for facility users (who are familiar with a facility) and another for design, construction, operations, and maintenance personnel. This latter group may represent a design/technical team that can address a very broad range of issues related not only to the owner's intent, but also to long-term performance of a facility.

In the case of an institution such as a college or university, design/technical team members are drawn from several departments such as: maintenance, construction, safety, energy management, communications, information technology, campus police, cafeteria services, and transportation, to name a few. In the case of a hospital, the POE survey should include many users—administrators, doctors, nurses, and most importantly, patients. Patients are seldom asked for their opinion and that runs counter to the lean philosophy. Other sources may be used for data gathering: the analysis of drawings, building observations through photography and video, and the analysis of operating cost. Maintenance records are one valuable source of operating cost and energy consumption data.

Based on the degree of need, a POE may be tailored to include such factors as general satisfaction, aesthetics, layout, air quality, privacy, durability of construction, maintainability of equipment, energy, and efficiency.

A candidate facility is selected for review; ideally, it should have been occupied for between one and two years, long enough not to be distracted by short-term final adjustments by the contractor, but not so long that shortcomings become accepted as part of "living with" the facility.

POE Procedures

- Invitations are sent to the team members for two distinctly different evaluation meetings:
  a. a facility user meeting, and
  b. a design/technical team meeting.

    The second group is especially critical as it is the best source of remedial action information. It is usually more convenient to have separate meetings on different occasions as the meeting agendas are somewhat different; the survey instruments, in particular, are different. The design/technical team meeting requires having team members acquainted with a facility they may be seeing for the first time, so more time has to be spent in providing guidance on the layout and features to be examined in detail.

- The facilitator leads the evaluation meetings. The design/technical team is instructed on the approach needed to make building observations and complete the survey documents.
- A general discussion is held on the background and conduct of the project. The owner's project staff and the architect/engineer (A/E) of record as well as other decision makers provide background on the inception of the project and its subsequent conduct.
- The POE participants walk through the entire facility in small groups and individually complete the surveys.
- Following the site visit the data from the design/technical team's surveys are collated and analyzed.
- The scored and narrative responses for both users' and design/technical surveys are combined with historical building performance information. Maintenance records are checked for operating costs, breakdowns, and malfunctions.
- A comprehensive POE report is prepared. The survey findings are reconciled with current standards and specifications to develop meaningful recommendations.
- The report is distributed to all involved parties.

Once a POE is completed and documented, it is critical that additional steps be taken to apply the lessons learned to the design and construction of future facilities. These steps represent the stages in the continuous improvement cycle shown in Figure 15.1, and are required for systematic improvements to occur. A meeting is conducted with selected decision makers, in which the survey findings and recommendations are presented.

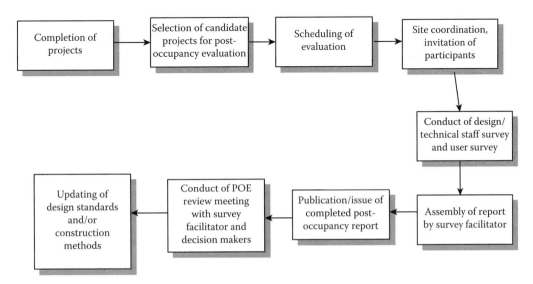

FIGURE 15.1
Flowchart of the post-occupancy process.

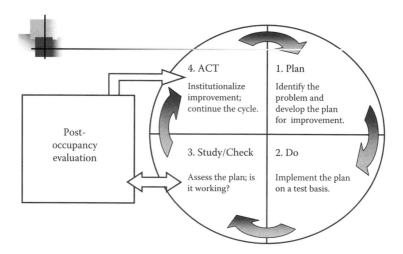

FIGURE 15.2
The Deming Cycle (PDSA Cycle) and Post-occupancy Evaluation. (Reprinted from Forbes, L., *Proceedings of the 2002 IIE Annual Conference 2002*, Orlando, FL, May 19–22, 2002. With permission.)

Approved changes are subsequently made in the appropriate reference documents, so that the specifications and construction procedures for future projects can have enforceable requirements to utilize identified best practices. These documents may include the master specifications, design criteria, and construction procedures manuals that many large organizations maintain for reference purposes.

As indicated in the foregoing figure, the post-occupancy process serves as the "Study/Check" phase of the PDSA cycle (Figure 15.2). The recommended changes are implemented in the "Act" phase. In subsequent iterations of designing and building a particular building or building feature, the use of the PDSA cycle enables the owners and designers to evaluate the impact of POE-based improvements.

Quality Score Calculations

The following equation provides for the determination of a "Quality Score" for each surveyed facility or project. The questionnaire responses are combined to obtain mean ratings (R) for each question. A weight (W) is assigned to reflect the importance of a particular objective. For example, under the heading of the objective "Function," respondents could be asked to rate the adequacy of a specific office space. The same question, when asked of different spaces may have different weights applied. The composite Quality Score would be based on the factors in the equation below: R = rating and W_{jk} = relationship of question j to objective k.

$$\text{Quality Score (QS}_k) = \frac{\displaystyle\sum_{j=1}^{n} R_j \times W_{jk}}{\displaystyle\sum_{j=1}^{n} W_{jk}}.$$

Industrial engineers or quality professionals can further enhance the POE process through life cycle analysis to assess building/facility performance on a wider scale that includes not only user satisfaction with everyday utilization, but also issues of operating cost, durability, and reliability. User satisfaction questionnaires provide valuable information on the perceptions of the client (construction owner) with respect to the finished product and, very importantly, the nature of the associated processes. A typical questionnaire would address contractor performance measures relating to such issues as timeliness, responsiveness, communication, empathy, cost, and construction quality. The performance of designers could be evaluated in a similar fashion.

Overall, the delivery of design and construction services can be significantly enhanced through the use of "lessons learned" and "best practices." The information collected through the POE application should also address how to design processes to improve design quality, cycle time, transfer of learning from past projects, and provide effective performance measurement systems for understanding, aligning, and improving performance at all levels. The quasi-experimental type of POE should be investigated for application where two or more facilities may be compared with each other.

Other POE Approaches

Several other POE approaches can be used within organizations, especially designers and contractors, to measure performance. The Construction Industry Institute (CII), headquartered in Austin, Texas, has produced important construction research findings that can be used very effectively in the POE process. Their research has looked into the performance of project teams that are used on certain types of construction projects where team membership changes during the transition from conception to completion. They have identified that project leadership behaviors have a great influence on project cost performance, and that the adoption of best practices influence schedule performance.

A POE could examine the extent to which the respective success factors had been incorporated in a project. Leadership behaviors, for example include: focus on objectives, team values and environment, empowerment and decision making, and external/internal relations. Best practices include: planning and project controls, scope management and project alignment, and quality.

Sample Questions

1. Focus on project objectives. Please respond to the following items, using a scale of 1 = not at all, to 7 = to a great extent.

   To what extent has your team leader

   a. Communicated team goals and objectives?

   b. Focused the team members?

   c. Aligned team member goals with the project goals and objectives?

2. Empowerment and decision making: Please respond to the following items, using a scale of 1 = not at all, to 7 = to a great extent

   To what extent has your team leader

   a. Allowed the team to make its own decisions?

   b. Supported the team's decisions?

   c. Promoted a spirit of empowerment among team members?

Client Satisfaction with Design and Construction Services

Many owners in both the public and private sectors have begun to use criteria other than low prices for selecting design and construction providers. In considering candidates for new business, these owners are increasingly referring to the nature of their interaction with designers and contractors on earlier projects. This past performance is rated and included in an evaluation of candidates for new work.

A performance evaluation questionnaire is an excellent adjunct to a POE. It addresses what is perhaps the most important part of the design and construction process; that is, the perceptions of the client (construction owner) with respect to the finished product and, very importantly, the nature of the associated processes. It goes without saying that the owner's level of satisfaction with the outcome of a project is an important measure of success. A typical questionnaire would address performance measures relating to such issues as timeliness, responsiveness, communication, empathy, cost, and quality. A typical survey could address these issues as follows.

Owners

Use the following scale to reflect how the following factors influence your satisfaction with a contractor organization. 1 = no importance to 5 = extreme importance.

1. Timeliness: Does the contractor (for miscellaneous jobs):
   a. Provide a fair estimate of cost and schedule?
   b. Give small jobs high priority?
   c. Complete jobs quickly, once they start?
   d. Respond promptly when asked for work status?
2. Communications: Does the contractor:
   a. Update me on jobs and their status?
   b. Explain each job before starting it?
   c. Notify me of and explain project delays?
   d. Follow up to ensure satisfactory completion?
3. Quality: Does the contractor:
   a. Emphasize the facility's performance characteristics?
   b. Strive to build in durability in the facility?
   c. Provide high quality workmanship, beyond code requirements?
   d. Treat quality as an essential element of owner satisfaction?

Other factors may be added to the foregoing lists as needed.

Best Practices to Derive the Best Results from POE

A number of best practices will help construction stakeholders to derive the best results from the use of POE. Utilize the services of an experienced professional such as a lean facilitator to interface with the involved parties to ascertain the organization's information needs and develop an appropriate survey instrument (questionnaire). The POE questionnaire should be validated to ensure that it asks the right questions and will be interpreted in a consistent manner in repeated use.

- Develop an organization-wide commitment to continuous improvement. A commitment to the lean construction philosophy would be ideal for harnessing lessons learned to deliver greater value and eliminate waste. Incorporate the Plan-Do-Study-Act (PDSA) cycle in the area of facilities construction activities. Top management must take the lead and encourage organization-wide participation in efforts to continuously learn and improve performance. Empower employees at all levels to recommend/implement improvements.

- Begin the journey toward adopting a new business model for facilities construction/management based on the principles of the Malcolm Baldrige National Quality Award (MBQNA) Criteria. Implement a training program that may start with conducting staff orientation sessions to learn about Baldrige principles.

- As a part of adopting the new business model, consider administering a self-assessment survey to staff involved with facilities design or construction. Voluntary and confidential staff responses to a battery of well-designed questions can provide senior management with the current state of the organization based on several criteria. These may range from employee morale to the effectiveness of the processes used to conduct business, as well as the extent to which some best practices are used.

- Devote the resources necessary to administer or participate in the POE activity. Develop the infrastructure necessary for a successful program; institute quality assurance reviews in the field and require that contractors address them, even if code compliance is otherwise achieved.

- Tailor the POE Program to the organization's needs; this will depend on whether your organization is an owner, architect, engineer, or contractor/builder. One size does not fit all.

- Develop quantitative performance measures and monitor them over time. Integrate the POE with the PDSA cycle. The findings from POE become inputs to the study phase of the cycle.

- Include the "voice of the customer" (the end user) in future projects as much as possible. Incorporate POE findings/recommendations in specifications, standards, and procedures promptly.

- Be willing to change.

Contractor Post-Mortems

Belair Contracting's Post-Mortem Process

A successful application of post-mortem studies to support lean construction is exemplified by Belair Contracting. Belair is a family owned site services company dating back to 1953. In 2005, the CFO, Tracy Dabrowski, worked with a consultant to introduce and institutionalize lean in Belair's three branches. Within 18 months the organization's management saw major improvements in their ability to predict success with a particular bid, monitor work in process (WIP) and not repeat previous mistakes.

Belair attributes a part of their success to conducting a mandatory post-mortem at the completion of each job. They will not close out a job until the post-mortem is held. They seek field staff's opinion of the job and whether it resulted in a win for both the customer and Belair. Using the principle of continuous improvement, Belair uses information from each completed project to update important checklists. These checklists serve as a guide to reduce the chances of repeating mistakes in future projects. (Belair's job close out process is displayed below in draft form. It has been slightly modified for explanatory purposes.)

Detailed Project Review/Post-Mortem Checklist and Report Card (Reproduced from Dabrowski, D., Company documents, February 28, 2008. With special permission of Belair Excavating.)

Date of review: Facilitator:

Purpose of Review:
   Budget Variance Mid-Job
   Change in Belair or GC Staff
   Change in Project Conditions or as Defined in Pre-Job Meetings
   Post-Mortem at 100% Completion

Project Information:
   Project #:   PM:
   Project Name: FS:
   GC: EST:

Describe Project:

Negotiated/Low Bid?

Other Business Units Involved:

	Original	to Date	Difference
Contract:	$0	$0	$0
Budget:	$0	$0	$0
	$0	$0	
Change Orders:	$0	$0	
CO Budget:	$0	$0	
	$0	$0	
Gross Profit:	$0	0%	$0

Evaluation:

The following questions explain positive and negative occurrences during the conduct or the project; this information facilitates a fact-based evaluation of the project.

   1. Did the scope of the job significantly change from the time it was bid versus actual mobilization?
   2. Were there challenges with the General Contractor on this job?

3. Were there challenges with the Subcontractor or a Vendor on this job either through Belair or others?
4. Did Belair stop the job for any reason? Identify these reasons.
5. Describe weather issues.
6. Describe site conditions.
7. List positives. (What did we do right?)
   1.
   2.
8. List negatives that led to detrimental results on this job. (What did we do wrong?)
   1.
   2.
9. List any safety issues or losses of property.
   1.
   2.
10. What did you learn from this exercise and what would you change in the future?
11. Is a Belair Process Change Required as a result of what we learned? Do we need to change our procedures?

The following questions are all pass/fail. NA is not an option. Any comments should have been addressed above. All questions are to be answered as of today, not in the future.

	Pass	Fail
1. PQM Completed Score _____	Y	N
2. Bid Hand Delivered	Y	N
3. All Contracts/Subcontracts "signed" Prior to Job Initiation	Y	N
4. Pre-Planning Meeting/Checklist Conducted and Signed Off	Y	N
5. On-Site Meeting Conducted Prior to Job Mobilization	Y	N
6. Equip/Materials Available and Delivered as Bid	Y	N
7. Were the Proper Personnel Available?	Y	N
8. Did Key Personnel Change During the Course of the Job?	N	Y
9. If Yes, Was a Formal Meeting Held With the Appropriate Individuals?	Y	N
10. EWO's Executed, Performed and Collected Upon	Y	N
11. Post-Mortem Conducted	Y	N
12. Formal Job Closeout Meeting Held	Y	N
13. Final Customer Survey Sent? Received Back?	Y	N
14. Final Bill to Include Retention Prepared and Collected	Y	N
15. Job Came in AT or Below Budget	Y	N
16. Any Accident Incident Reports Filed	N	Y
17. Does the FS Feel He Had Control to stop the Job/Responsible for the Job	Y	N

Overall evaluation (Figure 15.3) of project outcome (check the appropriate box)

Belair wins customer wins	Belair wins customer loses
Belair loses customer wins	Belair loses customer loses

FIGURE 15.3
Scoring of final project grade.

Action Items as a Result of this Detailed Job Review Meeting?

1.
2.

Upon Completion of Detailed Project Review/Post-Mortem

1. Please file all completed forms electronically under "Lean Processes and Tools/Detailed Project Reviews" by Job/Job # and Name. (PM/GM or DO)
2. Enter summary in the Belair Report Card for each job. If more than one Review is performed for a job, enter them all by date. (PMA by Local Office)
3. All projects that required a Detailed Project Review must be highlighted on the WIP by the PMA.

Submit copy to Local General Manager, and accounting with electronic WIP Reports.

Aligning the POE Process with Lean Construction

As described in the foregoing sections, the POE process captures extensive input from a variety of sources—surveys of various categories of users, maintenance records, warranty records, etc. Discussion in an open forum enables designers and builders to provide background information on issues such as site characteristics, programmatic needs, and their overall experience with the conduct of a project. The Last Planner® System (LPS) is built on a process of ongoing evaluation of job execution—both in the design and construction phases of a project. Through the weekly use of various metrics including PPC and reliability, waste can be detected and reduced.

The POE process to date has been mostly used on traditional projects with an emphasis on evaluating design parameters. It also suffers from the fact that it is usually conducted close to one year after occupancy. At that point, many memories have faded with regard to details of a project that could have tangible impacts, both positive and negative. In fact, there is a natural tendency to forget events that have had a negative impact, unless they are so big that they cannot be overlooked.

It is hereby proposed that the POE be aligned with Lean Construction by:

a. Having notes from weekly meetings assembled and made available for the POE meeting. This would start with design and continue through construction and completion.

b. Utilizing BIM to make information available in electronic media for improved access and analysis

c. Including in the POE documents data fields for lean statistics from the Last Planner, such as PPC averages, reasons for non completion (RNC), etc.

d. Qualitative data from Plus/Delta analysis of ongoing lean project meetings and reviews

e. Time frame information on lean activities such as Target Value Design (TVD) schedules and budgets

f. Project administration—RFIs, processing of payment requests

Many other capabilities can be built into the POE to support lean construction—once their value is established, they can easily be added. Above all, this information should be maintained in a database where it can be easily retrieved and made actionable for the next lean-based project.

Learning from Lean Construction Projects

As described in Chapter 5, Glenn Ballard (1999) suggests that companies should first understand themselves before rushing out to hire a consultant to make a lean transformation. He describes three possible roles for a consultant: (1) as a technical expert who tells people what to do to accomplish a specific goal; (2) as a temporary member of the organization, often for unpleasant jobs such as layoffs and reorganization; and (3) as a facilitative consultant that teaches an organization new skills. However, in the case of lean construction the people in the organization need to learn to think, see, and act in a lean mode.

Companies that understand and that have practiced self-transformation can benefit from the fresh ideas brought by a consultant. Those that have not tried to change from within may delude themselves by expecting consultants to bring about change extrinsically, when it needs to occur intrinsically. A lean consultant would benefit a company most by helping it to learn how to change, but a consultant is "always an outsider" and the people in the company must take the initiative to bring about change.

Ballard recommends that companies start with the LPS as it identifies improvement opportunities while freeing up resources to address them. The Five Big Ideas for Lean Project Delivery serve as a so-called True North and can be applied to virtually any large, complex project. They are described in detail in Chapter 3 and 7.

A lean facilitator has to serve as a change agent. Project work involves a lot of pressure and there are many technical issues to be addressed. The facilitator must be one who is free to focus on guiding people on a new process without getting caught up in the details of a project. The role calls for someone with passion who will hold people's feet to the fire when pressure builds.

It can be highly beneficial to the lean construction process, in fact, the owner is often in the strongest position to require lean behaviors, especially if the contract is written to include them. In the Sutter project (Chapter 7) the owner was part of the Core Group that performed contract administration, based on an Integrated Form of Agreement (IFOA). The owner's representative sat in on planning sessions. This compares dramatically with traditional projects where there is generally highly centralized planning, and extensive reporting to the owner of past events. This is like driving with a rear view mirror.

Another lean project, a petroleum refinery delivered by the lean production management method, did not have an IFOA, but the owner championed the adoption of lean in a formerly traditional project. The owner's representative became involved in the application of The Last Planner® System, and planning/review meetings at various levels: the master schedule, the look-ahead schedule, the weekly work plan, and the daily work plan. The lean facilitator on that project pointed out that this understanding helped the owner to avoid reverting to familiar behaviors when pressure mounted.

It is sometimes necessary to let people fail first in order for them to accept new ideas such as lean construction. In the refinery project, field supervisors were accustomed to starting work packages as soon as possible—past experience suggested that earlier completion and earlier progress payments would result. The PM, Dave Koester, insisted that the contractor should "Go slow to go fast." He pointed out that no work should be done until commitments could be made to have it completed by a specific date. As the project continued, the participants began to understand and accept the wisdom of mastering processes before becoming fixated on speed.

Collaborative design activities greatly influence the effectiveness of lean construction processes. Westbrook Air Conditioning and Plumbing pioneered the development of Integrated Project Delivery with a chiller plant installation in Orlando, Florida. In a departure from the typical design process for this type of project, there was a prioritization of function over form. For example, mechanical engineers (MEs) were included in designing column footers for the chiller plant facility. MEs would not normally participate in this aspect of the design, but collaboration with the structural engineers led to a sharing of information about the positioning of the equipment and the routing of 24 inch diameter chilled water lines. One column was moved 18", saving time and money for the overall project. In actual fact, the design cost was increased to re-engineer the layout. Trading higher design cost for lower overall project cost is one of the hallmarks of lean. The ME used 3-D modeling to configure this large piping to optimize the use of hangers from the building soffit; it also enabled prefabrication of the piping sections off-site. Once delivered to the site, these piping sections were rapidly installed and hung by prepositioned hangers. The steel erector rescheduled the installation of decking—this allowed the mechanical contractor to use a crane rigged through the steel frame. Prefabricated piping sections were connected quickly; an alternative method would have been much more expensive. Other benefits from the design phase collaboration included the installation of utility lines under the floor slab of the building, which was more cost effective. The GC made provisions for the electrical subcontractor that allowed them to lay one mile of conduit underground without needing to excavate.

Target value design (TVD) is effective for providing the owner's project requirements cost effectively. In addition to TVD, designing for lean operations ensures that space usage is optimized; the project scope meets the owner's needs without excess. Analytical support and performance measurement are necessary for the continuous improvement of construction processes. Measurements such as "Reasons for noncompletion" are critical, and enable improvement efforts to be targeted for the greatest impact.

With regard to deploying The Last Planner® System, it is advisable not to use the project estimate to back-calculate activity durations for plans. Project estimates are in reality optimistic forecasts of project activities made by people who are knowledgeable, but are not necessarily familiar with the prevailing field conditions. In fact, project supervisors are generally able to give more realistic estimates of activity durations. They are more accurate than we think they are.

Constraint analysis and assignment are essential for ensuring that work packages can be released to downstream crews without impediment. Lean project experience confirmed that identifying, listing, and assigning each constraint for resolution, regardless of size, resulted in all constraints being addressed or understood within a known and agreed time frame. Unmet expectations for solutions and continuation of constraining conditions were greatly reduced, especially since the responsible individual had the authority to agree when he would respond with resolution and answers. This approach allowed reasonable time for resolution and for accurate forecasts for completion of prerequisite work items. This proactive approach avoids situations where project team members erroneously assume that someone else is addressing the spoken needs or concerns and constraints linger unaddressed.

One very important lesson that was learned from several examples is that it is possible for a subcontractor to utilize lean approaches on a project, even if the other parties in the project do not represent a lean environment. Four subcontractors (Tweet/Garot, Belair Contracting, Grunau, and Superior Window Corp.) tell compelling stories about their lean journey and their transformation into lean organizations. The lean philosophy encourages project participants to collaborate as teams with a shared vision, such that their efforts optimize the overall project and not just their individual situation. Their experiences suggest that these subcontractors were more concerned with sharing the lean philosophy and in setting an example with high performance than with gaining profitability at the expense of others. These efforts may be paying off; as lean projects proliferate, known practitioners will be prime candidates for joining teams in such projects.

The Kaizen technique is short-term but very effective at identifying opportunities for improving construction operations. 5S is a simple but powerful technique that reduces clutter and diminishes or eliminates non-value-added time that is often spent in locating tools or equipment needed for construction tasks. As noted in Chapter 6, two specialty contractors documented savings as high as a hundred thousand dollars by disposing of excess inventory and streamlining their operations for better responsiveness.

A number of "people issues" need to be resolved with lean construction. Whereas lean projects have produced superior results with committed team members, it is difficult to coordinate their work with that of nonlean contractors. Reliable promising is an essential foundation for lean, and an uncommitted member cannot be relied upon to deliver as promised. In the first documented integrated project delivery (IPD) project (Matthews and Howell 2005), an uncommitted member was not heavily invested in the lean process, and reverted to traditional behaviors during the project. That member later dropped out by mutual consent. Even when faced with a condition whereby one or more stakeholders do not participate, it is still better than would occur in a traditional project.

The LPS is based on the principle of autonomation; that is, the empowerment of the people doing the work. This represents a paradigm shift for traditional project managers, and many are not prepared to make it. In one project, an experienced project manager was offended by this change in roles; he could not relate to empowering crews, and left the project.

Several integrated team issues need to be refined in order to reassure participants that they are structured for fairness. They include: insurance, bonding, job costing, accounting, collective warranty responsibility, and especially the distribution of gains and or/losses.

Enhancing The Lean Project Delivery System™ and IPD

Collectively, many lessons have been learned from a number of lean projects that have been executed in recent years. At the same time, the structure of lean methodologies has been evolving. For example, BIM technology has become more accessible, more capable, and more affordable. Sustainable construction has become a standard requirement as there is increasing interest in reducing the carbon footprint and in reducing operating costs. Above all it has become clear that having a multiskilled team working in close collaboration from project definition, design, and eventual construction is the best way to have a facility that not only meets the owner's requirements, but that exceeds them in terms of both initial costs and operating costs. With the passage of time, new approaches must be considered for governing the design and construction processes. A new model for the lean project delivery process is proposed. As shown in Figure 15.4, it comprises five overlapping trapeziums, a column on the left representing lean support from various professionals, and horizontal bars representing the project phases and activities that occur in parallel with them. It is adapted from The Lean Project Delivery System™ (Ballard 2000, 2008).

Importance of Champions

As mentioned throughout this book, lean requires transformation, both at the personal level and the organizational level.

> We cannot become what we need to be by remaining what we are.
>
> Max de Pree

This transformation has to take place in the hearts and minds of all stakeholders—owners, designers, contractors, subcontractors, and suppliers to name a few. This challenge is especially great in the design and construction environment, where people are creatures of habit, and often resist change. It falls on the lot of special individuals within stakeholder organizations to lead others toward a new way of doing things. To be effective, champions must be highly placed in their respective organizations, so they can get the ear of leaders and also have control over resources that support the transformation. The champion should be a member of the executive committee, or at lest a trusted direct report of a member of the executive committee. He or she should have enough influence to remove obstacles or provide resources without having to go higher in the organization. Clearly, the champion has to have enduring commitment to the lean concept, and a combination of the passion and motivational skills needed to engage others—at all levels—to embark on an unfamiliar course of action.

No one can "make people go lean". People have to be introduced to lean and taken to the point where they transition from curiosity to interest, and then to a desire to be a part of successful lean endeavors. While case studies abound on the results gained from lean implementation, they often understate the role of the champions in the respective organizations. Jay Berkowitz, President of Superior Window Corp has successfully implemented lean in his window manufacturing and installation company. Based on his personal experience, Berkowitz advocates that an organization needs a passionate "lightning rod" leader to be the combustible spark that drives the various phenomena involved

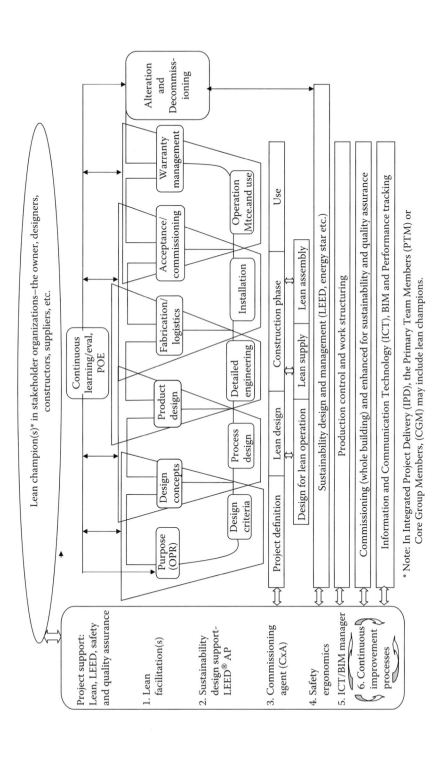

FIGURE 15.4

The enhanced lean project delivery process. (Adapted from Ballard, G., The Lean Project Delivery System™: An update. Lean Construction Journal 1–19, 2008; Ballard G. and Howell, G. Lean Project Management. Building Research and Information, 31 (2), 119–133, 2003.)

in going lean. While an organization needs a "hands-on" facilitator, a leader, essentially a lean champion, must be willing to stand up and represent the passion needed, otherwise the entire process quickly decays and becomes a giant waste of money and effort. In Berkowitz's words "Before your organization approaches LEAN, make sure they have the franchise player who has the passion around which they can build a team".

Observations

The effectiveness of lean project delivery methods is greatest when lean methodology is incorporated at the very beginning of the design process. The owner's project requirements (OPR) need to be defined at the project definition phase as they are an integral part of defining the purpose of a project, and later determining how well the owner's needs have been met. Sustainable construction involves designing and building a facility based on a system such as Leadership for Energy and Environmental Design (LEED). LEED considers many factors such as the location of a facility relative to the community in order to reduce transportation costs. It promotes the reuse of material from building demolition, and the recycling of various categories of building materials from off-site sources. LEED minimizes the use of potable water and seeks to maximize energy efficiency while improving indoor air quality for the benefit of building occupants.

Design based on these factors involves major long-term decisions and cost trade-offs on the part of project owners; initiatives such as LEED are feasible at project inception, and much less so at a later stage. The OPR should also be clarified relative to the configuration of a facility; the space requirements should be analyzed relative to the use of a facility for lean operations. Recent projects have been able to accommodate 90% of proposed operations in 70% of the space that was initially budgeted.

Commissioning should be expanded beyond its customary role in facility acceptance. As described in Chapter 10, commissioning is an important adjunct to sustainability initiatives; LEED standards require the support of a commissioning agent to ensure that both the design and execution comply with documented requirements.

Building information modeling (BIM) has been shown to enhance projects in several ways, especially when integrated with a project as early as the conceptual design phase. A Stanford study revealed improvements in a cross section of projects:

- Up to 40% elimination of unbudgeted change
- Cost estimation accuracy within 3%
- Up to 80% reduction in the time needed for a cost estimate
- Contract price savings of up to 10% through clash detection
- Up to 7% reduction in project duration.

Therefore BIM should be included in predesign team activities.

Lean construction has been shown to work best when the design has been planned to facilitate it. In an integrated team, construction representatives collaborate with designers to promote constructability. Because of the newness of lean construction, many projects have been designed in the traditional mode and lean has been adopted only during construction when the project has fallen behind schedule.

Table 15.1 describes the elements of an improved project delivery process. The process emphasizes the involvement and collaboration of specialists/experts in

TABLE 15.1

Enhancement of the Integrated Project Delivery Process

Project Phase	LEAN	LEED	BIM	Commissioning	Architectural/Engineering (A/E) Design
1 Predesign	Clarify owner's value proposition	LEED® AP support	Involve owner: verify OPR	Review OPR	A/E designers:
	Evaluate proposed facility use and size/configure for lean operations	Evaluate site feasibility with respect to sustainability standards		Review BOD	Clarify owner's value proposition
				Develop predesign commissioned outline	Lead design charrette with support of Lean, BIM, Commissioning, LEED facilitators
					Evaluate proposed facility use
					Clarify sustainability standards
2 Design	Target value design: provide best facility that meets owner's budget	Track incorporation of sustainability in design, g., LEED categories: sustainable sites, water efficiency, etc.	Better design options	CxA undertakes services with owner	A/E Designers:
	Collaboration between stakeholders to harness best ideas		Support design for constructability: "clash" detection	Perform 1 review	Perform design activities
	Guide process design		Provide evaluation of design alternatives	Provide VE and O&M recommendations	Target value design with aid of Lean facilitator, BIM/ICT manager
				Finalizes Cx plan for the project	Collaborate with CM/GC for constructability, process design

(Continued)

TABLE 15.1 (CONTINUED)

Enhancement of the Integrated Project Delivery Process

Project Phase	LEAN	LEED	BIM	Commissioning	Architectural/Engineering (A/E) Design
3 Construction	Require lean practices as a procurement strategy Use an integrated team: implement the IFOA Continuous learning and improvement through PPC tracking and analysis	Enhanced commissioning Closely monitor materials and equipment provided for the project for sustainability compliance	Support procurement with more accurate quantity take-offs Reduce conflicts between different trades, reduce change orders Better "real time" tracking of work	CxA leads Cx team, views contractor submittals Attends job site meetings	A/E Designers: Maintain technical support throughout construction
4 Acceptance	Conduct Commissioning for project acceptance	Meeting of LEED targets is a requirement for acceptance	Provide virtual start-up to verify system readiness Track/manage documentation of equipment	Verify functional performance tests Review test/balance activities	A/E Designers: Ensure finished facility meets owner's project requirements (OPR)
5 Postconstruction/warranty	Conduct post-occupancy evaluation	Monitor building performance in specific categories: energy usage, water usage, etc.	Track facility condition and performance of equipment, energy usage	Track warranty issues Perform semiannual, seasonal testing Revise/submit final Cx report	A/E Designers: Evaluate overall performance, owner satisfaction

several areas. The contractor executes the construction with the support of these professionals:

a. A lean facilitator: This individual provides orientation, training, measurement, and ongoing support with lean initiatives throughout the project.

b. A sustainability/LEED professional: Assists designers in determining the level of certification that is feasible on a given project. The LEED® AP verifies the points available based on the prevailing infrastructure.

c. An ICT/BIM manager: This professional uses informational technology including BIM to support the entire design and construction process, further facilitating lean supply and prefabrication in keeping with project requirements.

d. The commissioning agent (CxA): Plays a vital role in project design, ensuring that required systems are specified appropriately. The CxA provides quality assurance by verifying that functional performance tests confirm that design conditions are met by completed construction. The CxA also confirms sustainability requirements have been met, as required by LEED and other related standards.

e. The design team: Comprising architects and engineers as required orchestrates the project. This differs from traditional projects by their interaction and collaboration with the above mentioned professionals; these professionals are generally not included in nonlean projects.

f. The construction manager/general contractor (CM/GC) entity is not shown in the matrix; it is understood that they are required for the actual construction work as would be the case for any construction project. The proposed model requires their interaction with others to create the IPD team.

IPD is based on the adoption of lean principles and behaviors by a team that is bound together by a relational form of contract. While the commercial terms may range from GMP to Cost-Plus, IPD teams are guided toward overall project optimization by the relational contract. In some instances, it may contain a risk pool to protect participants against unexpected costs, while being able to harness overall project savings for everyone's financial benefit. Figure 15.4 illustrates an IPD process that has been expanded to incorporate the support of the professional described in the items above.

Enhanced Lean Project Delivery and IPD

As shown in Figure 15.4, Lean champions are represented at the top of the project delivery system, where they serve as drivers of the various lean processes. Ideally, each stakeholder organization should have a champion—owners, designers, constructors/contractors and suppliers. As IPD involves a relational contract such as the Integrated Form of Agreement (IFOA) or Consensus Docs 300, the champions should collaborate to align interests, objectives, and practices.

Professionals that oversee activities necessary for both lean and sustainable practices are:

- A lean facilitator
- A sustainability professional (e.g., a LEED® AP)

- An ICT/BIM manager to provide BIM support for design and construction activities, and performance information for project tracking.
- A commissioning agent to ensure that the OPR are met and that various systems function as designed (QA).
- The A/E team that traditionally carries out the design process.

It must be noted that these positions are not meant to be mutually exclusive; a LEED® AP could serve as both a sustainability professional and a commissioning agent (with certain conflict of interest restrictions). A talented lean facilitator could also provide ICT and BIM support, depending on the size of the project and its time requirements.

The alteration and decommissioning activities generally occur after many decades, hence they are separated from the typical design and construction cycle. Project definition is accompanied with clarifying owner's value proposition. If the owner wishes the project to emphasize sustainability, for example, this function could determine the LEED certification level required (silver, gold, etc.). The approach, lean design, extends the LPDS to consider the operational aspects of facility use. Designing for lean operations can identify a need for a smaller facility than traditional practice would suggest.

Two distinct phases, lean supply and lean assembly, are positioned to support the construction phase. These phases are not sequential but rather concurrent, as the construction process in a typical project would apply the Just-In-Time approach to the acquisition of materials and equipment. Lean supply would occur incrementally, with delivery to the site when needed, and prompt assembly in order to reduce waste caused by inventories and excessive movement of stockpiled materials and equipment.

Production control ranges from lean design through the end of the construction phase. Similarly, work structuring would be implemented with staff involved in design work, and subsequently with the providers of construction activity in the field. Sustainability ranges from the beginning of the project definition phase through use and to eventual alteration and commissioning. The sustainability professional determines the level of certification that is attainable for a specific project, and documents the specific actions of the project team that are needed to attain the certification. Some of the commissioning agent's actions are also included. Sustainable design is concerned with the reuse and or recycling of certain building components, even after many decades.

Project Definition Phase

In the needs and values determination, design professionals assist the owner/client in clarifying a value proposition; that is, the purpose of the project and the needs to be served. A design criteria document describes specific needs to be met, such as size, space proximities/adjacencies, and energy efficiency requirements. The conceptual design uses the design criteria and value proposition to define an outline design that serves as a starting point for the design phase.

Establishing Design Criteria

The design criteria define the owner's basis of design (BOD). It reflects the owner's needs and wants that must be satisfied by the design of the project. This may include the use of spaces, their sizes, finishes, and activities to be performed within them. The collaboration of the expanded project team could lead to enhancements of the design criteria, such as incorporating practices to reduce the carbon footprint, or reduce the spill-over of site lighting to adjacent areas.

Lean Design Phase

The lean design phase builds on the output from the project definition phase. With the enhanced model the expanded project team brings to the project the experience in their respective fields that improve the project in many ways. BIM tools enable extensive pre-planning and evaluation of alternatives proposed by the team—not just A/E designers and constructors, but with the sustainability and commissioning experts. In parallel with this bar is one labeled "Design for lean operations". It involves an analysis of the owner's requirements to establish the capacity and layout that facilitates lean operations. The construction phase follows the design phase and is linked to the bars titled "lean supply" and "lean assembly".

Lean Supply

Lean supply comprises detailed engineering, fabrication and logistics; it requires up-front product and process design to define what is needed and when it should be delivered. This is especially important with engineered to order components as utilized in EPC projects. Lean supply also includes reducing the lead time for project information requirements. The sustainability professional would dovetail lean supply activities with the acquisition of environmentally compliant materials. For example, LEED promotes the use of rapidly renewable wood products, minimizing the need to harvest lumber in virgin forests. Material acquisitions criteria favor shipments within a radius of 500 miles.

Lean Assembly

Lean assembly is practiced in the actual construction of a project, putting materials, systems and components in place to create a completed facility. As described in Chapter 4, in The Last Planner® System of production control (LPS), work structuring culminates in the form of schedules that represent specific project goals. Schedules are created for each phase of the project, beginning at the design phase and ending at project completion. The production control provided by the LPS deploys the activities necessary to accomplish those schedules. Production control and work structuring refer to the management of production throughout the project.

Commissioning

Commissioning is a significant departure from the prevailing lean project delivery system. That system includes commissioning as an acceptance-related activity at the end of the construction phase. While the activity is included in the enhanced model, commissioning is seen to play a larger role; it is shown in a horizontal bar at the bottom of the diagram. The commissioning process should start as early in a project as the pre-design phase, allowing a commissioning agent (CxA) to make vital input relating to constructability and sustainability. The CxA also provides quality assurance throughout both design and construction, limiting the possibility of costly errors that can leave an owner with a facility that does not perform as expected. As described in Chapter 10, commissioning includes the following:

1. Total building commissioning: Verifies attainment of owners' goals and needs with regard to all building features/systems, from predesign/planning through acceptance/occupancy and—post-occupancy.
2. Commissioning for equipment start-up and acceptance: The scope usually includes major MEP systems: HVAC, control systems, and electrical systems, fire suppression/alarm systems, final inspections, tests, and basic operating performance.
3. Commissioning for certification: LEED, Energy Star, Green Globes, and so on. The scope includes design reviews and construction inspections to certify performance goal attainment.
4. Commissioning for functional performance testing: Testing of selected building systems including the building envelope to verify that the respective performance expectations have been met.

Use

Use refers to a completed facility. Following successful commissioning, the facility should undergo a protracted operations and maintenance phase as it is used. The ICT/BIM function has a major role in tracking the performance of the entire building in the course of its use. Designers should explore the possibility of interfacing a BIM model for the entire building with a permanent system that is left in place after building completion to monitor system performance and warranty-related issues. A malfunction of a chiller, for example, should be recorded in a database together with diagnostic information. This approach represents continuous commissioning. If the performance of many building systems drifts out of specifications, then there could be undesirable increases in energy usage, or alternatively comfort conditions that fall short of users' expectations.

Alteration and Decommissioning

Alteration and decommissioning refers to a future activity when the facility may be repaired, renovated, or taken out of service. Learning loops refer to two levels of learning from the design and construction processes. Learning occurs in the near term through application of the LPS on a weekly basis to review PPC values and commitment reliability. At the completion of a facility, POE reviews the consequences of decisions made during the execution of a project. POE enables project participants to learn from the past. As The Lean Project Delivery System™ continues to evolve, other areas of expertise may be added in a framework of team collaboration, so that synergistic improvements can be made.

Continuous Learning/Evaluation/POE

This box represents a process of continuous learning at all stages of a project. The process includes weekly and other ongoing meeting where percent projects completed (PPC) values are reviewed and reasons for non-completion (RNC) are analyzed to improve forward planning. At the end of a project, this incremental learning information is integrated with the post occupancy evaluation (POE) process to close the "learning loops". This enhancement enables project teams to improve the entire supply chain in near "real time" as well as in future projects.

Work Structuring and Production Control

As defined in the LPDS, Ballard (2008) the purpose of work structuring is to make site operation flow in a reliable and quick manner while delivering value to the customer. Production Control governs the execution of plans and extends throughout a project. The Last Planner® System is central to the process of production planning and control.

Sustaining Lean Construction

Alan Mossman, Director of The Change Business, Ltd. UK, offers the following hypothesis about sustaining lean construction initiatives:

> "Implementation is most likely to succeed long-term where it is a conscious and consistent strategy voluntarily arrived at and systematically deployed by a senior level champion with the time and resources to manage and support over a number of years a bottom up "infection" of the organization using sound organization development/change management principles."

Mossman's advanced his hypothesis in a paper titled "Why isn't the UK construction industry going lean with gusto?", and came from years of involvement with lean construction initiatives as Director of the Lean Construction Institute in the United Kingdom.

Ed Anderson, principal/lean consultant with Anderson Technical Services, recommends the following strategies for sustaining lean construction initiatives (Ed Anderson, Project sustainability. Email in June 2009 to L. Forbes): First, everyone must understand that lean is a journey and not a quick fix. It is a journey of continuous improvement. Sustainability also requires that all stakeholders be assured that they will get *personal* benefits. "Take one for the team" (a personal loss) is the antithesis of what the LPDS is about. However, it must be understood that the *project* must benefit as opposed to individuals. This does not and cannot exclude individual benefits, but there must be an order to benefits. If an individual object and/or process does not benefit the *project*, then it should not be implemented. Allowing just one of these nonproject benefits, especially if imposed by a member of the management team, will eventually cause the entire enterprise to fail.

Sustainability requires that the owner of the result of the *project* drive the initiative. Owners who abdicate responsibility because they do not understand what the initiative is and how it can benefit all stakeholders will not only drive the effort to a standstill, but will actually make the system worse than it was before the initiative. Smart people will try to overcome this, but they will fail and move on to work for people who really mean it when they advocate a "win-win-win" environment.

There are two ways to go: (1) study history, or (2) reinvent the wheel. If one does not study and learn from history he or she is likely to reinvent the wheel, at best. Studying history will only get one so far, however. As Greg Howell (of LCI) likes to say "One must ride the bicycle in order to fully appreciate and understand the process and its implications." In Ed's words, "riding the bicycle is a great analogy because most of us went through that process. Think about how it went—Dad holding the bike up with us on it, running fast, and then letting us go, moments later to crash. During this short ride, our bodies experienced the physics of gyroscopic motion. We did not know how to respond to it, and we did not even call it by name."

"If we were not hurt too badly, we would let Dad push us off again. We went through a 'learning curve' and learned how to respond to gyroscopic motion. And, boy, what a thrill it was! Some of us were a little more lucky—we had a better management team; Dad put training wheels on the bike before we rode it. He did not even have to push us off. We were able to start at a standstill and slowly begin to 'feel' the gyroscopic motion." Ed continues: "Because no one explained what we were experiencing we did not even know what was occurring, but we learned to respond to and even use the gyroscopic motion to keep ourselves erect and on course. But when we tried to turn our first corner we experienced another phenomenon of gyroscopic motion. The bike did not want to turn, it wanted to go straight. So how did we learn to overcome this? We learned to lean into a curve (not an intuitive action) and the wheels were 'fooled' into going in another direction. So even with training wheels in place, we had to go through a process of continuous learning."

"Riding the bike" is an analogy Ed Anderson frequently uses when engaging people in learning/understanding the LPDS. What Ed does differently than what his Dad did, is to explain the "LPDS gyroscopic motion." Construction people need to understand why the system works the way it does, so they don't keep experiencing their own negative learning curve experience.

Ed sums up: "I also try to help people understand that the training wheels, while optional, can serve significant benefits. I represent what I do (mentoring/training) as their LPDS training wheels. I know that they go on the rear wheel, not the front; I know they must fit loosely; and finally I know when they must take the training wheels off in order to go to the next level of learning."

James Womack, co-author of *The Machine That Changed the World* (1990), is one of the great influences in today's lean movement. He notes a tendency for organizations to "backslide" to the old ways of working after making significant improvements with lean (J. Womack, e-mail sent May 30, 2007, Subject: The Problem of Sustainability). He notes that this is often the result of failing to connect the improvement efforts to the way the organization is managed. The most frequent cause is that of middle management resistance to change. Womack thinks that the root cause of that problem has two factors:

a. Confusion in the organization about priorities
b. The failure to make someone responsible for the performance of each so-called value stream.

As middle managers do their jobs they may not be equipped, or have the time to monitor gaps in performance between current levels and the level improvement efforts could make. In the lean environment this is called value stream management. Womack feels that fixing the root causes of poor performance requires the participation of everyone, *and a change in their behavior.* In answer to the question "What can we do about our sustainability problem?" Womack proposes an approach that would not require significant organizational changes; instead a manager in another area could be assigned to audit the performance of improvement initiatives as reflected in the horizontal flow of value along the value stream. That individual could serve as a liaison between responsible managers and top management to gain agreement on who must do what by when to achieve a sustainable leap in performance that will benefit the customer and the organization. He suggests that to prevent regression, someone needs to periodically clarify priorities for each value stream and identify the performance gap between what the customer needs and what the value stream is providing. The person taking responsibility then needs to engage everyone touching the value stream to carefully capture the current condition (the "current state") of the value stream that is causing the gap. The next step is to envision a better value stream and determine who will need to do what, by when, to bring it into being and move to a higher level of performance. Finally, the value stream leader needs to determine what will constitute evidence that the performance gap has been closed and collect the data to demonstrate this. Womack points out that this exercise is, of course, nothing but Dr. Deming's Plan-Do-Check-Act cycle conducted repetitively by the responsible person, ideally employing A3 analysis.

This approach implies that the organization has a degree of maturity in its lean journey, when a constituency of lean converts (and advocates) exists. This situation often occurs in the manufacturing environment, but very infrequently in the construction environment. Construction stakeholder organizations (i.e., designers, contractors, subcontractors, and owners) need to think beyond the needs of the immediate project and work toward a future state in the value stream that represents higher levels of performance, quality, safety, and customer satisfaction. By working with a champion within each stakeholder organization, a lean facilitator may be an effective catalyst. He or she can help stakeholders in acquiring the necessary knowledge and in adopting the beliefs and behaviors that are needed for "leanness."

Womack issues a caution that would serve the lean construction community well; Other organizations are different from Toyota and each other, hence experiments with value stream management methods are needed. If the sustainability problem is not addressed, he points out, the current surge of interest in lean—driven by the success of Toyota—may become just another episode in the long history of unsustainable management improvement campaigns.

The Way Forward

In the construction environment, the top managers of stakeholder organizations (i.e., owners, designers, construction managers/contractors and suppliers) should embrace the practice of continuous learning and improvement that is central to the lean construction philosophy. Above all, they should accept the fact that change takes time, and they should make the investment of the human and financial resources needed to make it occur.

> He who would learn to fly must first learn to walk and run and climb and dance; one cannot fly into flying. Friedrich Wilhelm Nietzsche

As studies have shown, various forms of waste in the construction supply chain account for as much as 30% of the construction dollar, there are many opportunities for improvement that will not only reduce construction costs and deliver value for owners, but also improve profitability for designers and constructors. Owners, in particular, need to recognize the superiority of lean methodologies such as lean project delivery and IPD in securing the foregoing objectives. They should seek to have their projects delivered through these methodologies, and follow the example set by large owners such as the Sutter Health Care System in California that garnered significant reductions in cost and schedule, with high levels of owner/user satisfaction. Forms of contracts have been developed to facilitate lean projects, such as the IFOA or Consensus Docs 300. The authors take the position that owners are in the strongest position to move the industry toward an adoption of lean, in a manner similar to the growing adoption of sustainability initiatives such as LEED.

Several positive outcomes will result from owners' leadership, including:

- Lean projects include prerequisite activities that involve team building and lean coaching/facilitation to ensure that participants understand and commit to lean beliefs and behaviors.

- Project teams have a vested interest in optimizing the project, as opposed to their respective "pieces"; some integrated agreements have a risk pool contingency that serves as a no-fault emergency fund, and also garners savings that are shared with the team.

- Stakeholders will see all project events, positive or negative, as learning opportunities as opposed to the tradition of placing blame.

- Autonomation will be encouraged; to empower the people in the field to secure reliable promises and execute work that CAN be done, instead of pushing crews to meet unrealistic deadlines.

This paradigm shift will not occur easily—it will be encouraged by repeated application of lean in projects and reinforced by ongoing measurement and reasons/root cause analysis that demonstrate that lean projects are in fact more successful and profitable for all stakeholders. Table 15.1 explains how various disciplines should collaborate, especially in the early project phases. Lean facilitators, sustainability/LEED professionals, BIM specialists, and commissioning agents should collaborate with the traditional architectural/engineering professionals. Together, they should not only design the product—the facility, but should influence the construction process to increase the probability of delivering a facility that provides the highest value to the owner/client.

The authors propose that all industry stakeholders should actively seek knowledge of lean construction and IPD through affiliation with the Lean Construction Institute (LCI). As a nonprofit organization, the LCI seeks to share knowledge with all industry stakeholders, owners, designers, constructors, and suppliers. LCI has been forming chapters in several cities around the US. They have also been promoting the concept of learning communities in which industry professionals share knowledge on lean concepts with each other, even if they are competitors. It seeks to unite them to engage actively in applying lean principles from design inception through the construction process; the benefits so derived can have a major impact at a national level by improving the competitiveness of the construction industry as a whole.

Questions for Discussion

1. What are the benefits of post-occupancy evaluation (POE)?
2. Who should participate in the process?
3. What can an owner learn from a POE?
4. What can a designer learn from a POE?
5. How does a so-called post-mortem study benefit a contractor?

6. How do post-occupancy studies relate to lean construction?
7. Who are the professionals that support traditional designers in the enhanced IPD process—what value do they add to a project?
8. Why does interest in new initiatives wane after initial implementation?
9. How can lean construction initiatives be sustained
   a. by designers?
   b. by contractors?
   c. by owners?

Appendix: Examples of "Lessons Learned" Recommendations from Post-Occupancy Evaluation of a New Educational Facility

Design Factors

- 1,460 student stations, 1,310 design capacity
- Gross square feet: 160,113 on a 17 acre single story campus plan.
- The design avoided dead-end halls and staircases to improve security.
- Open corridors reduced conditioned space by 15%.
- A thermal energy storage system (TES), ice production used to reduce energy cost.

Design/Construction Recommendations

In future facilities, investigate benefit of smaller footprint/2-story design. Confirm inclusion of review comments from specialists, maintenance, quality control staff, plant operations.

- Fast track review comments to avoid time delays and incorporate end users' comments.
- Ensure test/balance compliance before acceptance.
- Require contractors to take corrective action at no cost to the owner.
- Encourage designers, design builders to review problems in existing facilities and improve current designs.
- Involve all postconstruction stake holders in the design/review process.

Classroom Recommendations

- Stimulate learning through more generous use of windows.
- Provide adequate electrical outlets to anticipate expansion/overcrowding.
- Provide adequate (or complete) separation between classrooms and the administration area to minimize interruptions to the instructional process.
- Analyze instructional needs to provide adequate marker boards, computers, storage, etc.

- Minimize classroom cross talk with sound insulation. Distance quiet areas from noisy ones as much as possible.
- Conduct HVAC test and balance with computers in working mode. In computer labs the heat load can affect comfort conditions.

Restroom Recommendations

- Position staff restrooms to minimize travel distance.
- Anticipate over-enrollment in sizing restrooms; pay special attention to facilities for female students.
- Improve the privacy of locker rooms.
- Consider using electrically operated hand dryers to reduce paper towel use.

Auditorium Recommendations

- Consider designing auditorium lighting with adequate levels to promote dual use as classroom space.
- Design HVAC air distribution system to avoid high velocities/noise levels.
- Consider a variance to increase auditorium size.
- Consider wiring in public address connections with the option of enabling them.
- Provide adequate restroom facilities in the auditorium for stand-alone use.

Meal Preparation/Distribution Recommendations

- If at all possible, select kitchen equipment with additional capacity to anticipate over-enrollment.
- Configure spill-out areas to increase flexible space for eating.
- Provide space for expanding serving lines in cafeterias.
- Install vision panels in cafeteria doors to provide cafeteria staff with better control.
- Avoid adhered rubberized flooring. The seams tend to fail, causing trip hazards.

Security Recommendations

- Include traffic signalization provisions in overall project requirements.
- Address escalating school safety concerns with cameras or preinstalled conduits/infrastructure.
- Provide additional power outlets of various voltages to adequately provide for electronic surveillance equipment.
- Provide additional cooling in surveillance equipment locations to ensure comfortable conditions for staff, and stable operating temperatures for equipment.
- Utilize card access systems to control access to critical areas.
- Install swing gates with secret latches in administration office areas.
- Place fire alarm pull stations near doorways, avoiding swinging doors.

- Promote designs that facilitate visibility and supervision of entrances.
- Provide for emergency (911) access to playfields to promote rapid emergency vehicle response.
- Configure elevator systems to minimize or eliminate unauthorized access at ground level in order to maximize security at the upper floor levels.

Handicapped Accessibility Recommendations

- Design main counters for accessibility from both sides (36″ h × 36″ w).
- Ensure that accessible restrooms are labeled at 60″ above finished floor level.
- Ensure ADA accessibility for all sinks and water coolers.
- Install doors with a latch side clearance of 18″ to 24″.
- Provide accessible outdoor tables and benches.
- Design ramp handrails with 18″ extensions.
- Align ramps appropriately with parking spaces.

Furniture, Fixtures and Equipment (FF&E) Recommendations

- Consider ADA requirements when selecting FF&E (work stations, sink cabinets).
- Investigate products to promote greater longevity.
- Consider over-enrollment in allocating in-contract FF&E for physical education lockers.
- Align bookshelf selection with room for growth.

HVAC Recommendations

- Favorably consider specifying ice storage systems for future projects.
- Allow for over-enrollment of 30% or more when sizing systems to ensure that ice depletion does not occur, resulting in punitive electrical demand charges.
- Design chiller controls to alternate the starting sequence for longer life.
- Size air return openings/ducts to minimize noise levels.
- Serve fewer fire rated areas with each air handler (use more air handlers).

Quality Assurance Recommendations

- Institute the utilization of in-house journeymen as quality control staff to detect work quality deficiencies in contractors' work.
- Build into each project a requirement for work quality deficiencies to be addressed even if they meet the relevant codes, communicate this requirement at preconstruction meetings.
- Require detailed drawings of expansion joints and monitor construction to ensure proper execution.
- Maintain close control over contractors' material/equipment substitutions, especially in design/build projects.

General Recommendations

- Design custodial closet doors to swing outward, maximizing storage space.
- Design staff parking in anticipation of student over-enrollment.
- Make walkways and their roofs wider to better protect staff/students.
- Provide clear signage to improve way finding in the facility.

Examples of Actions Taken Following POE Recommendations:

1. Layout of air handlers in machine rooms improved to facilitate maintenance.
2. Walkway design revised to allow better rain protection.
3. Auditorium air conditioning deficiencies noted, high return air velocities create high noise levels. Design was modified to meet new noise criteria.
4. Auditorium seating, balcony views obstructed by guard rails, adjustments made to future specifications.
5. Running track deficiencies noted, improved specifications and construction methods instituted.
6. Ice storage HVAC system critiqued. specifications improved.
7. Electrical surge protection for electronic communications/security/alarm systems recommended.
8. Handicapped accessibility deficiencies noted in clinics, conference rooms, office spaces.
9. Emergency egress window deficiencies listed, operating levers inadequate.
10. Locations of security cameras and card access readers recommended.
11. Inadequate procedures identified with respect to programming and subsequent construction quality assurance.
12. Poor maintainability noted with clear-colored vinyl tile and textured wall paint finish.
13. Inward-swinging doors in custodial closets minimize storage space, specifications revised.
14. Specialized labs with large equipment provided with special access.
15. The flow of students through staff's parking lots was reduced.
16. Roof scuppers observed to spill on wall-mounted light fixtures, changes suggested.
17. Garage facility: Multiple roof levels to be avoided to obviate need for ladders.
18. Corridor lighting adjusted to light floors more than walls. Alarm panels to be located in supervised areas only. Expensive vehicle exhaust removal placed too low for effective operation.
19. Transmission repair shop inadequately vented, causing unsatisfactory working conditions.

References

Ballard, G. 1999. *The Challenge to Change.* From selected readings Lean Construction Institute.

Ballard G. and Howell, G. 2003. Lean Project Management. Building Research and Information, 31 (2), 119–133

Ballard, G. 2008. The Lean Project Delivery System™: An update. *Lean Construction Journal* 1–19.

Forbes, L. 2002. Continuous learning through quality-based post occupancy evaluation. *Proceedings of the 2002 IIE Annual Conference 2002,* Orlando, FL, May 19–22.

Matthews, O., and G. Howell. 2005. Integrated project delivery an example of relational contracting. *Lean Construction Journal* 2(1): 46–61.

Mossman, A. 2009. There really is another way, if only he could stop—for a moment and think of it—Why isn't the UK construction industry going lean with gusto?. *Lean Construction Journal* 24–36.

Preiser, W. F. E., Rabinowitz, H. Z. Rabinowitz, and E. T. White. 1998. Post Occupancy Evaluation, Van Nostrand Reinhold, New York.

Senge, P. M. 1990. *The fifth discipline: The art and practice of the learning organization.* New York: Doubleday.

Womack, J., D. Jones, and D. Roos. 1990. *The machine that changed the world.* New York: Harper Collins.

Bibliography

Abdel-Razek, R. 1997. How construction managers would like their performance to be evaluated. *ASCE Journal of Construction Engineering and Management* September.

Ballard G. 2000. The Lean Project Delivery System™. LCI White Paper-8, September 23, 2000 (Revision 1)

Davis, R. 1999. Creating high performance project teams. *Conference Presentations, Construction Project Improvement Conference,* Austin, TX, September 26–28.

Forbes, L. 2005. Industrial engineering applications in construction. *Handbook of Industrial Engineering,* ed. A. Badiru. 31–37. Boca Raton, New York: Taylor & Francis.

Garsden, B. 1995. Postconstruction evaluation. *Journal of Construction Engineering and Management* 121(1): 37–42.

Lichtig, W. A. 2005. Sutter health: Developing a contracting model to support lean project delivery. *Lean Construction Journal* 2(1, April): 105–12.

Williams, K., and L. Poltronieri. 1997. Compare and contrast: POEs of three outpatient dialysis units leads to new planning and design decisions. Proceedings of the International Conference and Exhibition on Health Care Facilities Planning, Design, and Construction.

Glossary of Lean Terms

Activity Definition Model: The ADM represents design tasks or construction processes and enables planners to examine scheduled tasks in detail and determine their readiness for execution. These tasks are released only if all constraints are removed, thereby improving performance.

A3 Report: The A3 report is a disciplined and rigorous system for implementing PDCA management. As used by Toyota, the report fits on a single sheet of A3 paper, measuring approximately 11×17 and documents problems and the available information about them in a manner that helps users to understand quickly. It focuses users on improving processes and solving problems.

Autonomation: Lean construction applies the principle of autonomation that is an important ingredient of the Toyota Production System (TPS); the people closest to the work are empowered to stop production if they determine that the upstream production is defective. The last planners and their crews fulfill this role when using The Last Planner® System .

Basis of Design: The basis of design (BOD) reflects the owner's needs and wants that must be satisfied by the design of a project. This may include the use of spaces, their sizes, finishes, and activities to be performed within them.

Benchmarking: This involves comparing actual or planned project practices to those of other projects to generate ideas for improvement and to provide a standard by which to measure performance.

Building Information Modeling: Building information modeling (BIM) is the process of generating and managing building data during its life cycle. It is a model-based technology linked with a database of project information. A BIM carries all information related to a facility, including its physical and functional characteristics and project life cycle information, in a series of smart objects.

Buffer: The means (capacity, inventory, time) used to cushion against the shock of variation in a process. Two types of buffers may be used to shield downstream construction processes from flow variation. Plan buffers are inventories of workable assignments. Schedule buffers are materials, tools, equipment, manpower, time, and so on.

Commissioning: From the design phase to a minimum of 1 year after construction, this is a systematic process of assurance by verification and documentation that all facility systems perform efficiently and in accordance with the design documentation and intent and meet the owner's operational needs.

Commissioning Agent: The CxA (or Authority) is a professional that serves as an objective advocate for the owner and is responsible for directing the commissioning team and process; coordinating, overseeing, and/or performing the commissioning testing; and reviewing the results of the system performance verification.

Commitment Planning: Commitment planning is a method for defining criteria that lead to the selection of quality assignments. It results in commitments to deliver that other production teams can rely on by following a prerequisite that only *sound* assignments should be made or accepted.

Commitment Reliability: Commitment reliability is made by producers (contractors, designers, etc.) in the design/construction supply chain in order to meet schedules. Commitment reliability is calculated by comparing completed work (DID) with planned work (WILL).

Constraint: A constraint is an obstacle that inhibits the execution of a task that is required in the look-ahead plan. It should also be beyond the control of the last planner.

Constraint Analysis: Foremen, project managers, schedulers, and other staff communicate to identify constraints and the reasons for their occurrence. After analysis, they are able to solve and remove the constraint and assign work tasks.

Consensus Docs 300: This is a standard form of contract for the construction industry that is based on an Integrated Form of Agreement (IFOA). Twenty-two leading construction associations came together and published a consensus set of contract documents in 2007 that they felt was fair to all parties.

Control Chart: A graphic display in the progression of a process. It is used to determine if the process is under control and progressing on time.

Continuous Flow: Producing and moving one item at a time (or a small and consistent batch of items) through a series of steps as continuously as possible, with each step making just what is requested by the next step. Also called one-piece flow or single-piece flow.

Continuous Improvement: Small improvements to reduce costs and ensure consistency in performance of products and services.

Cycle Time: How often a part or product is completed by a process, as timed by observation. This time includes operating time plus the time required to prepare, load, and unload. The appropriate calculation of cycle time may depend upon context. For example, if a paint process completes a batch of 22 parts every 5 minutes, the cycle time for the batch is 5 minutes. However, the cycle time for an individual part is 13.6 seconds (5 minutes × 60 seconds = 300 seconds, divided by 22 parts = 13.6 seconds).

Design Criteria: Design criteria for buildings define the owner's requirements, including those that may exceed applicable codes. They primarily serve as a performance-based guide to designers, to ensure that ensuing designs meet the functional needs and preferences of the owner. Design criteria list applicable codes and standards such as weather-related and earthquake codes as are relevant to the location. A criteria document typically describes the building systems and materials to be used. It may also define project-specific requirements such as building orientation, space descriptions, and adjacencies to meet owners' operating needs.

Design of Experiments: A statistical method that identifies those factors that might influence specific variables.

Energy Star: Energy Star is a reference system developed by the U.S. Environmental Protection Agency that promotes efficient energy usage in facilities and the systems and appliances that are used in them. Buildings that have earned the ENERGY STAR label use an average of almost 40% less energy than average buildings and emit 35% less carbon.

First Run Studies: These studies involve trying out new construction ideas on a pilot basis and applying the PDCA methodology to determine their success. First run studies also identify the best means, methods, and sequencing for a specific activity.

Fishbone Diagram: Also called the Ishikawa diagram or cause-and-effect diagram. The diagram illustrates how various causes and subcauses relate to creating potential problems. It may be used to determine the root of observed problems.

Five Ss: One of the guiding principles of Lean meant to achieve standardization and stability in the workplace through visual management. (1) **Sort:** Remove everything from the workplace that is not needed for current production. (2) **Set in order:** Arrange items so they are easy to find and return and locate items to minimize motion waste. (3) **Shine:** Keep everything clean and in top condition so that it is ready to be used. (4) **Standardize:** On-going maintenance of sort, set in order, and shine. (5) **Sustain:** Create the conditions or structures that will help sustain commitment.

Five Whys: A method of root cause analysis whereby the investigator repeatedly probes operators and other employees with the simple question "why?" until the problem is uncovered.

Flowcharting: A flow chart is any diagram that shows how various elements of a system relate (cause and effect diagrams, system or process flow charts).

Gemba: A Japanese term that means the place where things are actually happening. *Gembashugi*: shop floor oriented (Shugi = orientation or philosophy)

Hansei: A Japanese term that means critical self-reflection.

Heijunka: Creating basic stability in processes. To produce goods at a constant and predictable rate.

Hoshin Kanri: Japanese term for policy deployment. Its components are Hoshin, a compass, course, and policy; Kanri, management control.

Information and Communication Technology: The ICT encompasses computer hardware, software, and communications devices that allow the sharing and access of information between many users. In effect, it creates an "information village."

Integrated Form of Agreement: The IFOA is a legal agreement that seeks to align the commercial relationships of a construction project's design and construction participants that are assembled as a temporary production system. It also requires collaborating through planning, design, construction, and managing the project as a network of commitments. As a result it optimizes the project as a whole, rather than any particular piece. The agreement also calls for a combination of learning with action and promoting continuous improvement throughout the life of the project.

Integrated Project Delivery: The IPD is a relational contracting approach that aligns project objectives with the interests of key participants. It creates an organization able to apply the principles and practices of The Lean Project Delivery System™. The fundamental principle of IPD is the close collaboration of a team that is focused on optimizing the entire project as opposed to seeking the self-interest of their respective organizations.

IPD Team Performance Contingency: A provision in integrated project delivery for the sharing of costs and benefits based on the performance of the IPD team.

Jidoka: Providing machines and operators the ability to detect when an abnormal condition has occurred and immediately stop work. This enables operations to build in quality at each process and to separate men and machines for more efficient work. Jidoka is one of the two pillars of the Toyota Production System along with Just-In-Time. Jidoka is sometimes called autonomation, meaning automation with human intelligence.

Just-In-Time Production: A system of production that makes and delivers just what is needed, just when it is needed, and just in the amount needed. JIT and Jidoka are the two pillars of the Toyota Production System.

Kaizen: Continuous improvement of an entire value stream or an individual process to create more value with less waste. There are two levels of kaizen: (1) system or flow Kaizen focuses on the overall value stream and (2) process Kaizen focuses on individual processes.

Kanban: A signaling device that gives authorization and instructions for the production or withdrawal (conveyance) of items in a pull system. The term is Japanese for sign or signboard.

Last Planner System: The LPS is an important subset of The Lean Project Delivery System™ and is critical to its effective deployment. The LPS accommodates project variability and smooths workflow so that labor and material resources can be maximally productive. It uses lean methods to provide improved project control.

Lean Operations: Lean operations enable a facility owner/user to maximize the use of the built environment. Lean design should facilitate lean operations in a completed facility. A poorly laid out facility could be relatively expensive to operate—excessive travel by users from one point to another is one of the seven wastes documented in the Toyota Method.

Lean Production: A business system for organizing and managing product development, operations, suppliers, and customer relations that requires less human effort, less space, less capital, and less time to make products with fewer defects to precise customer desires, compared with the previous system of mass production. It was pioneered by Toyota after World War II.

Lean Project Delivery System: The LPDS is a lean technique that integrates phases to facilitate the design and delivery of construction projects. These phases are: (1) project definition, (2) lean design, (3) lean supply, (4) lean assembly, and (5) use. The LPDS is based on a close collaboration between the members of the project delivery team. They are bound by codes of conduct—both written and unwritten—that focus on the success of the overall project rather than their individual success

LEED: Leadership in Energy and Environmental Design was developed by the U.S. Green Building Council. It is a voluntary, consensus-based national standard for developing high-performance, sustainable buildings. The LEED® certification program is a feature-oriented rating system that awards buildings points for satisfying specified green building criteria. The six major environmental categories of review include: sustainable sites, water efficiency, energy and atmosphere, materials and resources, indoor environmental quality, and innovation and design.

Linguistic Action: The theory of linguistic action describes the purposeful ways in which people communicate—the extent to which they mean what they say. In the lean construction environment it relates to the conversational nature of design, planning, and coordination. The theory influences the securing of reliable promises that are critical to the success of The Last Planner® System and lean construction in general.

Muda: Seven classes of waste: over production, delays, transportation, processing, excess inventory, wasted motion, and defective parts.

Mura: Japanese for unevenness in the production system relating to man power or materials. Includes day-to-day variation in customer demand.

Muri: Overburden of employees and technologies having a direct impact on morale in a negative manner.

Obeya: In Japanese means simply "big room." At Toyota it has become a major project management tool, used especially in product development, to enhance effective and timely communication. Similar in concept to traditional "war rooms," an

Obeya will contain highly visual charts and graphs depicting program timing, milestones and progress to date, and countermeasures to existing timing or technical problems.

Owner's Project Requirements (OPR): (See Design Criteria)

Pacemaker Process: Any process along a value stream that sets the pace for the entire stream. (The pacemaker process should not be confused with a bottleneck process, which necessarily constrains downstream processes due to a lack of capacity.)

Pareto Chart: A diagram that is used to display the relative importance in a number of factors. In evaluating problems, it shows the types of problems and the frequency of their occurrence or cost impact based on the so-called 80/20 rule. The chart enables users to readily prioritize needed corrective action.

Plan, Do, Check, Act: An improvement cycle based on the scientific method of proposing a change in a process, implementing the change, measuring the results, and taking appropriate action. The PDCA cycle has four stages: Plan, Do, Check, and Act.

Poka Yoke: Japanese term that means mistake (or error) proofing. A poka yoke device is one that prevents incorrect parts from being made or assembled, or easily identifies a flaw or error. Error-proofing is a manufacturing technique of preventing errors by designing the manufacturing process, equipment, and tools so that an operation literally cannot be performed incorrectly.

Percent Plans Complete: In lean construction, work accomplishment is recorded as a graphical plot of PPC; it shows the percentage of the assigned plans (i.e., commitments that were completed, fractional completion is not considered). Some lean construction practitioners refer to PPC as commitment reliability.

Prevention Over Inspection: A well-known mantra emphasized by proponents of quality. Prior to the emergence of the quality movement, the main focus of quality was on inspection. Both research and experience pointed to the fact that the net cost of inspecting is so high that it is better to spend money on preventing problems. "Quality must be planned in not inspected in."

Production Lead Time: Also called **Throughput Time** and **Total Product Cycle Time**. The time required for a product to move all the way through a process from start to finish. At the plant level this is often termed door-to-door time. The concept can also be applied to the time required for a design to progress from start to finish in product development or for a product to proceed from raw materials all the way to the customer.

Productivity: Productivity is the measure of how well resources are brought together and utilized for accomplishing a set of goals. Productivity is measured as the ratio of outputs to inputs. In the construction environment it may be represented as the constant-in-place value divided by inputs such as the dollar value of material and labor.

Quality Assurance: Planned and systematic actions to help assure that project components are being designed and constructed in accordance with applicable standards and contract documents.

Quality Control: The review of project services, construction work, management, and documentation for compliance with contractual and regulatory obligations and accepted industry practices.

Quality Function Deployment: The QFD is defined as "a technique to deploy customer requirements into design characteristics and deploy them into subsystems, components, materials, and production processes." The QFD may be applied to the design and planning of construction projects; the built facilities meet users' needs more closely than is derived through practitioners' experience alone.

Reasons for Noncompletion: In conjunction with lean-based construction projects, incomplete work plans are studied each week to determine the root causes or RNC. At each weekly meeting time is devoted to learning why certain tasks were not completed, in order to improve the effectiveness of future work plans.

Relational Contracting: Relational contracting is a transaction or contracting mechanism that apportions responsibilities and benefits of the contract fairly and transparently, based on trust and partnership between the parties. It provides a more efficient and effective system for construction delivery in projects that require close collaboration for execution. The relationship between the parties transcends the exchange of goods and services and displays the attributes of a community with shared values and trust-based interaction.

Reliability: The extent to which a commitment is fulfilled.

Reverse Phase Scheduling: The RPS involves working backward from the deliverables of the completed project, determining what needs to be done at each stage to complete the preceding activity. It is a detailed work plan that specifies the hand offs between trades for each project phase (i.e., what SHOULD be done at each phase). In lean construction, the reverse phase schedule is developed through conversations with the primary subcontractors, including their last planners, typically with sticky notes on a display board.

Sensei: A Japanese title used to refer to or address teachers, professors, professionals such as lawyers and doctors, politicians, clergymen, and other figures of authority. The word is also used to show respect to someone who has achieved a certain level of mastery in an art form or some other skill.

Set-Based Design: A design management approach that defers design decisions to the "last responsible moment" to allow for the evaluation of alternatives that improve constructability. Toyota has used this technique to design new models in a much shorter time than the industry standard. They have learned that the best solutions are hybrids of original design options.

Six S: Six S is an enhancement of the 5S methodology to include an emphasis on safety. Like 5S it also promotes Lean through visual management to achieve standardization and stability in the workplace

Statistical Sampling: Involves choosing part of a population of interest for inspection.

Supply Chain: The term encompasses all the activities that lead to having an end user provided with a product or service—the chain is comparable to a network that provides a conduit for flows in both directions, such as materials, information, funds, paper, and people.

Sustainability: The ability of a society to operate indefinitely into the future without depleting its resources. Sustainability includes concepts of green building design and construction, reuse and recycling of materials, reduced use of material and energy resources for building construction and operation, water conservation, and responsible stewardship of the surrounding environment.

Takt Time: The available production time divided by customer demand. For example, if a widget factory operates 480 minutes per day and customers demand 240 widgets per day, takt time is 2 minutes. Similarly, if customers want two new products per month, takt time is 2 weeks. The purpose of takt time is to precisely match production with demand. It provides the heartbeat of a lean production system.

Target Value Design: The TVD involves designing to a specific estimate instead of estimating based on a detailed design. It seeks to address the problem that affects many projects—various design disciplines work from a common schematic design to do design development in their areas of expertise.

Team Dynamics: Influences problem solving, communications skills, conflict resolution.

Total Quality Management: A philosophy that encourages companies and their employees to focus on finding ways to continuously improve the quality of their business practices and products.

Toyota Production System: The production system developed by Toyota Motor Corporation to provide best quality, lowest cost, and shortest lead time through the elimination of waste. TPS is comprised of two pillars, JIT production and Jidoka. TPS is maintained and improved through iterations of standardized work and Kaizen, following the scientific method of the plan-do-check-act cycle.

Trend Analysis: Involves using mathematical techniques to forecast future outcomes based on results. It is often used to monitor technical and cost/schedule performance.

U.S. Green Building Council: This is a non-profit membership organization whose vision is "a sustainable built environment within a generation," Its membership includes corporations, builders, universities, government agencies, and other non-profit organizations.

Value-Added and Non-value-Added Work Activities: Value-added activities provide desirable outcomes for a customer; that is, they meet the customer's value proposition. Conversely, non-value-added activities do not contribute to meeting a customer's expectations. For example, searching for a tool to perform a task does not get the task done and does not provide value.

Value Stream: All of the actions, both value-creating and non-value-creating, required to bring a product from concept to launch and from order to delivery. These include actions to process information from the customer and actions to transform the product on its way to the customer.

Value Stream Mapping: A simple diagram of every step involved in the material and information flows needed to bring a product from order to delivery. A current-state map follows a product's path from order to delivery to determine the current conditions. A future-state map shows the opportunities for improvement identified in the current-state map to achieve a higher level of performance at some future point.

Variation: The divergence of a process from an intended plan. The lean philosophy seeks to reduce process variation in order to obtain reliable flow through a process.

Visual Management: Using visual displays for better communication. It works in conjunction with empowerment so that anyone, not just a supervisor, can detect process anomalies and take prompt corrective action.

Waste: Any activity that consumes resources but creates no value for the customer. There are seven wastes as identified in Lean: overproduction, excessive inventory, unnecessary conveyance, overprocessing, excessive motion, waiting, and corrections (of defects).

Workable Backlog: Is used to describe assignments that have met all quality criteria, but may need other prerequisite work to be done before they can be started. Workable backlog enables crews to continue working productively if there is a constraint that prevents the completion of an item on the weekly work plan.

Work Structuring: The process of defining and organizing activities into logical, practical groups of work. It is a process of subdividing work so that the pieces are different from one production unit to the next to promote flow and throughput, and to have work organized and executed to benefit the project as a whole.

Work Package: A compilation of information regarding the availability of material, equipment, resources, and other pertinent information that is necessary to perform a specific scope of work. The work package is coordinated with other interdependent work packages and requirements.

Index